Principles of Physical Chemistry

David H. Mansfield, B.Sc.
Senior Teacher and Head of the Chemistry Department,
The Harvey Grammar School, Folkestone, Kent

Heinemann Educational Books
London

Heinemann Educational Books Ltd
22 Bedford Square, London WC1B 3HH

LONDON EDINBURGH MELBOURNE AUCKLAND
HONG KONG SINGAPORE KUALA LUMPUR NEW DELHI
IBADAN NAIROBI JOHANNESBURG
EXETER (NH) KINGSTON PORT OF SPAIN

ISBN 0 435 65563 9

First published 1978
Reprinted with corrections 1979
Reprinted with corrections 1981

Filmset in 'Monophoto' Times 10 on 11 pt by
Richard Clay (The Chaucer Press), Ltd, Bungay, Suffolk
and printed in Great Britain by
Fletcher & Son Ltd, Norwich

Principles of Physical Chemistry

Other books by David Mansfield

Heinemann Experimental Chemistry Series

2. Electrochemistry

10. Radiochemistry

Published by McGraw-Hill

Chemistry by Experiment and Understanding (*with John R. Gerrish*)

Published by Longman (*for the Nuffield Foundation*)

Drugs and Medicines (Revised Nuffield Chemistry)

Preface

MODERN chemistry syllabus reforms at all levels place an increasing emphasis on Physical Chemistry to provide a conceptual framework within which the more factual material of Organic and Inorganic Chemistry can be systematized. This book is intended primarily to meet the needs of students preparing for G.C.E. A and S level examinations and equivalent qualifications, such as Ordinary National Certificates and Diplomas. It also covers the work required for the entrance and award papers of the Oxford, Cambridge, and London colleges. Students on introductory courses at University and other Colleges of Higher Education should also find the text useful.

There is considerably more material than is likely to be required for any particular examination syllabus and the student must make an appropriate selection under the guidance of his teacher. There is now greater divergence than ever before between individual syllabuses at intermediate level, and some traditional topics have necessarily been truncated or abandoned to make way for relevant new material. Novelty itself has not been the criterion for selection, however, and sufficient traditional material is retained here to emphasize the historical development of the subject and the debt each generation of innovators owes to workers of the past. Where appropriate this material has been restated in present-day terms to assist understanding. In addition to the historical perspective, the international nature of scientific research is stressed so that the development of important concepts can be seen to result from the combined efforts of many workers in different lands and times. Every effort has been made to show the relevance of physical chemistry in a wide range of other disciplines such as biology, engineering, geology, archaeology, food science, and in everyday life.

The mathematical demands have been carefully considered. It is recognized that many students whose interest is mainly in the biological sciences will not be studying mathematics as a main subject. The great majority of these, however, will be taking a 'mathematics for science' course. In those sections of the book covering the basic A and S level material, only mathematics familiar to such students has been employed. The use of some elementary calculus has not been avoided in the chapters on Chemical Kinetics and Chemical Thermodynamics as its use can add so enormously to understanding, even at intermediate level. The student on a modern biology course in Higher Education is certain to encounter these concepts and it is thus advantageous that a satisfactory treatment is given here.

This is not a practical book, but the results of many experiments are discussed to provide evidence for the ideas and concepts introduced. Wherever possible, experiments that can be performed in teaching laboratories at intermediate level have been chosen and sufficient experimental detail given to permit their reconstruction. Many of the results quoted have been obtained by my students. Suitable experimental work is contained in the *Heinemann Experimental Chemistry Series*. A selection of problems, some from recent public examinations, are given at the end of each chapter together with a short bibliography where this is appropriate.

I would like to thank the staff of Heinemann Educational Books, particularly Hamish MacGibbon and Graham Taylor, for their help, advice, and encouragement. I would welcome constructive criticism and suggestions to improve the text in future editions.

Folkestone, 1978 David H. Mansfield

Contents

PREFACE		v
ACKNOWLEDGEMENTS		ix
Chapter		
1	Gases and the Kinetic Theory of Matter	1
2	Atomic Structure	15
3	Spectroscopy and Analysis	34
4	Thermochemistry	45
5	Structure and Bonding	60
6	The Equilibrium Law	82
7	Vapour Pressure	94
8	The Ionic Theory	108
9	Ionic Equilibria	124
10	Rates of Reaction—Chemical Kinetics	145
11	Chemical Thermodynamics	167
12	The Periodic Classification	184
13	Surface Chemistry and Colloidal Systems	194
14	Nuclear and Radiochemistry	201
TABLE OF RELATIVE ATOMIC MASSES		219
UNITS AND SYMBOLS		220
INDEX		223

Acknowledgements

THE author gratefully records the help and advice given by:

The late Sir Lawrence Bragg
Professor Sir George Porter, F.R.S., Director of the Royal Institution of London
Dr. M. Goggins, Renal Unit, Kent and Canterbury Hospital
Dr. B. H. Robinson and Dr. J. F. J. Todd of the Chemistry Department, University of Kent at Canterbury
Dr. M. Gilbert, Department of Materials Science, University of Loughborough

The author and publishers are grateful to the following for permission to reproduce photographs and other illustrations:

A.E.I. (Scientific) Ltd.
Esso Petroleum Company Ltd.
Evans Electroselenium, Halstead
Griffin and George Ltd.
Philip Harris Ltd.
Pye-Unicam of Cambridge
Shell Petroleum Company Ltd.
The Director of the Science Museum, South Kensington, London for the following items from the photographic library:
Crown copyright, Science Museum, London, negative numbers 213/50, 214/50, 4589, and 7067
Photographs, Science Museum, London, negative numbers 88/46, 89/46, 83/65, and 87/46
Lent to the Science Museum, London, by the late Sir J. J. Thomson, negative number 375
United Kingdom Atomic Energy Authority.

Permission to reproduce questions from their past papers has been kindly granted by the secretaries of the following examination bodies:

Cambridge Colleges' Joint Examination
Oxford Colleges Admissions Office
Associated Examining Board (AEB)
University of London Schools Examinations Council
Oxford Local Examinations
Oxford and Cambridge Schools Examination Board
Joint Matriculation Board
University of Cambridge Local Examinations Syndicate
Welsh Joint Education Committee (WJEC)

To my students at Slough, Braintree, and Folkestone, whose enthusiasm and skill in probing into difficult areas has sustained and encouraged my interest and enthusiasm for continued enquiry into the difficult but rewarding search for a consistent and coherent systematization of the physical world.

1. Gases and the Kinetic Theory of Matter

TRADITIONALLY, matter has been recognized to exist in three states: gaseous, liquid, and solid. Although some systems (such as helium at temperatures close to the absolute zero of temperature, and the so-called 'liquid-crystals') seem to demand the recognition of additional states of matter, it is possible to interpret the vast majority of observations made on material systems in terms of this traditional tripartite division. In this chapter we shall be concerned with the gaseous state and in particular with the development of the Kinetic Theory of Matter which grew mainly out of the investigations made on the behaviour and properties of gases.

The effect of pressure on the volume of gases

This effect can be conveniently studied in the apparatus illustrated in Figure 1.1. By altering the relative heights of the two limbs of the apparatus the volume of the fixed mass of gas trapped in the left-hand tube of the assembly can be recorded under a range of pressures. It will be observed that the temperature is kept constant throughout the experiment.

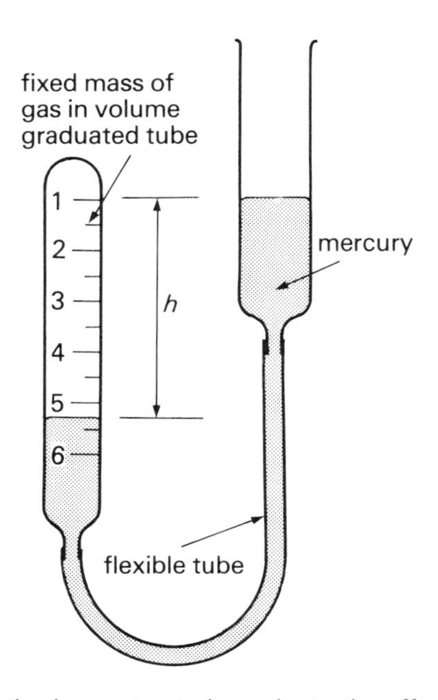

Figure 1.1 Apparatus to investigate the effect of pressure on the volume of a gas at constant temperature

Pressure of gas/mmHg
$$= \text{atmospheric pressure/mmHg} + h/\text{mmHg}$$

If the results are plotted graphically, the results shown in Figure 1.2 are obtained. Investigations of this type were carried out by Robert Boyle who published his results in 1662. His conclusions are stated in *Boyle's Law*:

The volume of a fixed mass of gas (V) *is inversely proportional to the pressure* (p), *provided the temperature remains constant.* Stated mathematically:

$$V \propto 1/p \quad \text{or} \quad pV = \text{constant}$$

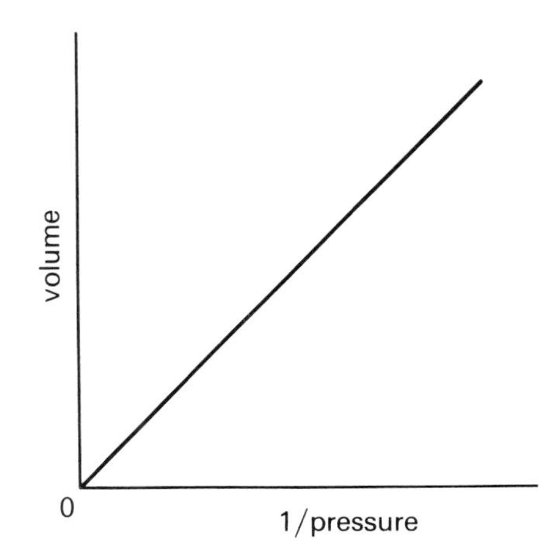

Figure 1.2 Variation of volume with pressure of a gas at constant temperature

The effect of temperature on the volume of gases

The apparatus illustrated in Figure 1.3 can be used to investigate this effect. Corresponding values of volume and temperature are noted as the water is heated electrically. The height of the right-hand limb of the apparatus is adjusted so that the mercury level is identical in both sides of the apparatus, so ensuring that all the volume measurements are made under atmospheric pressure. The results are plotted graphically as in Figure 1.4.

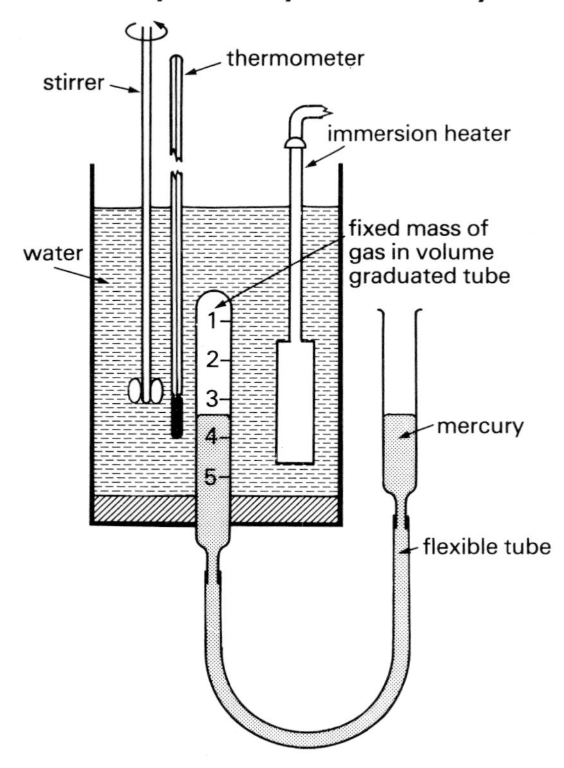

Figure 1.3 Apparatus to investigate the effect of temperature on the volume of a gas at constant pressure

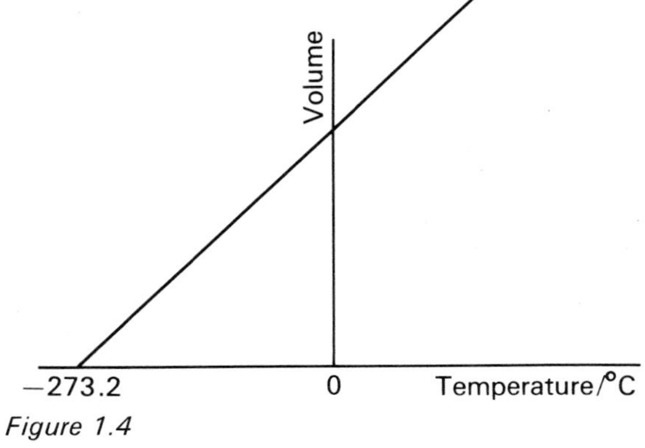

Figure 1.4

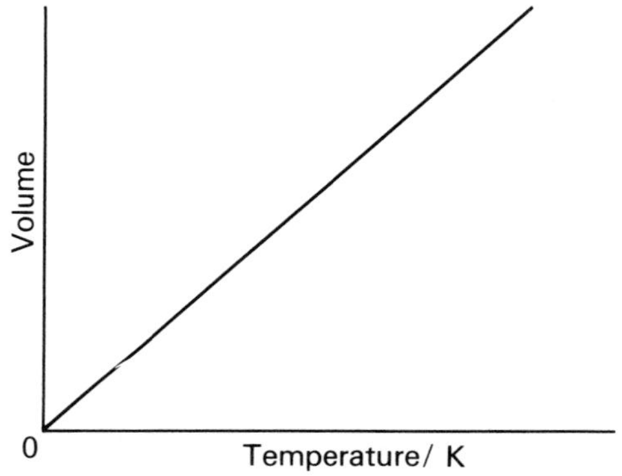

Volume
0
Temperature/ K

Figure 1.5

If the graph is extrapolated to find the temperature at which the volume would be zero, this is found to be −273.2 °C. This finding is independent of the nature and quantity of the gas employed. It is convenient to define a new temperature scale on which this temperature is taken as zero, each unit of the scale being identical with the Celsius scale. It is known as the Kelvin scale and its unit is the Kelvin. When the experimental results are replotted with the temperatures in Kelvin, the graph in Figure 1.5 is obtained. This effect was investigated by Jacques Charles in 1787 and his conclusion, in modern wording, is contained in the following statement of *Charles' Law*:

The volume of a fixed mass of gas (V) is directly proportional to its Kelvin temperature (T), provided the pressure remains constant. Stated mathematically:

$$V \propto T \quad \text{or} \quad V/T = \text{constant}$$

Ideal and real gases

More accurate measurements of these effects show that real gases obey the laws of Boyle and Charles with only limited accuracy. The importance of these deviations from the gas laws will be discussed later in this chapter. A useful concept is that of the 'ideal gas'. Such a gas is one that obeys the gas laws with complete accuracy.

The general gas equation

This can be obtained by combining the mathematical statements of Boyle's and Charles' laws:

$$\frac{pV}{T} = k \quad \text{or} \quad pV = kT \quad \text{(where } k \text{ is a constant)}$$

The value of k depends on the amount of gas considered, and for the special case of 1 mole of gas the symbol R is employed for the constant, i.e.

$$pV = RT \text{ (for 1 mole of gas)}$$

Hence for n moles of gas:

$$pV = nRT$$

or

$$pV = \frac{m}{M}RT$$

(where m is the mass and M the molar mass). This can be transposed to give:

$$M = \frac{m}{V} \times \frac{RT}{p}$$

$$= \rho \times \frac{RT}{p}$$

(where ρ is the density of the gas)

Thus, given a value for the Gas Constant, R, if the density of a gas is known under stated conditions of temperature and pressure, its molar mass can be calculated.

Standard temperature and pressure (s.t.p.)

It is convenient to define a standard temperature and pressure for use when quoting the volumes and densities of gases. The pressure of 1 atmosphere (760 mmHg; 101.3 kPa) and the temperature of 0 °C (273.2 K) have been accepted for this purpose.

The determination of gas densities

The classical methods of gas density determination associated with the names of Victor Meyer, Regnault, and Dumas have now been superseded by the direct weighing technique described below, the use of gas syringes heated in a suitable furnace, and the gas micro-balance which will be discussed later in this chapter. Details of the older methods will be found in more traditional textbooks, such as *Physical Chemistry* by A. J. Mee (London: Heinemann Educational Books, 1962), or *Advanced Level Physical Chemistry* by A. Holderness (revised by J. Lazonby) (London: Heinemann Educational Books, 1976).

Gas densities by direct weighing

The gas-tight flask with fittings as illustrated in Figure 1.6 is connected via tap B to a vacuum system and exhausted with tap A closed and tap B

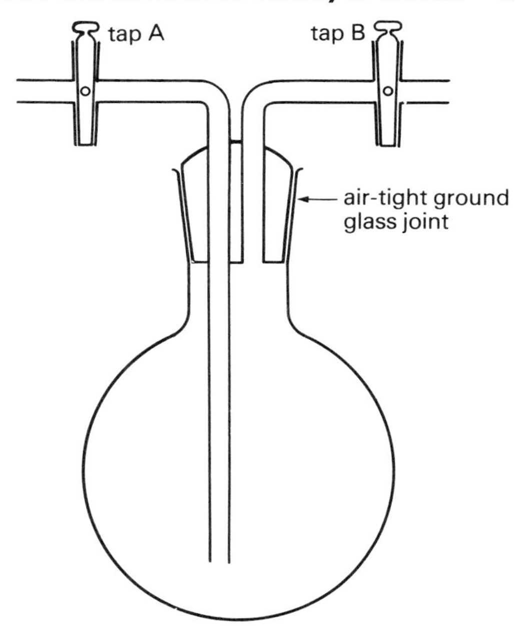

Figure 1.6 Apparatus for the direct weighing of gases

open. Tap B is closed, the flask is disconnected and weighed. A suitable source of dry gas is connected to tap A and the flask filled with both taps open. The taps are closed and the flask is disconnected and weighed. The difference in weighings gives the mass of gas contained in the flask. The flask is filled with water and weighed once more. The mass of water it will hold is found by subtraction and, knowing that 1 cm^3 of water has a mass of 1 g, the volume capacity of the flask obtained. Dividing the mass of gas by its volume gives the density under the laboratory conditions of temperature and pressure. If the density of gas at s.t.p. is required, the volume can be corrected to its value under these conditions by applying the laws of Boyle and Charles.

From the results of such experiments, if the relative molar mass of the gas is known, the volume occupied by one mole of gas can be calculated. This quantity is known as the molar volume of the gas.

It will be seen that the molar volumes of these real gases, at s.t.p., is almost constant. The accepted value for this constant for an ideal gas is 22.41 $dm^3\ mol^{-1}$ and it will be seen that real gases have constants approaching this value with reasonable accuracy. A further discussion of deviation from ideality will be given later in the chapter.

Table 1.1 The molar volumes of some gases

Gas	*Relative molar mass*	*Density at s.t.p./g dm^{-3}*	*Molar volume at s.t.p./ dm^3 mol^{-1}*
Carbon monoxide	28.01	1.251	22.40
Carbon dioxide	44.01	1.977	22.26
Chlorine	70.90	3.215	22.06
Helium	4.00	0.178	22.46
Hydrogen	2.02	0.090	22.42
Nitrogen	28.01	1.251	22.40
Oxygen	32.00	1.429	22.39
Sulphur dioxide	64.06	2.927	21.89

It has been shown previously that for 1 mole of gas

$$R = \frac{pV}{T}$$

Substituting suitable values in this equation enables us to evaluate the constant, R:

$$R = \frac{(1\ \text{atmosphere}) \times (22.41\ \text{dm}^3)}{(273.2\ \text{K})}$$
$$= 0.08205\ \text{atmosphere dm}^{-3}\ \text{K}^{-1}$$

As discussed in the introduction, the atmosphere is not the SI unit of pressure and thus this value of R is not in SI units. It is a form of expressing R, however, which is convenient for calculations using the gas equation.

The determination of relative molecular masses (formerly called molecular weights)

The relative molecular masses of gases can be determined by the direct weighing method described earlier. An elegant technique to supplement this was developed in 1910 by W. Ramsey and R. Whytlaw-Gray to determine the relative molecular mass of radon gas. This is suitable for use when very small quantities of gas are available; the apparatus can be as small as 2 × 8 cm and is known as the microbalance. A diagrammatic illustration of the apparatus is shown in Figure 1.7. The balance is first evacuated and the quartz bulb falls as the buoyancy force of the air displaced no longer applies. A gas of known density is then slowly admitted until the pointer reaches a predetermined position on the scale and the pressure noted on the manometer. The apparatus is evacuated again and

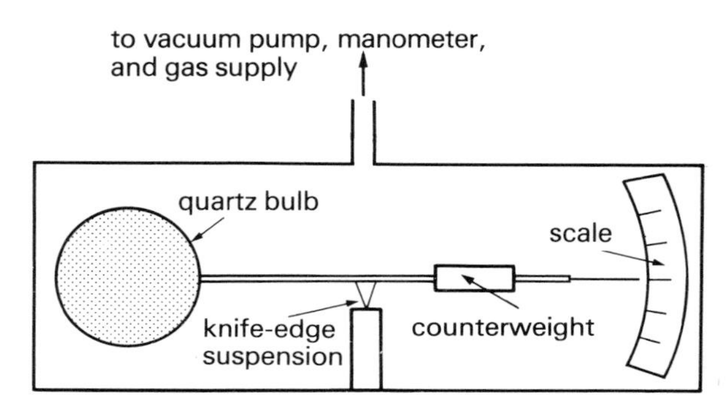

Figure 1.7 The microbalance

the gas of unknown density is admitted slowly until the pointer reaches the same position on the scale; the pressure is again noted.

Now for the standard gas of known density:

$$M(\text{s}) = \rho(\text{s}) \times \frac{RT}{p_\text{s}}$$

and for the gas of unknown density:

$$M(\text{u}) = \rho(\text{u}) \times \frac{RT}{p_\text{u}}$$

When the two pressures were recorded, the buoyancy force on the quartz bulb was equal and hence $\rho(\text{u}) = \rho(\text{s})$ we can thus combine these two equations:

$$\frac{M(\text{s})}{M(\text{u})} = \frac{p_\text{u}}{p_\text{s}}$$

Hence as $M(\text{s})$, p_u and p_s are known, substitution allows the calculation of $M(\text{u})$.

Relative molecular masses of liquids

The relative molecular masses of liquids can be determined if a known mass is vaporized and its volume determined under known conditions of temperature and pressure. A convenient apparatus for this purpose is shown in Figure 1.8. A gas syringe is contained in a suitable furnace which maintains it at a steady temperature at least 10 °C above the boiling point of the liquid under test. This furnace may be electrically heated or the vapour of a suitable liquid, usually water, may be passed continuously through it. The nozzle of the gas syringe is fitted with a self-sealing rubber cap. The apparatus is left until thermal equilibrium is reached between the furnace interior and the 15 cm^3 or so of air the syringe contains initially. A smaller syringe fitted with a hypodermic needle is filled with the liquid under test, some of which is injected into the gas syringe and the small syringe is reweighed to give the mass of liquid injected. The liquid vaporizes in the gas syringe and its volume at the temperature of the furnace and under laboratory pressure can be recorded. This procedure is repeated using different masses of liquid and the results plotted graphically. Graphs of the type shown in Figure 1.9 are obtained. These results were obtained at a temperature of 100 °C (373 K) and 756 mmHg (756/760 atmosphere).

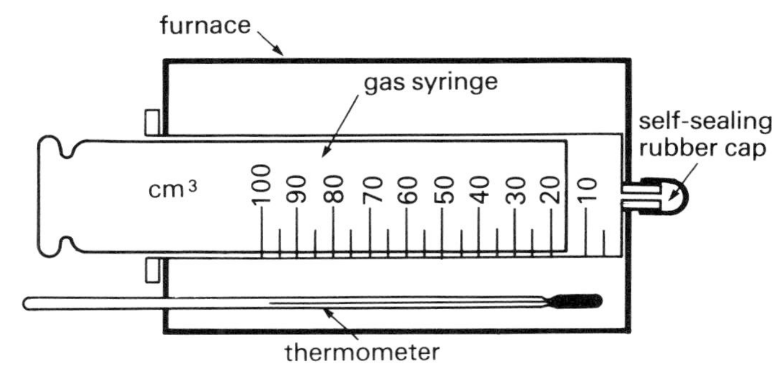

Figure 1.8 Apparatus to determine relative molecular masses of liquids

The reciprocal of the slope of each curve is the density of the vapour under the temperature and pressure at which the measurements were made. Hence the density of propanone vapour under these conditions is 0.0019 g cm^{-3}, or 1.9 g dm^{-3}. We can substitute this value in the equation:

$$M = \rho \times \frac{RT}{p}$$

Figure 1.9 Graph from which the densities of vapours can be deduced

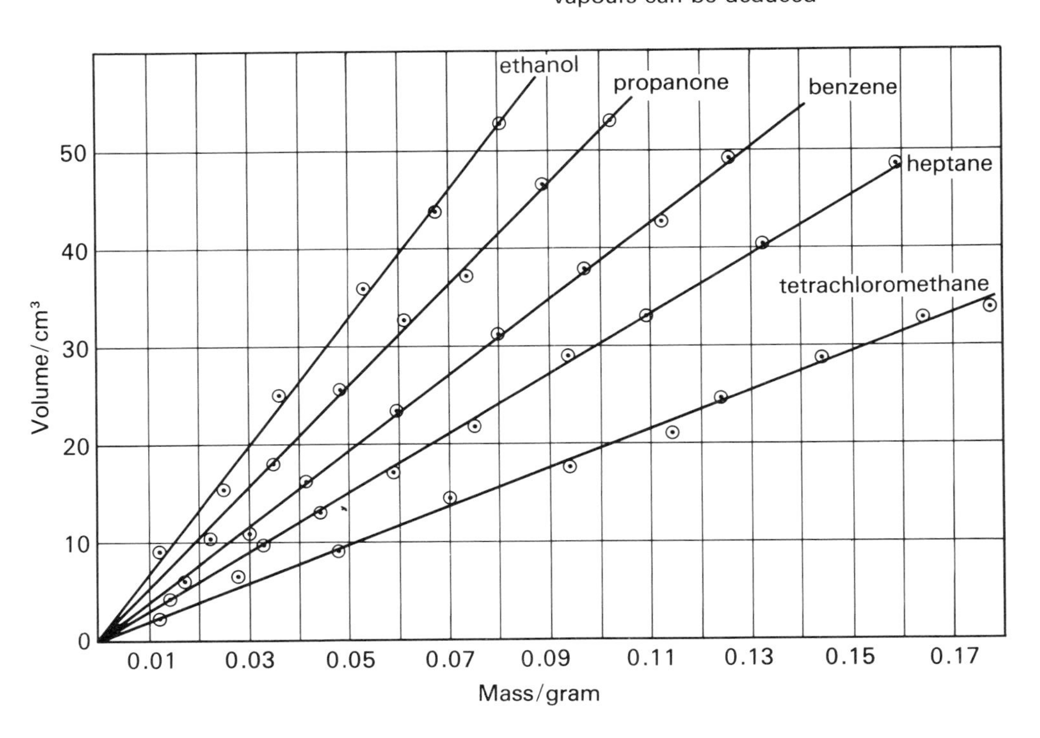

remembering that the units used for other quantities are consistent with those for R.

$$M = (1.9\,\text{g}\,\text{dm}^{-3}) \times \frac{(0.082\,\text{atm}\,\text{dm}^3\,\text{K}^{-1}) \times (373\,\text{K})}{(756/760\ \text{atm})}$$

$$= 1.9 \times 0.082 \times 373 \times 760/756\ \text{g}$$

$$= 58\ \text{g}$$

thus $M_r = 58$

Alternatively, the SI value of R can be used. When this is done, it is not necessary to insert units as they will be consistent by definition. If the SI value is used, however, the pressure and the density must be inserted in their basic SI units.

$$760\ \text{mmHg} = 101\,300\ \text{Pa}$$

thus $756\ \text{mmHg} = 100\,800\ \text{Pa} = 1.01 \times 10^5\ \text{Pa}$

$$1.90\ \text{g dm}^{-3} = 1.90\ \text{g dm}^{-3} \times 10^{-3}\ \text{kg g}^{-1} \times 10^6\ \text{dm}^3\ \text{m}^{-3}$$

$$= 1.90 \times 10^3\ \text{kg m}^{-3}$$

Substitution in the general gas equation gives:

$$M = \frac{1.90 \times 10^3 \times 8.31 \times 373}{1.01 \times 10^5}$$

$$= 58$$

If such a series of results is not available an individual pair of values of mass and volume may be substituted in the equation; however, the averaging of errors available by using the graphical method is lost. An alternative method of carrying out these calculations is to correct the volume of vapour to s.t.p. and then find the mass of vapour that would occupy 22.4 dm^3 under these conditions. This mass is, of course, the relative molar mass.

The composition of water

Water was regarded as an element until towards the end of the eighteenth century when Joseph Priestley observed the formation of water when mixtures of air and hydrogen were exploded by electrical detonation in a closed vessel. Henry Cavendish, in 1781, measured the volume proportions in which hydrogen and oxygen combine during the formation of water and found them to be 201 : 100.

Jean Dumas, in 1842, investigated the gravimetric composition of water by passing carefully purified hydrogen over heated black copper oxide and measuring the loss of mass of the copper oxide due to removal of oxygen and the mass of the water formed. He found that the gravimetric proportions in which hydrogen and oxygen combined to form water was 7.98 : 1. Little was known at that time about the proportions in which atoms combined either with themselves in pure elements or with the atoms of other elements in compounds. Hence it was assumed that, in terms of John Dalton's Atomic Theory of 1802, that one atom of hydrogen combined with one atom of oxygen to form one 'compound-atom' (in modern language one molecule) of water. The formula of water was thus declared to be HO and the atomic mass of oxygen (on the scale H = 1) to be 8.

It was difficult to reconcile these conclusions with the volumetric investigations made earlier; particularly as Dalton, in an extension of his Atomic Theory, postulated that the volumes occupied by the atoms of all gases are equal when measured under the same conditions of temperature and pressure. A modification to Dalton's postulate by Amadeo Avogadro of Italy brought the solution to such problems.

Avogadro's hypothesis states: *equal volumes of all gases, measured under the same conditions of temperature and pressure, contain the same number of* **molecules**. Thus the observed facts of both the volumetric and gravimetric composition of water could be explained. We would nowadays write the equation for the formation of water as:

$$2H_2(g) + O_2(g) = 2H_2O(l)$$

The difficulty of applying Dalton's theory disappears when instead of having to consider the ultimate particles in gaseous elements to be *atoms*, Avogadro's interpretation allows us to regard the ultimate particles in such gaseous elements as hydrogen and oxygen to be *diatomic molecules*. Avogadro's hypothesis is also consistent with the observation that one mole of gases have a volume of about 22.4 dm^3 at s.t.p., for one mole of any gas contains the same number of molecules. This number is known as the Avogadro or Loschmidt number, and given the symbol L. Methods for determining the Avogadro number will be discussed in later chapters.

The volume proportions in which gases react

This aspect of reactions between gases was studied by many investigators in the eighteenth and nineteenth centuries. It is possible to duplicate their experiments today, using gas-syringe techniques, with an ease that these investigators would envy.

Ammonia gas is known to be composed of nitrogen and hydrogen. If a fixed volume of ammonia is passed between two gas syringes over a heated iron catalyst, it is virtually decomposed into its elements. The volume of the nitrogen/hydrogen mixture so formed is double that of the original ammonia. If this mixture is passed over heated copper oxide, the hydrogen in it combines with oxygen from the solid to form water which on cooling occupies negligible volume. The residual gas, which is nitrogen, is found to have a volume of one-third that of the nitrogen/hydrogen mixture. Thus

two volumes of ammonia → one volume of nitrogen + three volumes of hydrogen

Assuming the diatomicity of nitrogen and hydrogen, this is consistent with the equation:

$$2NH_3(g) \rightarrow N_2(g) + 3H_2(g)$$

From a study of the results obtained by his own investigations and those of others, the French chemist Joseph Gay-Lussac published his Law of Volume Combinations in Gases. This can be stated in the form: *when gases combine, the volume proportions in which they do so are in simple numerical ratios to each other and to those of any gaseous products. All volumes must be measured under the same conditions of temperature and pressure.*

Determination of relative atomic masses (formerly called atomic weights)

In 1858 the Italian chemist Stanislao Cannizaro demonstrated how the work of Avogadro particularly could form the basis,of a powerful method of deducing relative atomic masses. His method can be applied only to those elements that form a reasonably wide range of volatile compounds and was most useful in fixing the relative atomic masses of the non-metallic elements such as carbon. Examination of the data in Table 1.2 will make his method clear.

Table 1.2 Cannizzaro's method applied to carbon

Compound	*Relative molar mass*	*% carbon*	*Mass of carbon in one mole/g*
Carbon dioxide	44	23.7	12
Carbon monoxide	28	42.9	12
Cyanogen	52	46.2	24
Ethane	30	80	24
Ethene	28	85.7	24
Ethyne	26	92.3	24
Methane	16	75	12
Propane	44	81.8	36
Benzene	78	92.3	72
Butane	58	82.7	48
Trichloromethane	119.5	10	12

Since the smallest mass of carbon in one mole of any of the above compounds is 12 and the other masses of carbon per mole are whole number multiples of this, the relative atomic mass of carbon is likely to be 12.

The use of the mass spectrometer has, in recent years, rendered these classical methods of determining relative molar masses and atomic masses obsolete. The principle of the mass spectrometer is discussed in the next chapter.

The kinetic theory of matter

In 1827, the English botanist Robert Brown was observing a suspension of pollen grains in water through a microscope. He was interested to note that the pollen grains exhibited constant and random motion. This type of motion, now known as Brownian motion, was also observed when other relatively small particles were suspended in a liquid or a gas. An explanation of this effect was suggested in 1863 by Wiener. He explained the random motion as due to the continual bombardment of the particles by the molecules of the material in which they were suspended. The motion of the invisible molecules is thus made manifest by its effect on the suspended particles.

Out of such observations the Kinetic Theory, the theory of molecules in motion, gradually emerged. The postulates on which the quantitative interpretation of the Kinetic Theory is based are:

(*a*) gases consist of large assemblies of molecules in which the distance between the molecules is very much greater than their diameters;
(*b*) there are no attractive or repulsive forces between the molecules;
(*c*) the molecules are in a rapid state of random motion in all directions;
(*d*) the temperature of the gas is a function of the average kinetic energy of the molecules.

It is possible to apply the principles of Newtonian mechanics to these postulates to obtain an equation relating the motion of the molecules to the measurable parameters volume, pressure, and temperature.

In mechanics a distinction is made between speed and velocity. The speed of a body is the rate of change of position with time without specification of direction; it is said to be a *scalar* quantity. Velocity is the rate of change of position in a specific direction; it is said to be a *vector* quantity. Hence the average velocity of the molecules in a sample of gas is zero. (If they are regarded as travelling in every possible direction, the velocity of any one molecule being balanced on average by that of another molecule travelling at the same speed and in a direction 180° to it.) For this reason it is necessary to define the quantity *root mean square velocity*, given the symbol $\overline{c^2}$. Consider n molecules which at a given instant have velocities c_1, c_2, $c_3, \ldots c_n$. The root mean square velocity is given by:

$$\overline{c^2} = \sqrt{[c_1{}^2 + c_2{}^2 + c_3{}^2 + \ldots + c_n{}^2] \times \frac{1}{n}}$$

This device makes all the velocity values positive by squaring the negative values, and hence the average is not zero.

Let these n molecules be contained in a cubic vessel of side l metres and let the mass of each molecule be m grams. The root mean square velocity will be in m s^{-1}. Consider one of these molecules, of velocity c_1, whose motion and velocity is represented by the line OA on Figure 1.10. Its velocity can be resolved into the three components x, y, and z represented by the lines Ox, Oy, and Oz which are parallel to the walls of the container.

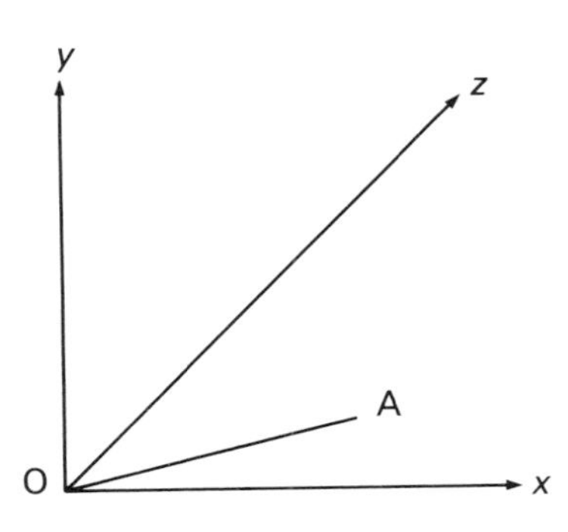

Figure 1.10 The motion of a molecule in a cubical container

Thus

$$c_1{}^2 = x^2 + y^2 + z^2$$

Consider the component of the motion of the molecule along the axis Ox. On collision with one wall of the containing vessel, which it strikes normally, it is reflected in a direction 180° to its original motion by this collision and it suffers a change of momentum:

$$mx - (-mx) = 2mx$$

Travelling x metres in one second, the molecule makes one impact with a wall for each distance of l metres travelled, and it suffers x/l changes of direction per second. The change of momentum per second is therefore:

$$\frac{x}{l} \times 2mx = \frac{2mx^2}{l}$$

Similarly its change of momentum per second along the y and z axes is given by $2my^2/l$ and $2mz^2/l$. Thus the total change of momentum per second of the molecule is:

$$\frac{2m}{l}(x^2 + y^2 + z^2) = \frac{2mc_1{}^2}{l}$$

According to Newton's second law of motion, the force exerted by the impact of a body on a surface is equal to the rate of change of momentum of that body. Thus the force exerted on the walls of the containing cube by our molecule is $2mc_1{}^2/l$. The total force exerted by the n molecules is:

$$\frac{2m}{l}(c_1{}^2 + c_2{}^2 + c_3{}^2 + \ldots + c_n{}^2) = \frac{2nm\overline{c^2}}{l}$$

(Collisions between molecules do not result in change of total momentum and so can be neglected in this argument.)

Pressure is force per unit area and the total area of the six faces of our cube is $6l^2$, hence the total pressure of the gas is:

$$\frac{2nm\overline{c^2}}{l} \times \frac{1}{6l^2} = \frac{2nm\overline{c^2}}{6l^3}$$

and as l^3 is the volume of the container, this becomes:

$$pV = \tfrac{1}{3}nm\overline{c^2}$$

or for one mole of gas containing Avogadro's number of molecules (L),

$$pV = RT = \tfrac{1}{3}Lm\overline{c^2}$$

We can use this relationship to demonstrate the connection between the temperature and the kinetic energy of the molecules,

$$RT = \tfrac{1}{3}L m\bar{c}^2 = \tfrac{2}{3} \times \frac{L m\overline{c^2}}{2}$$

But $Lm\bar{c}^2/2$ is the total kinetic energy (K.E.) of the molecules.
Hence,

$$RT = \tfrac{2}{3}\ \text{K.E.}$$

$$T = \tfrac{2}{3}\ \text{K.E.} \times \frac{1}{R}$$

Thus the total, and therefore the average, kinetic energy of the molecules is directly proportional to the temperature.

Root mean square velocities of gaseous molecules

We can use the equation $RT = \frac{1}{3}Lm\overline{c^2}$ to calculate the root mean square velocities of gases under known temperature conditions.

Table 1.3 Root mean square velocities of some gases at 0 °C

Gas	*Root mean square velocity at 0 °C/m s⁻¹*
Argon	410
Carbon dioxide	390
Hydrogen	1838
Nitrogen	493
Oxygen	460

The atomicities of gases

In earlier work in this chapter the diatomicity of many common elemental gases has been assumed. It is now possible to present evidence in support of this assumption.

(1) *Evidence from Gay-Lussac's law and Avogadro's hypothesis* When the combination of hydrogen and chlorine to form hydrogen chloride is investigated volumetrically, it is found that:

1 volume of hydrogen + 1 volume of chlorine →
2 volumes of hydrogen chloride

Applying Avogadro's hypothesis,

n molecules of hydrogen + n molecules of chlorine → $2n$ molecules of hydrogen chloride

Thus

1 molecule of hydrogen + 1 molecule of chlorine →
2 molecules of hydrogen chloride

Each molecule of hydrogen chloride must thus contain *half* a molecule of hydrogen and thus the hydrogen molecule must contain an even number of hydrogen atoms. Additional evidence that this even number is two can be deduced from the observation that hydrogen chloride dissolves in water to form a monobasic acid (i.e. each mole of acid contains one mole of replaceable hydrogen).

(2) *Evidence from the Kinetic Theory* It can be shown, as a consequence of the kinetic theory, that the ratio

$$\gamma = \frac{\text{specific heat of a gas at constant pressure}}{\text{specific heat of the gas at constant volume}}$$

is, theoretically, 1.67 for a monatomic gas and 1.40 for a diatomic gas. The measured value for this ratio for hydrogen is 1.40.

Deducing the gas laws from the kinetic theory

Boyle's law is directly deducible from the kinetic theory as $Lm\overline{c^2}$ is a constant at any one temperature and hence:

$$pV = \tfrac{1}{3}Lm\overline{c^2} = \text{constant}$$

Charles' law is involved in deducing the direct relationship between the temperature of a gas and the average kinetic energy of the molecules and this demonstrates the fundamental nature of this law.

Two other gas laws, originally established on an empirical basis, can be deduced from the kinetic theory of gases.

(*a*) *Dalton's law of partial pressures* We have seen that, in terms of the kinetic theory, the pressure of a gas can be regarded as due to the forces exerted on the walls of the containing vessel due to their change of momentum on collision. Thus if a mixture of different types of molecule is contained in the vessel, each component of the mixture contributes its partial pressure to the total pressure independently of the contribution of the other components. Hence the total pressure (p_{total}) is equal to the sum of the partial pressures of the components ($p_1, p_2, p_3, \ldots, p_n$).

$$p_{\text{total}} = p_1 + p_2 + p_3 + \cdots + p_n$$

(*b*) *Graham's law of diffusion (or effusion)* Diffusion in a gas is the tendency of that gas to fill the entire space available to it. If a gas is held in a container with an outlet into an environment at lower pressure, because of diffusion the gas will pass from the container to the outside until the pressures inside and outside the container are equalized. This process is defined as effusion. Applying the quantitative statement of the kinetic theory in the form:

$$pV = \tfrac{1}{3}Lm\overline{c^2}$$

wee see that $\overline{c^2} \propto 1/m$ and as the rate of diffusion of a gas is proportional to the average speed of the molecules,

$$\text{rate of diffusion (or effusion)} \propto \frac{1}{\sqrt{\text{molar mass}}}$$

This relationship was first established experimentally by Thomas Graham in 1829. Experimental evidence for the law can be obtained using the apparatus shown in Figure 1.11. A supply of dry gas is attached to the apparatus via the three-way tap as shown and used to sweep air out of the gas syringe. When this has been accomplished, the tap is adjusted to connect the syringe to the lower tube and the time taken for a suitable volume of gas to effuse from the apparatus via the pin-hole aperture is noted. The experiment is then repeated for other gases.

Specimen results are shown in Figure 1.12. The linear plot, passing through the origin, shows experimental agreement with the theoretical prediction.

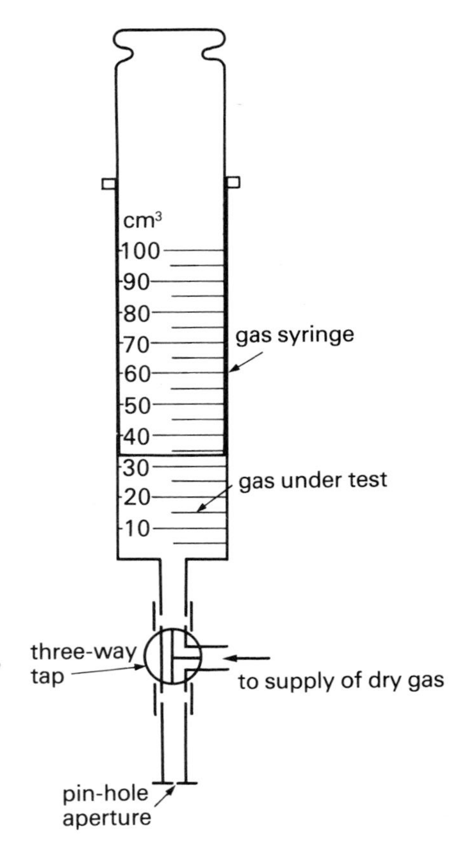

Figure 1.11 Apparatus to investigate the rates of effusion of gases

Real gases and deviations from the gas laws

From the mathematical statement of Boyle's law, pV = constant, we should expect a graph of the product of pressure and volume, for a fixed mass of gas at constant temperature, plotted against pressure to be linear and parallel to the x axis. In Figure 1.13, this plot is made for sulphur dioxide and it can be seen that considerable deviation from this expectation occurs. This is due to the fact that real gases do not show behaviour in complete accord with the postulates of the kinetic theory in two important respects:

(1) There is not zero attraction between the molecules of the gas, the bulk of the gas molecules exert forces on those on the outside of the gas having the effect of pulling these latter molecules away from the wall of the container. The effective pressure is thus less than the ideal pressure postulated by the kinetic theory. This 'internal pressure' demands that a correction term be added to the measured pressure. The forces responsible for the internal pressure of a gas are known as van der Waals' forces, a name whose significance will become apparent later.

(2) The volumes of the molecules are not completely negligible in comparison with the total volume of the gas. The effective volume of gas is thus the measured volume less a correction term for the volumes of the molecules.

In 1873, Johannes van der Waals proposed the following modification to the gas equation to take account of these considerations:

$$\left(p + \frac{a}{V^2}\right)(V - b) = RT$$

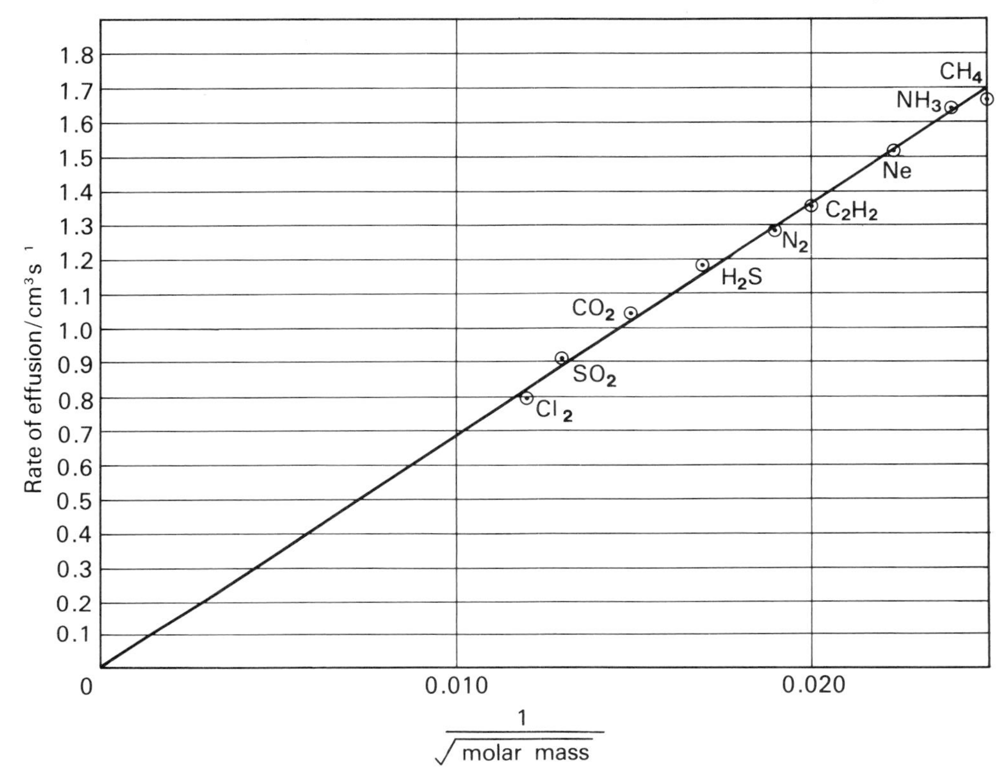

Figure 1.12 Rates of effusion of some gases

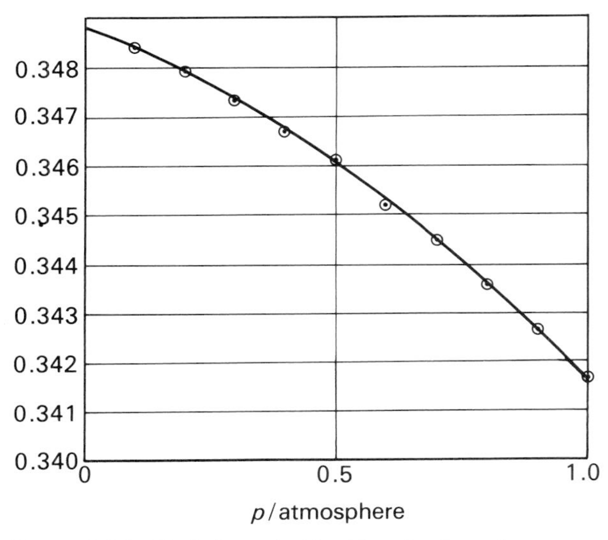

Figure 1.13 Variation of pV with p for 1 g of sulphur dioxide at 0 °C

Van der Waals' equation is one of several equations of state for real gases.

It will be appreciated that these two effects have their greatest influence at high pressures and that real gases behave most ideally at low pressures. In the graphical data for 1 g of sulphur dioxide in Figure 1.13, therefore, the value for the product pV closest to that which would be observed at all pressures if the gas were ideal, is the value obtained by extrapolation to zero pressure. This is the basis of the so-called 'limiting densities' method for the accurate determination of molar masses.

Molar masses of gases by limiting densities

Using the extrapolated value of 0.3488 atm dm^3 for pV at zero pressure from Figure 1.13, we can substitute this and other appropriate data into the gas equation to compute an accurate value for the molar mass of sulphur dioxide:

$$pV = \frac{m}{M}RT$$

$$(0.3488 \text{ atm dm}^3) = \frac{1\text{g}}{M}(0.08205 \text{ atm dm}^3 \text{ K}^{-1}) \times (273.2 \text{ K})$$

thus,

$$M = \frac{0.08205 \times 273.2}{0.3488}\text{ g}$$
$$= 64.27 \text{ g}$$

The relative molar mass of sulphur dioxide, from these data, is 64.27.

Molar masses of gases from van der Waals' equation

The coefficients a and b in van der Waals' equation can be calculated from experimental data not directly involving gas density measurements. If such values are known for a particular gas together with accurate data concerning its density, the molar mass can be calculated with considerable precision. The worked example below should make the principle clear.

Baxter and Starkweather determined the density of nitrogen as 0.4167 g dm^{-3} at 253.3 mmHg and 0 °C. The ratio of density/pressure was found not to vary detectably at pressures up to one atmosphere. The van der Waals' coefficients for nitrogen are: $a = 1.390$ atm dm^6 mol^{-2} and $b = 0.03913$ dm^3 mol^{-1}.

For n moles of gas, van der Waals' equation takes the form:

$$\left(p + \frac{n^2a}{V^2}\right)(V - nb) = nRT$$

Substituting the values given (the units have been omitted for clarity),

$$\left(\frac{253.3}{760} + \left[\frac{0.4167}{M}\right]^2 \times 1.390\right) \times \left(1 - \frac{0.4167}{M} \times 0.03919\right) = \frac{0.4167}{M} \times 0.08205 \times 273.2$$

$$\left(0.3333 + \frac{0.2414}{M^2}\right) \times \left(1 - \frac{0.01633}{M}\right) = \frac{9.341}{M}$$

$$0.3333 + 0.2414/M^2 - 0.005443/M - 0.003942/M^3 = 9.341/M$$

$$0.3333M^3 - 9.341M^2 + 0.2414M - 0.003942 = 0$$

Using the Newton–Raphson method of iteration the three roots of this cubic equation were computed. The value of 28.00 for one of these roots is the only acceptable value for the molar mass of nitrogen, the other two roots being fractional.

The movement of molecules

The motion of a molecule can be regarded as having translational, rotation, and vibrational components. These three types of motion are illustrated for a diatomic molecule in Figure 1.14.

The distribution of molecular energies

An elegant method for the direct determination of the distribution of molecular translational energies is illustrated in Figure 1.15. A sample of tin is heated in the furnace to a sufficiently high temperature to atomize it. A stream of tin atoms emerges from the furnace exit and passes up the apparatus in a collimated pencil-like stream after passing through the collimating screens. The apparatus is evacuated to eliminate the collisions between air molecules and tin atoms which would otherwise cause the latter to be deflected from a straight line path. On meeting the rotating drum, the stream of tin atoms is only admitted when the slit points directly at the holes in the collimating screens. When

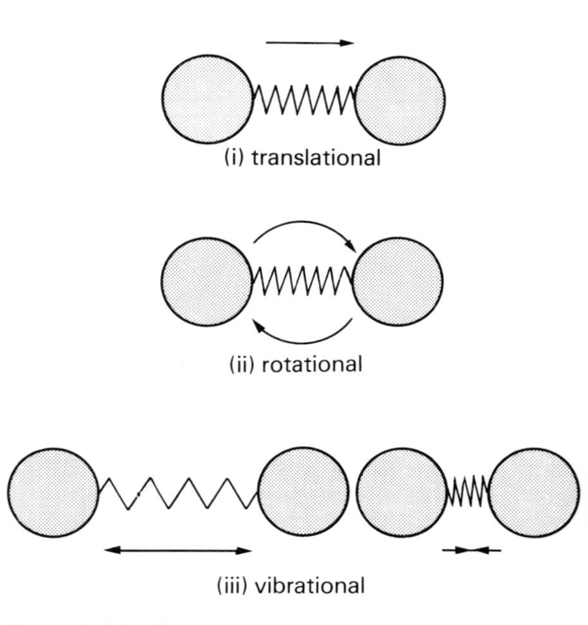

Figure 1.14 The three components of molecular motion

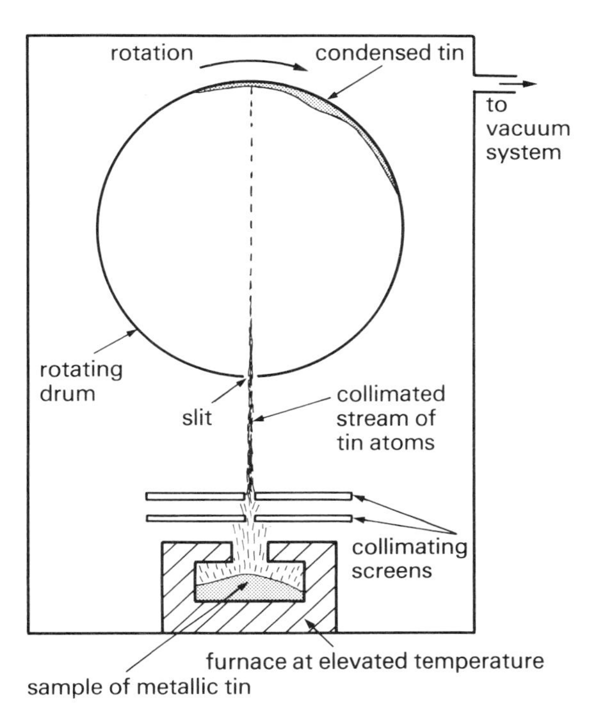

Figure 1.15 Zartmann's apparatus for investigating the distribution of molecular velocities

the tin enters the drum it travels across the vertical diameter. The atoms travelling fastest reach the back of the drum, on which they condense, first; the slowest atoms reach the back of the drum later and thus condense at a different place. After several revolutions of the drum, the thickness of the metal condensed at different points can be measured and from the dimensions of the apparatus and the speed of rotation of the drum, the fraction of the molecules with each speed can be calculated. The experiment is repeated with the furnace at other temperatures and the results plotted to give graphs of the type shown in Figure 1.16.

By applying statistical techniques to the kinetic theory, Boltzmann in 1886 was able to deduce the distribution of molecular velocities in a sample of gas without recourse to experimental measurements. His conclusion can be stated in the form:

$$\text{Fraction of molecules having energy } E \text{ or greater} = \exp(-E/RT)$$

The further importance of this relationship will be discussed in Chapter 10.

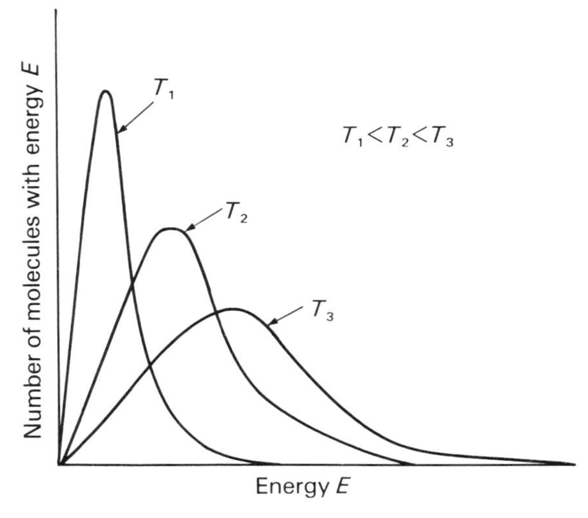

Figure 1.16 Distribution of molecular velocities

Suggestions for further reading

KNIGHT, D. M. *Classical Scientific Papers – Chemistry* (London: Mills and Boon, 1968)

JEANS, J. H. *The Kinetic Theory of Gases* (Cambridge: Cambridge University Press, 1952)

Problems

1. Convert the following volumes of gases to s.t.p.

(i) 27.5 cm^3 of carbon dioxide at 18 °C and 756 mmHg pressure
(ii) 82 cm^3 of nitrogen at 21 °C and 763 mmHg pressure
(iii) 112 dm^3 of hydrogen at 19 °C and 20 atmospheres pressure
(iv) 58 cm^3 of methane at 100 °C and 733 mmHg pressure.

2. The following results were obtained in a direct weighing experiment:

Mass of flask + fittings, evacuated = 183.257 g

Mass of flask + fittings, + gas X = 187.942 g

Mass of flask + fittings, + water = 987.560 g

The experiment was carried out at 23.0 °C and 733 mmHg pressure. Calculate the relative molar mass of gas X.

3. In an experiment using the microbalance of Ramsey and Whytlaw-Gray, carbon dioxide (relative molar mass 44) was first introduced into the balance and the point indicated on the scale noted when the pressure of the gas was 238 mmHg. The apparatus was re-evacuated and gas Y was admitted until the pointer returned to the same position on the scale. This occurred with the pressure of gas Y at 308 mmHg. Calculate the relative molar mass of Y.
4. Use the data in Figure 1.9 to calculate the relative molar masses of (*a*) ethanol, (*b*) benzene, and (*c*) tetrachloromethane.
5. In the table below, data concerning some typical and volatile compounds of an element Z are given.

Compound	*Relative molar mass*	*Percentage Z in the compound*
A	28	57.2
B	64	50.0
C	30	53.3
D	44	36.4
E	108	74.1
F	60	53.3

What value for the relative atomic mass of Z do these results indicate?

6. In an effusion experiment, a certain volume of methane (relative molar mass 16) is found to take 23 seconds to effuse. How long will the same volume of hydrogen chloride (relative molar mass 36.5) take to diffuse under the same conditions?
7. Stating the assumptions you make, estimate the pastial pressures, in mmHg, of (*a*) nitrogen, (*b*) oxygen, and (*c*) carbon dioxide in air at room temperature and pressure.
8. Using the method of limiting densities, calculate a value for the relative molecular mass of chloromethane from the data given below:

Pressure/atmosphere	1	$\frac{2}{3}$	$\frac{1}{2}$	$\frac{1}{3}$	$\frac{1}{4}$
Density/g dm^{-3} at 0 °C	2.307	1.526	1.140	0.7571	0.5666

9. Summarize the postulates or assumptions of the kinetic molecular theory of gases.

   Show how the ready compressibility of gases and the ease of mixing of gases are explainable using these postulates.

   Give two important ways in which real gases deviate from ideal behaviour.

   What is meant by *mean molecular velocity* and *root mean square velocity*. Explain the significance of the latter in connection with Graham's Law of Diffusion.

   Calculate the temperature at which hydrogen has the same root mean square velocity as oxygen has at 273 K, both being at the same pressure. (WJEC 'A')
10. Describe qualitatively in terms of kinetic theory the essential differences between the gaseous, liquid, and solid states.

    Outline one method by which the rates of diffusion of gases can be compared.

    In a diffusion apparatus, a volume of ethanoic (acetic) acid vapour diffused in 7 minutes and, under the same conditions, the same volume of oxygen diffused in 4 minutes. What conclusions can you draw from this information about the particles present in the ethanoic (acetic) acid vapour? (Cambridge 'A')
11. Many attempts have been made to modify the ideal gas law $pV = nRT$ to represent the behaviour of real gases. Explain the physical significance of the terms introduced in the van der Waals' equation of state, and describe how you would attempt to verify this equation experimentally.

    Two flasks of equal volume are connected by a narrow tube (of negligible volume). Initially both flasks are at 27 °C and contain 1.4 moles of an ideal gas, the pressure being 1 atmosphere. One of the flasks is then immersed in a bath at 127 °C while the other is kept at 27 °C. Assuming the volume does not change, calculate the final pressure and the number of moles of gas in each flask.

    (Cambridge Entrance and Awards)

2. Atomic Structure

Before Dalton

SPECULATIONS concerning the possible atomic nature of matter are found in the writings of Hindu and Egyptian philosophers living many centuries before the birth of Christ. In the fifth century B.C. the Greek philosopher Leucippus and his student Democritus strongly advocated the concept that matter was ultimately composed of tiny individual particles, and coined the word 'atmos' for them. In the third century B.C. these ideas were opposed by Aristotle, a thinker of enormous breadth and depth on subjects as diverse as art, astronomy, ethics, mechanics, and politics. Aristotle's thinking proved so influential on the philosophers of the early Christian church that his teachings became mandatory in all academic institutions throughout Christendom until the end of the Middle Ages. The atomic concept was thus lost to man's thought until the Renaissance in the fourteenth century A.D. The revival of the atomic hypothesis can be traced in the writings of Robert Boyle (1627–1691) and Sir Isaac Newton (1642–1727). In Newton's celebrated treatise *Opticks* we read: 'It seems probable to me that God ... formed matter in solid, massy, hard, impenetrable, movable particles ... so very hard as never to wear or break to pieces; no ordinary power being able to divide what God himself made one in the first creation.' Newton's speculation was not founded directly on experimental evidence, the need for which had been advocated strongly by the monk Roger Bacon (1214–1294) who wrote: 'There are two modes of investigation, through argument and through experiment.... Argument does not suffice but experiment does.... Armed with experiment and calculation, science must not be content with facts.... It wants to find out the laws, the causes.'

The first statement of an atomic theory based on experimental evidence was that of John Dalton, an English Quaker, published in 1807. The experimental observations that led Dalton to his statement were the Laws of Chemical Combination, which summarized the then known facts about the proportions by mass in which elements combine to form compounds. These laws are:

1. *The Law of Conservation of Mass* which states that: *no change in the total mass of all the substances involved is observed when a chemical reaction occurs.* By carrying out reactions in sealed glass tubes, Landolt between 1893 and 1908 was able to show that this law held true to 1 part in 10 000 000. (An important amendment to this law will be discussed in Chapter 4.)

2. *The Law of Constant Composition* (Proust, 1802) which states that: *a given chemical compound always contains the same elements in the same proportions by mass no matter how it is prepared.* The law was shown to operate to great accuracy in a series of experiments by Stas in 1860. He prepared and analysed silver chloride from highly purified starting materials in four different ways. He showed that 100 g of silver gave 132.843, 132.848, 132.842, and 132.849 g of silver chloride. The divergence between the greatest and the least of these results is less than 1 part in 20 000.

3. *The Law of Multiple Proportions* (Dalton, 1803) which states that: *when two elements combine to form more than one compound, if a fixed mass of one of these elements is taken, the masses of the other element that combines with it are in a simple numerical ratio.* Copper, for example, forms two oxides. In black copper oxide 4 g of copper are combined with 1 g of oxygen and in red copper oxide 8 g of copper are combined with 1 g of oxygen.

4. *The Law of Reciprocal Proportions* (Richter, 1792) which states that: *the proportions by mass with which two elements separately combine with a third element are in a simple numerical ratio with the proportions by mass with which they combine with each other.* This law is illustrated in Table 2.1 with data for hydrogen, carbon, and oxygen.

Table 2.1

Compound	*Mass of hydrogen/g*	*Mass of carbon/g*	*Mass of oxygen/g*
Methane	1	4	–
Carbon dioxide	–	3	8
Water	1	–	8

Dalton's Atomic Theory

The theory can be stated in modern terms as follows:

1. Atoms are the ultimate, discrete particles of matter which are not subdivided during the course of chemical changes. They cannot be created or destroyed.
2. All atoms of any one element are identical in all respects, including mass, and different from the atoms of other elements.
3. Compounds are formed by the combination of atoms of different elements in simple, integral, numerical proportions forming the 'compound-atoms' (or molecules) of such compounds.

The explanation of the Laws of Chemical Combination in terms of the Atomic Theory can now be discussed.

1. *The Law of Conservation of Mass.* Because atoms can neither be created nor destroyed and they are not subdivided during chemical change, no loss in mass is observed during chemical changes, which are essentially a regrouping of the atoms which make up the reactants to produce the products.
2. *Law of Constant Composition.* If a sample of a given compound is taken, it will consist of a certain number of molecules all of which are identical. In the case of silver chloride, therefore,

$$\frac{\text{Mass of chlorine in sample}}{\text{Mass of silver in sample}} = \frac{\text{Mass of 1 atom of chlorine}}{\text{Mass of 1 atom of silver}} = \frac{35.45}{107.9}$$

3. *Law of Multiple Proportions.* The formulae of the two oxides of copper discussed previously are written (in terms of the Atomic Theory):

   Black: CuO Red: Cu_2O

   Thus, with one atomic mass of oxygen, there is one atomic mass of copper combined in the black oxide and two atomic masses of copper combined in the red oxide.
4. *Law of Reciprocal Proportions.* The formulae of the compounds discussed previously are written (in terms of the Atomic Theory):

   Methane: CH_4 Carbon dioxide: CO_2
   Water: H_2O

   Thus the mass proportions observed are in simple numerical proportions because atoms combine in simple integral proportions in compound formation.

As will be discussed later, Dalton's Atomic Theory requires some modification to accommodate more recent discoveries.

The electrical nature of the atom

In 1791, the Italian physician, Luigi Galvani made the accidental discovery that a freshly dissected frog's leg gave a sharp convulsion when brought into contact with a discharging static electricity machine. It was later discovered that the electrical machine was not essential and that the convulsion could be achieved by joining the nerve and muscle via a metallic strip. The effect was more pronounced when two different metals were used to complete the circuit. Because of his biological inclinations, Galvani thought the phenomenon to be characteristic only of living materials and named it 'animal electricity'.

Figure 2.1 The 'pile' invented by Alessandro Volta

In 1800, Alessandro Volta, Professor of Natural Philosophy in the University of Pavia, showed that a 'pile', consisting of alternate discs of silver and zinc separated by cloth soaked in brine, acted as a source of current electricity. Volta's invention was speedily copied in scientific laboratories all over Europe, including London's Royal Institution. It was used there by Sir Humphry Davy, his assistant Michael Faraday, and others in the earliest experiments in electrochemistry. In the same year as Volta's invention, Nicholson and Carlyle decomposed water into its elements, and around 1806 Davy first isolated sodium and potassium from their molten hydroxides by electrolysis.

The idea that atoms were held together in compounds by the attractions between opposite charges rapidly gained acceptance. In his *Experimental Researches into Electricity* Faraday wrote: 'Although we know nothing of what an atom is, yet we cannot resist forming some idea of a small particle, which represents it to the mind: and although we are in equal, if not greater, ignorance of electricity, so as to be unable to say whether it is a particular matter or matters or mere motion of ordinary matter, or some kind of third power or agent, yet there is an immensity of facts which justifies us in believing that the atoms of matter are in some way endowed or associated with electrical powers, to which they owe their most striking qualities and among them their mutual chemical affinity.' Faraday's work in electrochemistry is discussed more completely in Chapter 8.

Electrical discharge through gases

In 1748, William Watson described the discharge of a static electricity machine through a gas held in a glass tube at low pressure: 'It was a most delightful spectacle when the room was darkened, to see the electricity in its passage: to be able to observe not, as in the open air, its brushes or pencils of rays an inch or two in length, but here the coruscations were of the whole length of the tube between the plates, that is to say, thirty-two inches.'

After Hans Geissler's invention of the much more efficient mercury vacuum pump, Julius Plücker in 1858 was able to observe the deflection of glow by a magnetic field. In 1869 Hittorf found that a shadow was cast by an opaque object placed between the walls of the tube and the cathode, and suggested that the fluorescence was caused by rays travelling in straight lines from the cathode. Sir William Crooke in 1879 developed the 'Maltese cross' tube shown in Figure 2.2 in which rays from the cathode caused a characteristic shadow to fall on the front of the tube. He also built his 'railway' tube shown in Figure 2.3, in which the cathode rays were made to turn the paddles of the object in the middle of the tube, causing it to travel along the glass rails. From these observations Crooke postulated the cathode rays to consist of a stream of negatively charged particles repelled from the cathode.

Herman von Helmholtz in a lecture to the Chemical Society of London in 1881 declared: 'If we accept the hypothesis that the elementary substances

Figure 2.2 Crooke's Maltese cross tube

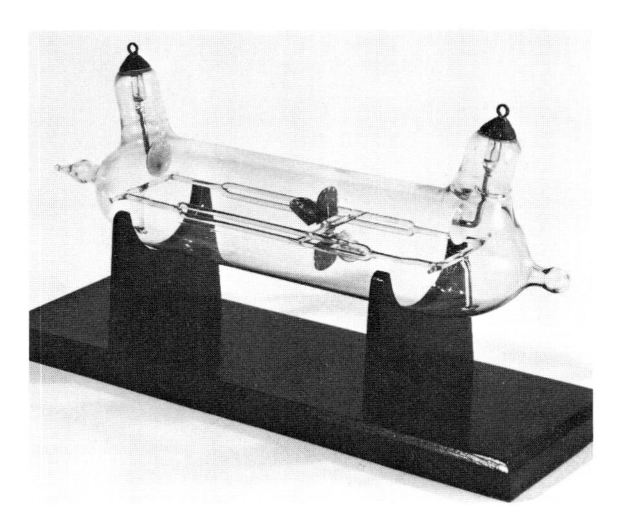

Figure 2.3 Crooke's 'railway' tube

are composed of atoms, we cannot avoid concluding that electricity also, positive as well as negative, is divided into definite elementary portions which behave like atoms of electricity.'

The notion that the atoms of gas in discharge tubes were being broken up by the electrical discharge meant that scientists had to think again about Dalton's 'indivisible atoms'.

The charge to mass ratio of cathode particles

In 1897 Sir J. J. Thomson used the apparatus shown in Figure 2.4 and diagrammatically in Figure 2.5 to determine the ratio of charge to mass for cathode particles. Rays from the cathode are accelerated towards the anode, some passing through the slit, and thence through the slit in the second metal plate. The particles pass down the tube between the deflecting plates before striking the rounded end of the tube, on which they form a fluorescent spot. When a potential difference is applied across the deflecting plates, the fluorescent spot moves in a direction depending upon the polarity of the deflecting plates. It can be shown that the deflection caused by the applied potential, y, is related to the electrostatic field strength, E, mass, m, charge, e, and velocity, v, of the cathode particles and to distance x by the equation:

$$\frac{e}{m} = \frac{2yv^2}{x^2E}$$

All the quantities on the right-hand side of the equation can be readily measured or calculated, except the velocity, v. This can be eliminated from the equation by applying a magnetic field at right angles to both the electrostatic field and direction of motion of the cathode particles. The magnetic coils Thomson used can be seen in the left-hand corner of Figure 2.4. The magnetic field strength, B, is now adjusted until the deflected cathode particles are returned to their initial position, when the electrostatic and magnetic fields will be exactly balanced. Under these conditions it can be shown that $v = E/B$, and if this is substituted into the previous equation we have:

$$\frac{e}{m} = \frac{2yE}{x^2B^2}$$

The ratio e/m can thus be calculated and the modern accepted value for this constant is 1.76×10^{11} C kg^{-1}. Thomson showed that the cathode particles had the same value for this con-

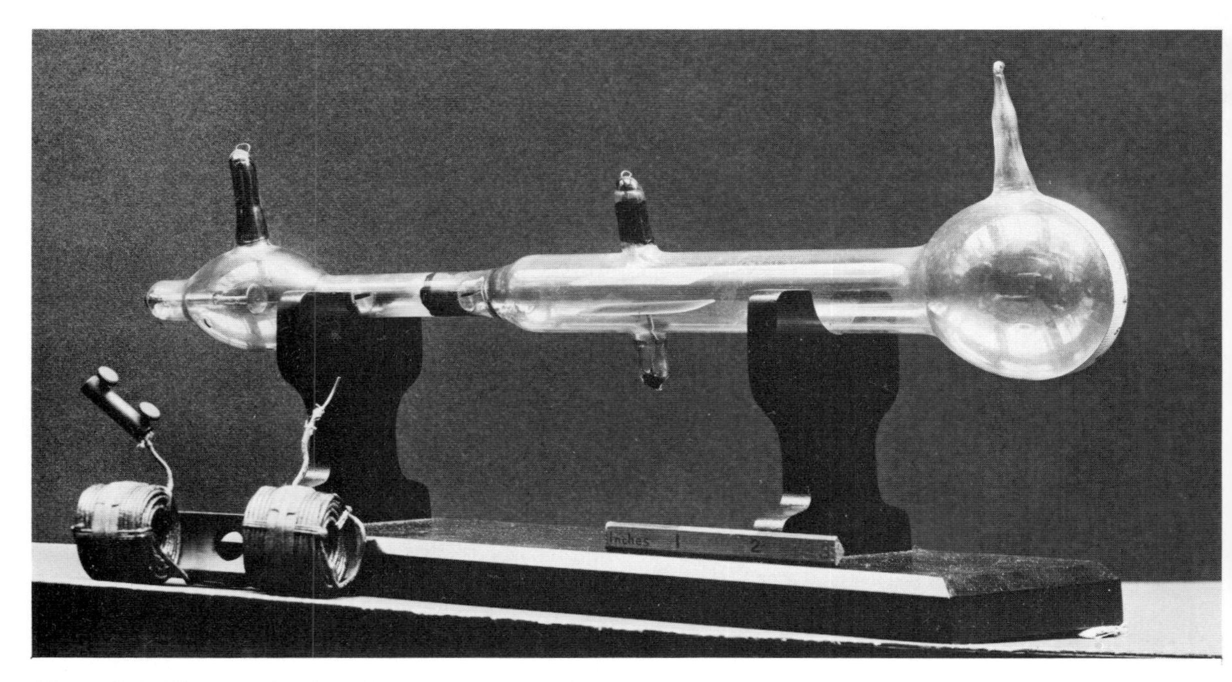

Figure 2.4 Thomson's *e*/*m* tube

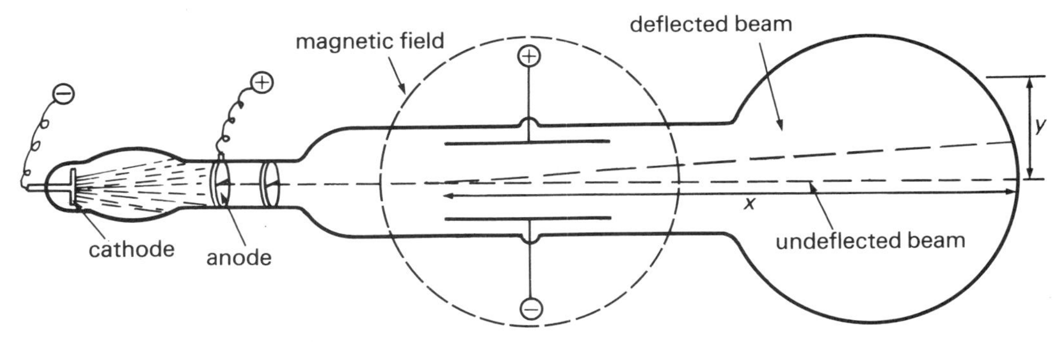

Figure 2.5 Thomson's e/m tube (diagrammatic)

stant no matter what gas was contained in the tube and that they were therefore likely to be the constituents of all types of atoms. Stoney introduced the name *electron* for these particles.

The charge e on the electron

This quantity was determined by Robert Millikan in 1909 using the apparatus shown in Figure 2.6. As the oil droplets enter the chamber they are ionized by bombardment with X-rays. If a suitable potential difference is then applied across the electrodes it is possible just to arrest the fall of the charged droplet, causing it to balance motionless between the plates; the gravitational and electrostatic forces acting are then equal. Writing an equation in which these forces are balanced and substituting the experimental results enables the only unknown quantity, the charge on the oil droplet, to be calculated. In each case the charge on the oil droplet is found to be a small integral multiple of the smallest charge observed. Millikan suggested that this smallest charge is the charge on one electron, other oil droplets having acquired multiple charges. The accepted value from such experiments for the charge on an electron is 1.60×10^{-19} C.

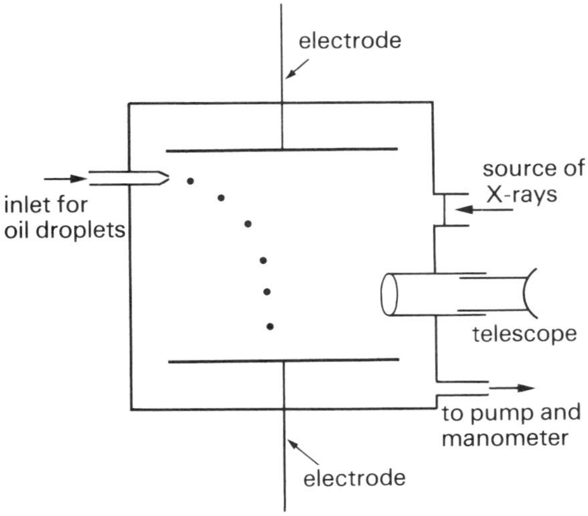

Figure 2.6 Millikan's apparatus for the charge on the electron (diagrammatic)

The mass of the electron

If we combine the values of e/m from Thomson's experiment with the value of e from Millikan's, we can obtain the mass of an electron. Converting this value from the usual units of mass to the scale of a hydrogen atom having a mass of 1 unit, we obtain a value for the mass of the electron of 1/1836.

Thomson's model of atomic structure

Thomson was now in a position to suggest a new model for atomic structure that would account for the advances in experimental knowledge since the time of Dalton. His model pictured the atom as a positive spongy material in which the small negative electrons are embedded. In gas discharge tubes he pictured these atoms as being torn apart by the applied electrical potential, the negative electrons travelling from cathode to anode and the residual positive ions from anode to cathode. This model is often known as the 'plum-pudding' model of the atom, the electrons being likened to the currants embedded in the floury mass of pudding. Thomson was later to turn his attention to the positive ions, and we shall take up this aspect of his work later in the chapter (page 23).

The discovery of X-rays

Using a discharge tube of the type illustrated in Figure 2.7, Wilhelm Roentgen in 1895 discovered that a very penetrating radiation, capable of passing through the wrappings around a photographic plate and causing exposure, was formed when cathode rays were made to strike a suitably shaped anode target. He discovered that these 'Roentgen' rays, later to be known as *X-rays*, passed more readily through some objects than others. In particular, he noticed that if a human hand was placed between the tube and the wrapped photographic plate, an image was obtained which showed the less penetrable bone in relief against the fleshier tissues. Because of the great potential for his discovery in surgery, he rapidly wrote up his findings and sent copies to a selection of the most eminent scientists of Europe, enclosing sample photographs.

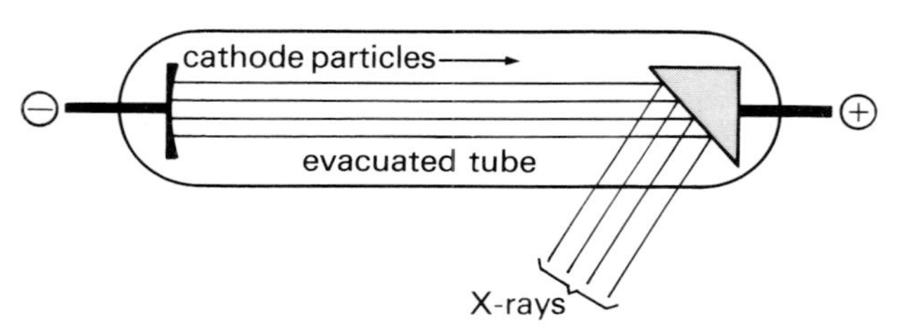

Figure 2.7 A tube for the production of X-rays

The discovery of radioactivity

One recipient of Roentgen's communication was the celebrated French scientist, Henri Poincaré. Early in 1896 Poincaré gave a report of Roentgen's discovery to the Academy of Sciences in Paris. A member of his audience, Henri Becquerel, was particularly interested to hear that the X-rays caused the sides of the discharge tube to fluoresce at the point of exit. Becquerel had been interested in the properties of salts of the element uranium which were known to fluoresce under the stimulation of the ultra-violet radiation in sunlight. He thought that the fluorescence might be connected with the ability to expose a wrapped photographic plate.

To investigate this he sprinkled a uranium salt on a wrapped photographic plate and left the assembly on a window-sill in bright sunlight. At the end of the day the experiment was taken indoors and the plate developed. To Becquerel's satisfaction the film had been exposed by the uranium salt, leaving a sharp outline of its presence. Becquerel repeated the experiment many times, sometimes using a solution of the salt, sometimes with a metal object placed between the salt and the plate. In each case exposure of the plate occurred.

Figure 2.9 shows a photograph of the type obtained by Becquerel; in this example a perforated metal sheet was placed between the salt and the wrapped plate.

One day during the course of such an experiment, the sky clouded over and Becquerel placed

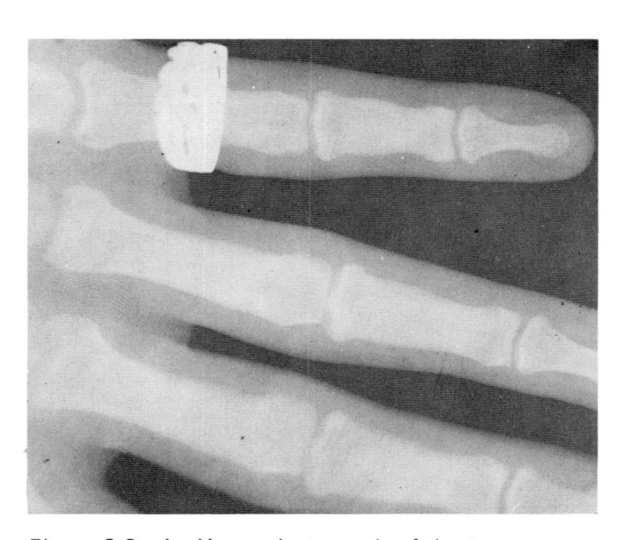

Figure 2.8 An X-ray photograph of the type obtained by Roentgen

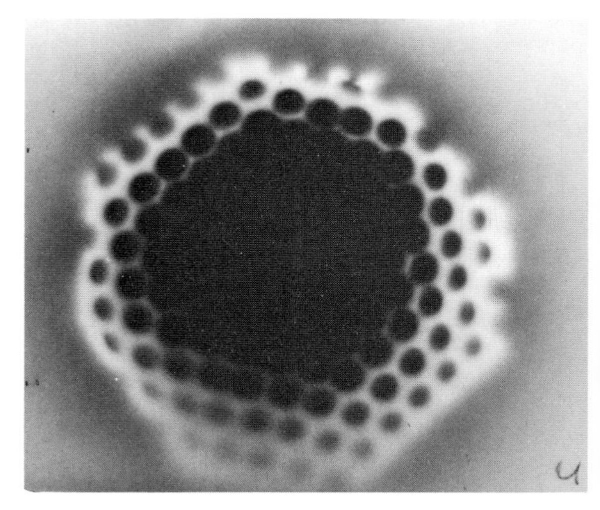

Figure 2.9 A photograph of the type obtained by Becquerel

the plate with the uranium salt in a drawer. Although he could not have expected the experiment to give a positive result, for some reason Becquerel developed that plate and was amazed to find that the exposure was as great as previously. He next tried the experiment with sunlight completely excluded, and when fogging occurred as before he realized that the effect had nothing to do with the action of sunlight, nor with the fluorescence of the salt. The uranium compound was giving off some previously undiscovered radiation capable of penetrating the wrappings and exposing the plate quite spontaneously. Becquerel had discovered radioactivity.

All over Europe scientists set about the thorough examination of this intriguing new phenomenon. Marie and Pierre Curie shortly afterwards isolated two previously unknown and powerfully radioactive elements, polonium and radium, from the mineral pitchblende. The new science of radiochemistry was born. In Chapter 14 there is a more complete discussion of the properties of radioactive materials; we shall be concerned here only with those aspects of radioactivity that impinge directly on a discussion of atomic structure.

Types of radioactive radiation

Towards the end of the nineteenth century, a series of experiments indicated that the radiation from radium comprised three different components. These are depicted diagrammatically in Figure 2.10. As the radiation passes through the electrostatic field, it is split into a component attracted to the negative plate (named *α-radiation*), a component attracted to the positive plate (named *β-radiation*), and a component undeflected by either plate (named *γ-radiation*). A similar effect is observed when the radiation is passed between the poles of a magnet. The properties of these types of radiation will be considered more fully in Chapter 14; it is sufficient for our present purpose to realize that α-particles carry a positive charge and are about four times as heavy as hydrogen atoms.

The discovery of the atomic nucleus

Hans Geiger and Ernest Marsden, working under the direction of Ernest Rutherford at the University of Manchester in 1910, carried out experiments in which a thin pencil of α-particles were used to bombard samples of metal foil only a few atoms thick. The experiments were carried out in an evacuated chamber so that the α-particles were

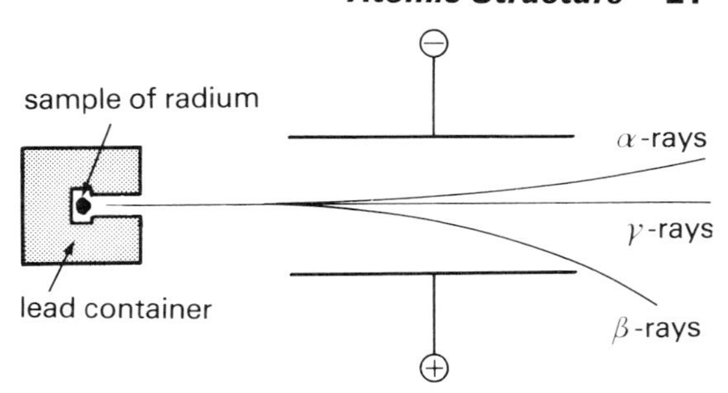

Figure 2.10 The components of radiation from radium

not absorbed by air molecules. The eventual position of the α-particles was detected by allowing them to fall on a zinc sulphide screen which scintillated whenever struck by a particle. In a lecture at the University of Cambridge in 1936, Lord Rutherford gave an account of this work:

> 'In the early days I had observed the scattering of α-particles and Dr. Geiger in my laboratory had examined it in detail. He found in thin pieces of heavy metal that the scattering was small, of the order one degree. One day Geiger came to me and said, "Don't you think that young Marsden, whom I am training in radioactive methods, ought to begin a small research?" Now I had thought that too, so I said, "Why not let him see if any α-particles can be scattered through a large angle?" I may tell you in confidence that I did not believe that they would be, since we knew that the α-particle was a very fast massive particle, with a great deal of energy, and you could show that if the scattering was the accumulated effect of a number of small scatterings, the chance of an α-particle's being scattered backwards was very small. Then I remember two or three days later Geiger coming to me in great excitement and saying "We have been able to get some of the α-particles coming backwards . . ." It was quite the most incredible event that has ever happened to me in my life. It was almost as incredible as if you fired a 15-inch shell at a piece of tissue paper and it came back and hit you.
>
> 'On consideration I realized that this scattering backwards must be the result of a single collision, and when I made calculations I saw it was impossible to get anything of that order of magnitude unless you took a system in which the greater part of the mass of the atom was concentrated in a minute nucleus . . .'

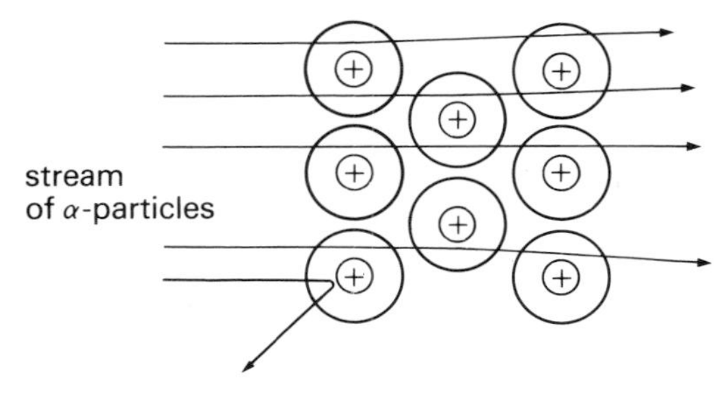

Figure 2.11 Deflection of α-particles by the nuclei of atoms of gold foil

In 1911 Rutherford proposed the nuclear theory of the atom based on this experimental evidence. He calculated that atoms must have nuclei of some 10^{-15} metres diameter. As atoms have diameters of the order 10^{-10} metres, the diameter of the atom is about 10^5 times that of the nucleus. Rutherford himself stated that the size of atoms to their nuclei is as the dome of St. Paul's cathedral to a man's clenched fist. This tiny nucleus contains the total positive charge associated with the atom and nearly all of its mass.

As a consequence of this we can see that the nucleus of an atom must be extremely dense. We can calculate the density of the nuclei of gold atoms knowing that its atomic mass is 197 and that this consists of 6.023×10^{23} atoms.

Mass of one gold atom =

$$\frac{197}{6.024 \times 10^{23}} \text{ g} = 3.28 \times 10^{-22} \text{ g}$$

Volume of 1 gold nucleus =

$$\tfrac{4}{3}\pi r^3 \approx 3 \times 10^{-45} \text{ m}^3$$

Assuming that the total mass of the gold atom is concentrated in the nucleus (and rounding off the value to 3×10^{-22} g),

$$\text{density of gold nucleus} = \frac{3 \times 10^{-22} \text{ g}}{3 \times 10^{-45} \text{ m}^3} - 10^{23} \text{ g m}^{-3} = 10^{17} \text{ g dm}^{-3}.$$

The significance of this becomes even more apparent if we consider the mass of the earth (5×10^{26} g) and calculate the volume it would occupy if it was comprised solely of such nuclei:

Volume of 5×10^{26} g of gold nuclei =

$$\frac{5 \times 10^{26} \text{ g}}{10^{23} \text{ g m}^{-3}} = 5 \times 10^3 \text{ m}^3$$

This volume is of the same order of magnitude as that of a large swimming pool.

The constitution of the nucleus

At this stage it was realized that the nucleus must contain positively charged particles, and these were given the name *protons*. Some difficulty, however, was experienced in deciding how many protons were contained in the nucleus of each element. When the elements are listed in order of relative atomic mass, their ordinal number in this list being known as the *atomic number*, it was observed that successive elements did not show equal differences of atomic mass. Additionally, several elements had atomic masses which were far from integral. Clearly it was not possible to equate each unit of relative atomic mass with the presence of a proton in the nucleus. As we shall see a little later (page 25), this problem was resolved with the postulation that nuclei also contained uncharged particles called *neutrons*. However, we shall first consider some important evidence that the atomic numbers of the elements are of more fundamental importance.

X-rays and atomic number

Charles Barkla observed in 1911 that in addition to the background, or white, X-radiation emitted by elements on bombardment with cathode rays, there were two pronounced X-ray lines, called the K and L lines, for each element that were characteristic of the particular element from which the anode was made. In 1913 Henry Moseley discovered that the frequency, v, of each line was related to the atomic number of the element, Z, by the equation

$$\sqrt{v} = a(Z - b)$$

the series of K lines for all the elements examined having the same value for the constants a and b. The frequencies of the L lines similarly fitting the equation but with different numerical values of a and b. The results of Moseley's analysis are shown in Figure 2.12.

This suggested that the atomic number of an element was related to some more fundamental property of the atoms of that element. The conviction gradually grew that the atomic number was in fact the number of positive charges, i.e. the number of protons, in the nucleus.

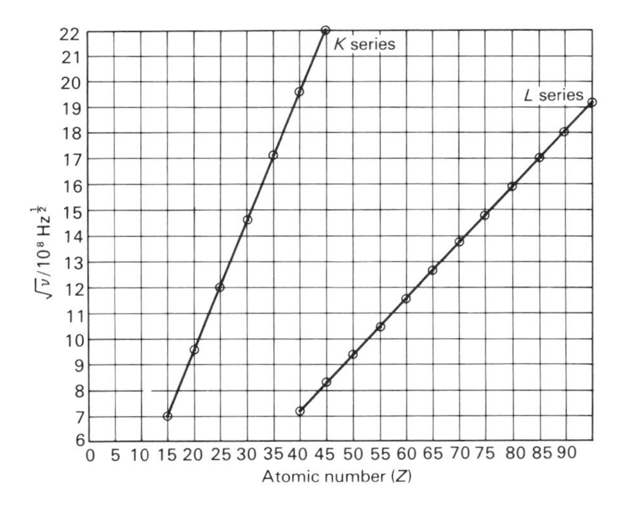

Figure 2.12 Relationship between the characteristic X-ray frequencies of elements and their atomic number

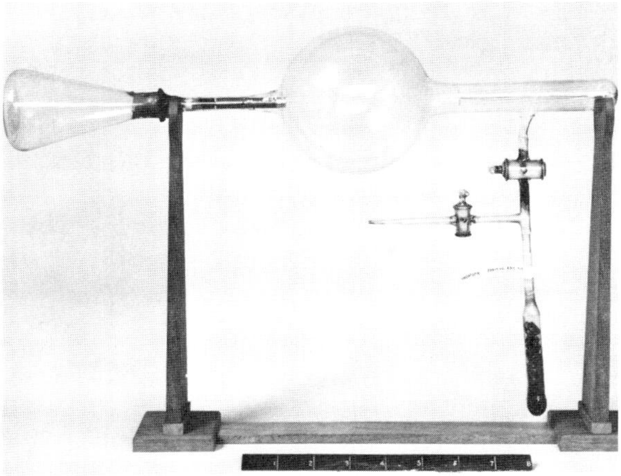

Thomson's positive ray apparatus

Having thoroughly investigated the properties of the cathode rays produced in gas discharge tubes, Thomson turned his attention to the positive particles travelling from the anode to the cathode. He recognized these as the positive ions formed when one or more electrons are ripped from the electrically neutral atoms by the electrical potential. In 1912 he used the apparatus illustrated in Figure 2.13 to investigate the path taken by these positive ions when subjected to the influence of a magnetic field and an electrostatic field in parallel. The eventual position of the ions was detected on a photographic plate placed inside the apparatus against the flat end of the conical flask. From an analysis of the forces acting on the ions during their passage through the tube he was able to show that all ions of a given mass to charge ratio would strike the photographic plate at some position on a parabolic curve. The exact position of any ion on that curve was shown to depend on its velocity. In distinction from the case of cathode rays, he found that the mass to charge ratios observed in this apparatus depended on the particular gas in the tube. While investigating the positive ions formed by gaseous neon, Thomson found that in addition to a strong parabolic image corresponding to a mass/charge ratio of 20, a weaker parabola also occurred corresponding to a mass/charge ratio of 22. He concluded from this that neon contains atoms of two types, having masses of 20 and 22 respectively. The existence of atoms of the same element having different masses was thus established, and for such entities the name *isotopes* was coined by Frederick Soddy from the Greek meaning 'the same place' (in the Periodic Table of the Elements).

Figure 2.13 Thomson's positive ray apparatus

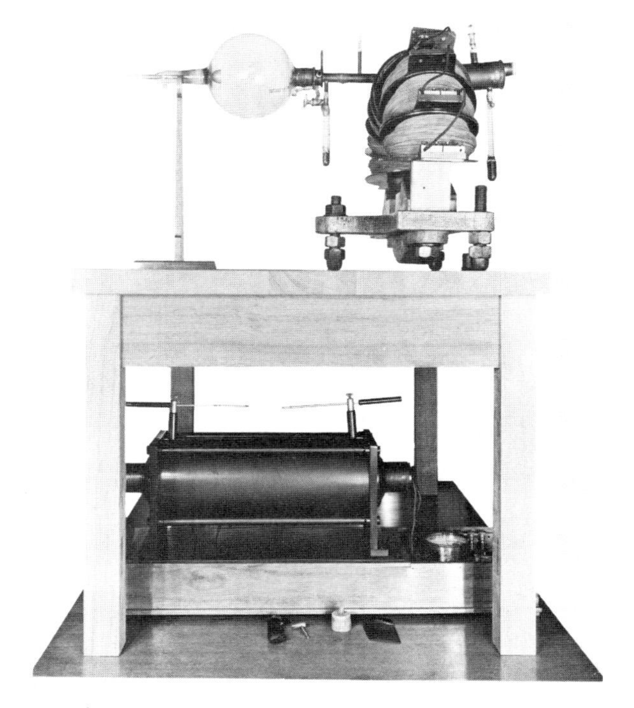

Figure 2.14 Aston's mass spectrograph

The work on the analysis of positive rays was developed by A. J. Dempster, K. T. Bainbridge, F. W. Aston, and others. In Aston's apparatus, shown in Figure 2.14 and known as a *mass spectrograph*, the magnetic and electrostatic fields are so arranged that ions of any one mass/charge ratio are brought to a point focus regardless of their velocity. Out of such developments the modern mass spectrometer eventually emerged.

The mass spectrometer

The schematic diagram of a modern mass spectrometer is shown in Figure 2.15. The material for analysis enters via the injection point and passes through a heated inlet chamber to vaporize any liquids or solids. As the sample passes between the anode and heated cathode of the ionization chamber, it is bombarded by electrons freed from the cathode and travelling upwards to the anode across its path. Ionization of the sample occurs:

$$X(g) + e^- \text{ (fast)} \rightarrow X^+(g) + 2e^- \text{ (slow)}$$

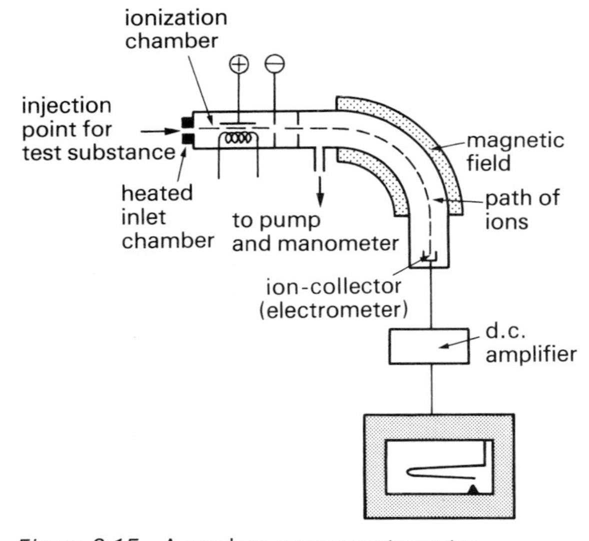

Figure 2.15 A modern mass spectrometer

(Multiple charged ions can be formed if the accelerating potential of the electrons is sufficiently high. This effect is discussed in more detail in Chapter 14; here we shall be concerned only with single positively charged ions, whose mass/charge ratio is thus equal to their mass.) The ions so formed are accelerated towards the perforated cathode, and a stream of these particles passes along the instrument until deflected by the magnetic field. The magnetic field strength is adjusted so that only ions of a particular mass are focused into the ion-collector. The movement of charge in the ion-collector constitutes an electric current; this is amplified and the signal is fed to a recording meter. This meter effectively plots ion current (proportional to the rate at which ions enter the collector and hence to the number of such ions) against time. The magnetic field strength is continuously varied and during the course of such a 'scan' the ions of each mass/charge ratio are focused in turn into the ion-collector.

An ion of mass m and charge e travelling with velocity v will be deflected along a path of radius r by a magnetic field strength B, i.e.

$$mv = Ber \qquad \mathbf{1}$$

Thus all ions with the same *momentum* arrive at the collector for a given value of B. If V is the accelerating voltage of the ions, their potential and kinetic energies can be equated:

$$mv^2/2 = Ve \qquad \mathbf{2}$$

Combining equations 1 and 2 yields:

$$\frac{m}{e} = \frac{B^2r^2}{2V}$$

Thus the ratio m/e can be deduced when the appropriate values of B, V, and r are known.

(In order that the magnetic field strength B can be varied, an electromagnet must be employed. This increases the cost of the instrument, but is preferable to using a fixed magnetic field and varying the ion accelerating voltage V because this latter procedure alters the efficiency with which ions are extracted from the ionization chamber.)

The mass spectrum of neon is shown in Figure 2.16. Modern instruments show that there is a third, much less abundant isotope not known to J. J. Thomson. The sensitivity of the recording meter has been adjusted to accommodate the differing proportions of the three isotopes.

Determination of relative atomic masses by mass spectrometry

The relative atomic mass of an element is the mean of the isotopic masses weighted to allow for the relative abundances of the isotopes. The peak heights of the mass spectrometer readout (adjusted to the sensitivity setting employed) can be taken as a measure of the abundance of each isotope. (The area under each peak should strictly be used for this purpose.) The method of calculation is illus-

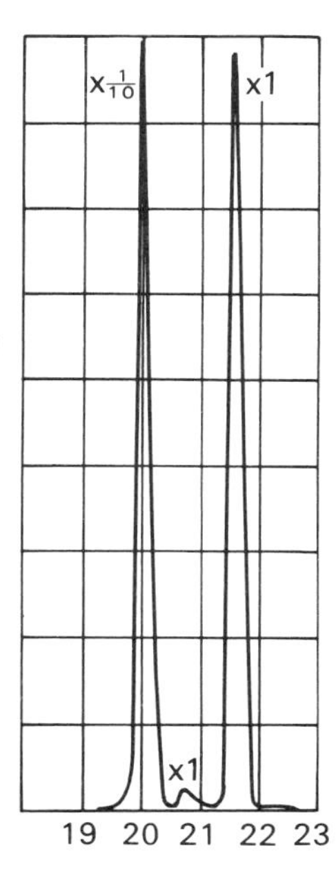

Figure 2.16 The mass spectrum of neon

trated here using the results for neon from Figure 2.16:

11.2 × 22 =	246.4	(weighted mass of Ne-22)
0.2 × 21 =	4.2	(weighted mass of Ne-21)
114 × 20 =	2280	(weighted mass of Ne-20)
125.4	2530.6	

Average mass of neon atoms (relative atomic mass)

$$= \frac{2531}{125.4} = 20.18$$

The neutron

Moseley's work had established the concept that the number of protons in the nucleus of the atoms of each element is characteristic of that element and numerically equal to its atomic number. To explain the existence of isotopes, Rutherford postulated the existence of a second type of particle in the nucleus of atoms, having one unit of mass (on the hydrogen scale) but carrying no charge. For these particles he coined the name *neutrons*. In the case of neon which has an atomic number of 10 and hence 10 protons in the nucleus of each atom, the isotopes of mass numbers 20, 21, and 22 have 10, 11, and 12 neutrons respectively in their nuclei.

The properties of the three fundamental particles of which atoms are composed are summarized in Table 2.2.

Table 2.2 Properties of the fundamental particles

Particle	*Mass* (H = 1)	*Charge*
Electron	1/1836	−
Proton	1	+
Neutron	1	0

Because it carries no charge, the neutron is more difficult to detect than the electron or proton. In 1930 Bothe and Becker reported that when certain light elements (such as beryllium and boron) are bombarded with α-particles, a hitherto unknown and highly penetrating radiation is emitted. In 1932 Irene and Frederic Joliot-Curie observed that high-speed protons are displaced from a block of paraffin wax placed in the path of this new radiation. James Chadwick explained these observations by postulating that the new radiation consists of neutrons, and thus the first experimental evidence of the neutron's existence was recorded.

The nature and arrangement of the electrons

In our historical survey we have reached a model of atomic structure in which the atom is pictured as consisting of a minute nucleus composed of protons and neutrons which accounts for nearly all the mass of the atom and its total positive charge. Outside this nucleus the electrons are arranged, but the manner of this arrangement has yet to be discussed. This is a subject of particular interest to chemists because during chemical reactions it is the outer parts of atoms that come into contact; the electronic structure is thus likely to dictate the type of chemistry exhibited by the atoms of each element. Before considering the problem of the electronic structures of atoms, however, we must first consider some theories concerning the *nature* of electrons themselves.

From the results of experiments with the 'Maltese cross' and 'railway' tubes described previously, many scientists regarded electrons as *minute particles*. When it was shown that a stream of electrons exhibited properties such as diffraction and interference associated with wave motions, other scientists held the view that the electron existed as a *wave motion*. (Diffraction of electrons is

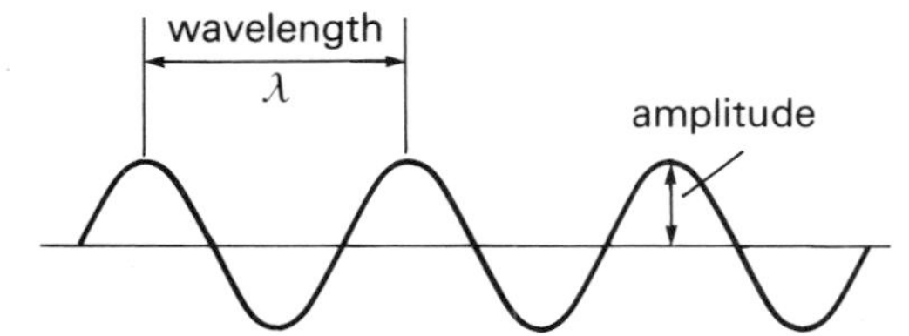

Figure 2.17 The wave motion of an electron

considered more thoroughly in Chapter 5.) If electrons are to be regarded as a form of wave motion, they must have a wavelength, λ, and a frequency, ν, as pictured in Figure 2.17. These quantities are related to the velocity, c, of the waves by the equation:

$$\lambda \times \nu = c$$

This dichotomy was resolved in 1924 by Louis de Broglie, who proposed that any moving particle has wave properties associated with it. He deduced the relationship between the mass, m, the velocity, c, and the wavelength, λ, of a particle in motion to be:

$$\lambda = \frac{h}{mc}$$

(where h is a constant known as Planck's constant). We are not aware of the wave properties of the moving macroscopic objects of everyday existence because their associated wavelengths are so short that they escape our attention. (For example, a golf ball of mass 45 g travelling with velocity 30 $m\ s^{-1}$ has a wavelength of less than 5×10^{-34} metres.)

Evidence from emission spectroscopy

An important clue to the extranuclear arrangement of the electrons came from the study of the light emitted by gaseous atoms of elements when suitably excited. A spectrograph suitable for analysing and photographing such emissions is shown in Figure 2.18. As the light passes down the instrument it is split up into its component colours by the system of lenses and prism. The spectrum finally produced is photographed.

The observation that such spectra are not continuous but consist of lines with particular frequencies and wavelengths required explanation. Scientists had further difficulty in explaining why

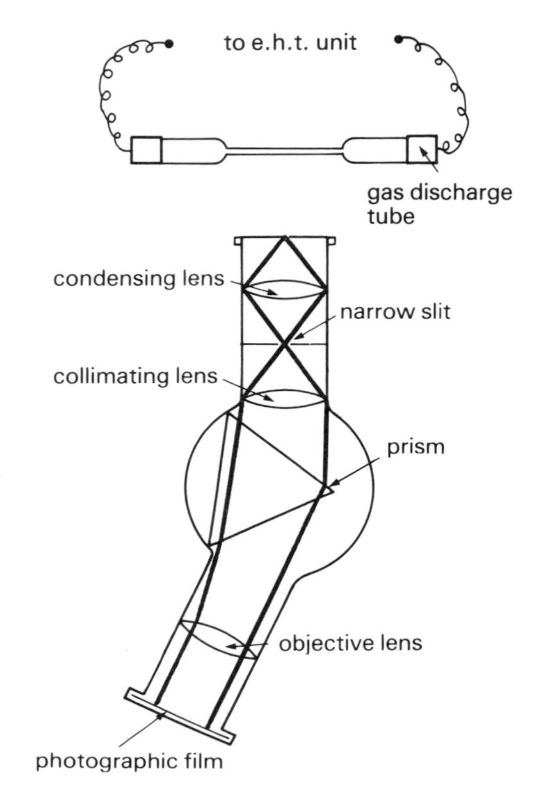

Figure 2.18 Spectrograph for photographing emission spectra

the negative electrons travelling around the positive nucleus do not spiral inwards towards the centre of the atom and eventually become entrapped in it.

The solution to these problems came in 1913 when the great Danish physicist, Niels Bohr, proposed his quantum theory of the atom. He postulated that electrons can exist outside the nucleus only in certain well-defined *orbits*; the closer any orbit is to the nucleus, the lower is its energy level. An electron, on passing from an orbit of higher energy to one of lower, must therefore emit a definite quantity of energy. In 1901 Max Planck had published his quantum theory of energy based on a study of the energy emitted when so-called 'black body' sources of radiant heat cool down. Planck's theory established that energy changes are *quantized*, i.e. the energy emitted or absorbed must be done so in discrete packets, and cannot change continuously as had been previously supposed. He showed that the quantum of energy associated with a light wave is proportional to its frequency. Thus,

$$E = h\nu$$

(h, the proportionality constant, is known as Planck's constant.)

Bohr showed that Planck's theory would explain the line spectra observed when excited atoms cool. As electrons fall from an orbit of higher energy to an orbit of lower energy, a definite quantity or quantum of energy must be lost. According to Planck's relationship, this energy must correspond to a definite frequency and thus appear as a line in the appropriate part of the spectrum.

The emission spectrum of hydrogen

Not unexpectedly, the simplest of all line spectra is that of the hydrogen atom. In 1885 J. J. Balmer showed that the wavelengths of certain lines in this spectrum could be expressed by the equation:

$$\lambda = k\frac{n^2}{n^2 - 4}$$

where k is a constant and n is a series of integers having values 3, 4, 5, 6, etc., each value corresponding to a particular line. The series of lines defined by this equation is known as the Balmer series. It was later shown that Balmer's equation was a special case of a more general relationship (the Rydberg equation) between the reciprocal of the wavelengths of the spectral lines:

$$\frac{1}{\lambda} = R_H\left(\frac{1}{m^2} - \frac{1}{n^2}\right)$$

where R_H, Rydberg's constant for the hydrogen atom, equals 1.097×10^7 m^{-1}. For the Balmer series, m equals 2. Using photographic film of extended sensitivity, it is possible to obtain spectra of hydrogen extending beyond the limits of visible light. Four further series of lines were thus discovered and named after the scientists responsible. These series are given in Table 2.3.

Table 2.3 The line series in the spectrum of atomic hydrogen

Series	*m*	*n*	*Region observed*
Lyman	1	2, 3, 4, . . .	ultraviolet
Balmer	2	3, 4, 5, . . .	visible
Paschen	3	4, 5, 6, . . .	infrared
Brackett	4	5, 6, 7, . . .	far infrared
Pfund	5	6, 7, 8, . . .	far infrared

From Planck's relationship we can deduce that the emissions in the ultraviolet, having the greatest frequencies, are associated with the highest energies. This equation can be used to calculate the energies associated with specific electron transitions; in Figure 2.19 the transitions responsible for the Lyman, Balmer, and Paschen series are represented in such a way that the spacings between the energy levels are proportional to the energy changes involved.

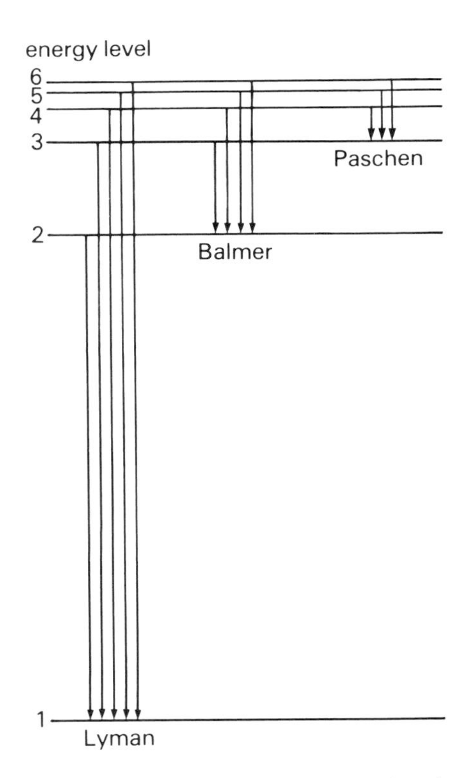

Figure 2.19 Electron transitions associated with lines in the Lyman, Balmer, and Paschen series

The Lyman series

As the lines in this series correspond to electron transitions between higher energy levels of the hydrogen atom and the orbit of lowest energy, it can be used to determine the amount of energy required to remove an electron from this lowest energy level (when the atom is said to be in the *ground state*) to the outermost part of the atom (when the atom can be said to be *ionized*). This quantity of energy is known as the *ionization energy of hydrogen*. Table 2.4 lists the frequencies of the lines in the Lyman series.

Table 2.4

$\nu/10^{15}$ Hz	$\Delta\nu/10^{15}$ Hz
2.466	
	0.457
2.923	
	0.160
3.083	
	0.074
3.157	
	0.040
3.197	
	0.024
3.221	
	0.016
3.237	
	0.011
3.248	

Inspection of these figures shows that the frequencies converge towards a maximum as the distances between the energy levels becomes smaller at the outer fringes of the atom. The convergence limit of the frequencies thus represents the movement of the electron from outside of the atom to the lowest energy level within the atom; the energy associated with this transition is thus equal and opposite to the ionization energy of hydrogen. We can determine the convergence limit by plotting differences between frequencies, $\Delta\nu$, against either the higher or lower of the two frequencies and extrapolating to find the frequency when $\Delta\nu = 0$. This has been plotted in Figure 2.20; the limiting frequency is found to be 3.275×10^{15} Hz. Planck's relationship can then be used to calculate the ionization energy.

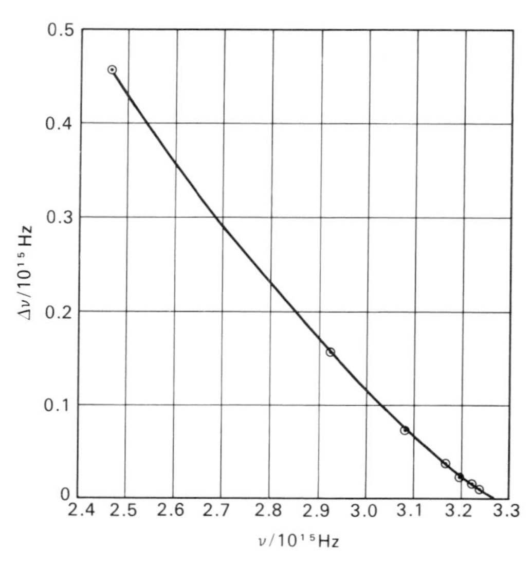

Figure 2.20 Finding the convergence limit of the frequencies in the Lyman series

Ionization energy of hydrogen,

$$E = h\nu$$
$$= (3.99 \times 10^{-13}\ \text{kJ mol}^{-1}\ \text{Hz}^{-1}) \times (3.275 \times 10^{15}\ \text{Hz})$$
$$= 1310\ \text{kJ mol}^{-1}$$

Systems with more than one electron

The Rydberg equation (page 27) is found to fit only single electron systems, such as the hydrogen atom and ions such as He^+ and Li^{2+}, but for each of these a different numerical value for the constant is observed. With polyelectron systems the analysis of their spectra is more difficult; it is found, however, that in each case the observed lines can be divided into four categories which are given the names 'sharp', 'principal', 'diffuse', and 'fundamental'. From analysis of these lines it is possible to deduce values for the energies of further ionization, the first ionization energy being the energy to remove a first electron from the ground state of the gaseous atom to infinity, the second ionization energy being that required to remove a second electron from the monopositively charged ion to infinity, and so on.

Mass spectrometry can also yield information concerning ionization energies. This is discussed in Chapter 14.

A further method of determining ionization energies, which is rather more limited in application, depends upon the use of thermionic valves known as *thyratrons*.

Ionization energies by electron impact

Using the circuit shown in Figure 2.21, the voltage across the valve is increased in regular increments. Electrons freed from the heated cathode travel to the anode and thus constitute an electrical current. The corresponding values of voltage and current are noted. At first the valve is found to obey Ohm's law (i.e. the current is proportional to the voltage). At a critical voltage, however (depending on the particular gas in the tube), there is a sudden escalation of current. Typical results are plotted in Figure 2.22 for a helium-filled thyratron. This escalation of current can be explained in terms of an increase of conducting particles being produced at the critical voltage due to ionization of the gas.

$$He(g) + e^-(g)\ (\text{fast}) \rightarrow He^+(g) + 2e^-(g)\ (\text{slow})$$

At voltages below this critical value the electrons do not have sufficient energy to ionize the gaseous atoms. When the critical voltage of the gas

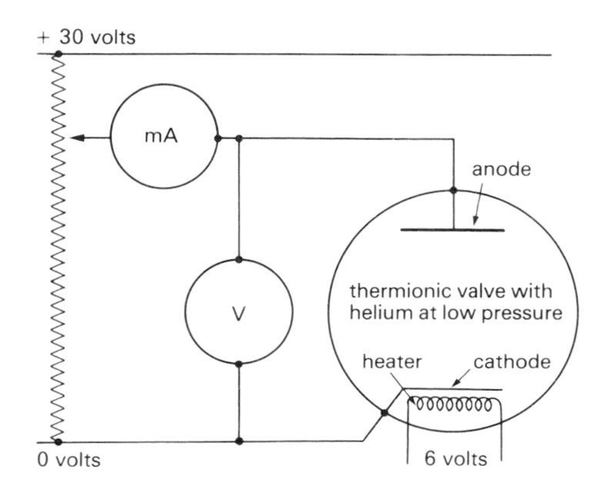

Figure 2.21 Determination of ionization energies by electron impact

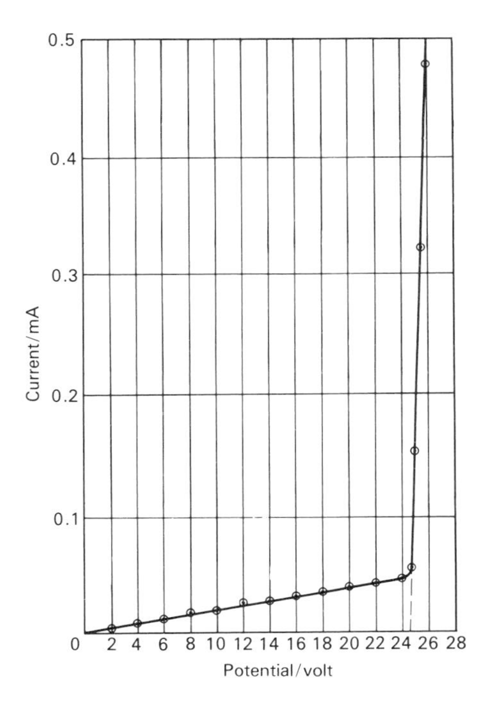

Figure 2.22 Variation of current with voltage using a helium filled thyratron valve

is reached the valve may 'strike' and the spectral colour of the element it contains emitted. The voltage at which this first occurs is known as the *first ionization potential* of the gas. From Figure 2.22 we can see that this constant has a value of 24.6 volts for helium. We can convert this ionization potential into ionization energy:

$$\text{Potential/volt} = \frac{\text{Energy/joule}}{\text{Charge/coulomb}} \quad \text{(by definition)}$$

Thus,

$$\text{Energy} = \text{Potential} \times \text{Charge}$$

Now the charge carried by one mole of helium ions is the Avogadro constant times the charge on one electron. Substituting suitable values and the previously determined ionization potential in this equation yields:

First ionization energy of helium =

$(24.6\ \text{V}) \times (6.02 \times 10^{23}\ \text{mol}^{-1}) \times (1.60 \times 10^{-19}\ \text{C})$
$= 2360\ \text{kJ mol}^{-1}$.

Patterns in ionization energies

The pattern that emerges when First Ionization Energies of elements are plotted against Atomic Number is discussed in Chapter 12 (page 188). Useful information concerning the arrangement of the electrons for a particular element can also be obtained by plotting the successive ionization energies. These have been plotted for magnesium in Figure 2.23; a logarithmic scale is necessary to accommodate the wide range of numerical values of the twelve successive ionization energies. We should expect successive ionization energies to increase in magnitude because each additional electron must be removed against a progressively increasing net positive charge on the magnesium ion. Inspection of the data in Figure 2.23 shows this to be the case; however, the lack of regularity between certain successive values requires explanation.

We are now in a position to propose a model of the electronic structure of magnesium which is consistent with this pattern.

The electronic configuration of magnesium

Present day interpretations of electronic configurations regard the 'energy levels' of electrons as positions in the atom at which there is the greatest *probability* of finding electrons and not, as previously, a definite orbit. For this reason the term *orbital* is now preferred.

Figure 2.24 shows the proposed model for the *ground state* of the magnesium atom. Each orbital can hold a maximum of two electrons and is represented by a box, the horizontal position of which indicates its energy level. Each arrow represents an electron; the significance of the direction of these is discussed later. The following interpretation

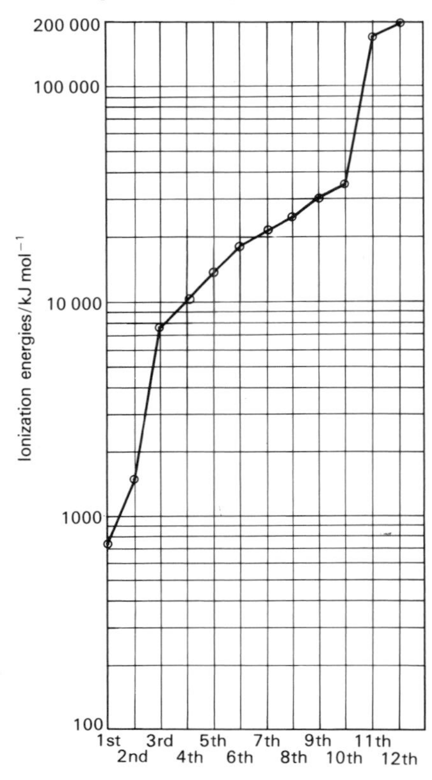

Figure 2.23 The successive ionization energies of magnesium

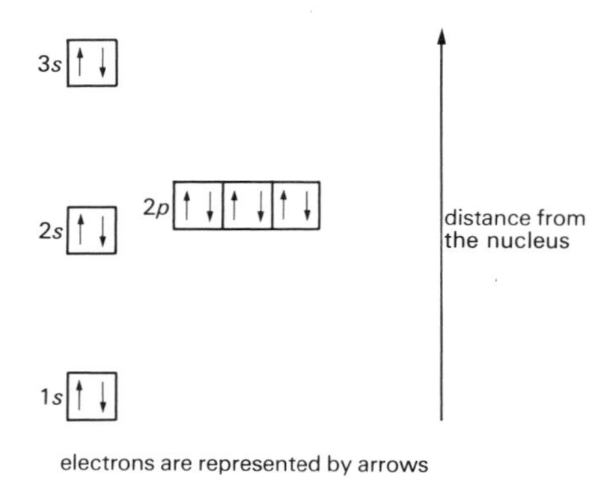

Figure 2.24 The 'ground state' electronic configuration of magnesium

should be read in conjunction with this, and also with Figure 2.23. The 1st and 2nd ionization energies correspond to the removal of the two electrons in the orbital represented by the box labelled 3*s*. The jump between the values for 2nd and 3rd ionization energies is due to the associated electron (in box 2*p*) being deeper within the atom and thus nearer the nucleus. The fairly regular increase between the 3rd and 10th ionization energies is a consequence of the 2*s* and all the 2*p* orbitals involved being at approximately the same energy level. Looked at in more detail, this sequence shows a small jump between the 5th and 6th ionization energies showing that one electron is removed from each of the 2*p* orbitals before any one of them loses both its electrons. The jump between ionization energies 8 and 9 is due to the removal of electrons from the 2*s* orbital which is slightly nearer the nucleus. The jump between the 10th and 11th ionization energies is due to the necessity of probing deeper still into the atom to remove the 11th and 12th electrons.

The principal energy levels (or shells) of atoms are designated by the numbers 1, 2, 3, etc. (or the letters K, L, M, etc.). The letters *s* and *p* refer to the 'sharp' and 'principal' categories of spectral lines with which the orbitals are thought to be associated. Atoms with more electrons than magnesium may also have electrons in *d* and *f* orbitals, associated with the 'diffuse' and 'fundamental' lines in their spectrum. The very large atoms beyond Lawrencium made artificially will have further orbital types.

Principles governing the electronic build-up of atoms in their ground states

1. The orbitals of lowest energy are filled first with electrons.
2. When more than one orbital has the same energy level, no pairing of electrons occurs until each contains one electron. This is a statement of *Hund's rule.*
3. Each electron in an atom must be distinguishable from the others. This is a statement of the *Pauli exclusion principle.*
4. No orbital can contain more than two electrons; these are distinguishable from each other as they are said to have *opposite spins.* (This distinction is represented in Figure 2.24 by showing electrons as arrows pointing in opposite directions.)

The main energy levels, or shells

These are sometimes referred to as the principal quantum numbers. The maximum number of electrons each can contain is $2n^2$, where n is the ordinal number of the shell.

The orbitals

Every shell contains one *s* orbital, and this contains two electrons of opposite spins when full. Each shell, except for the first or *K* shell, contains three *p*-orbitals all with the same energy. Each shell, except for the first and second, or *K* and *L*, shells contain five *d*-orbitals. Each shell after the third, or *M*, shell contains seven *f*-orbitals. The order in which the orbitals are filled with electrons is summarized in Figure 2.25. Many chemical data books contain tables showing the detailed application of this principle to each of the elements.

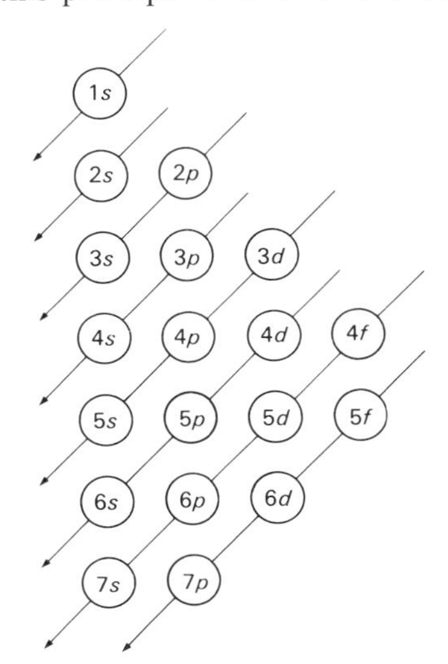

Figure 2.25 Order of occupancy of orbitals

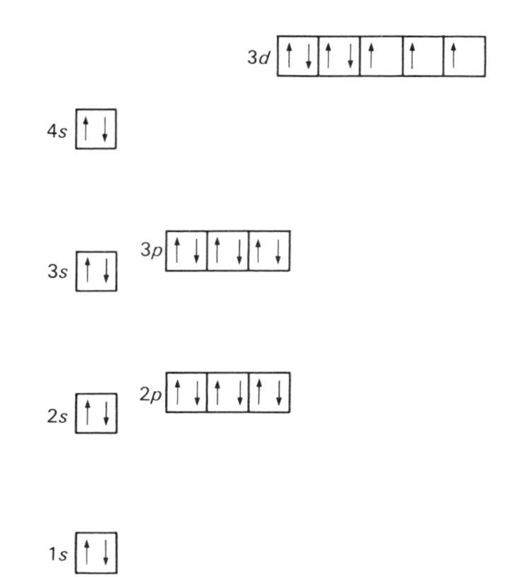

Figure 2.26 The *ground state* electron configuration of cobalt

A notation for electron structure

The electronic structure of magnesium shown in Figure 2.24 is conveniently notated as:

$$1s^2,\ 2s^2,\ 2p^6,\ 3s^2$$

This notation can be used to describe the electronic structure of all the other elements. In the case of cobalt, with the electronic structure shown in Figure 2.26, this is:

$$1s^2,\ 2s^2,\ 2p^6,\ 3s^2,\ 3p^6,\ 3d^7,\ 4s^2$$

This can be abbreviated to

$$[\text{Argon}],\ 3d^7,\ 4s^2$$

meaning that cobalt has the electronic configuration of argon with the addition of the electrons specified in the 3*d* and 4*s* orbitals. When this type of abbreviation is used, it is the immediately previous noble gas that must be named.

The relationship between electronic structures of the elements and the Periodic Table is discussed in Chapter 12.

Developments of Bohr's theory of the hydrogen atom

Although it was successful in explaining the simple line spectrum of hydrogen, Bohr's theory was not adequate when the fine structure of the spectrum was analysed using a high resolution spectrograph. It also proved inadequate for explaining the splitting observed to occur in the spectral lines when the discharging gases were subjected to a magnetic or an electrostatic field. The first of these effects was observed by Pieter Zeeman in 1896 and the second by Johannes Stark in 1911.

Then in 1916 Arnold Sommerfeld introduced his extension of the Bohr theory, which proved capable of avoiding these difficulties. This involved the introduction of the concept of elliptical orbits and a set of arbitrary rules known as 'selection principles', the sole justification for these being purely pragmatic; by application of these rules it was possible to avoid predicting the existence of a plethora of lines not actually observed in the spectrum. By the mid 1920s however, the Bohr–Sommerfeld theory had become too artificial and complex for its retention to be justified.

It was replaced by the wave mechanics of Erwin Schrödinger and the quantum mechanics of Werner Heisenberg. Of these two approaches, Schrödinger's has proved the more influential. Starting from de Broglie's postulates about the wave motion associated with particles of matter in motion, Schrödinger deduced the following equation concerning the wave motion of the electron in the hydrogen atom.

$$\frac{\partial^2\psi}{\partial x^2} + \frac{\partial^2\psi}{\partial y^2} + \frac{\partial^2\psi}{\partial z^2} + \frac{8\pi^2 m}{h^2}(E - V)\psi = 0$$

where x, y, and z are the Cartesian coordinates of the electron; m, E, and V are the mass, total energy, and potential energy respectively of the electron; h is Planck's constant. The quantity ψ is called the *wave function*, and ψ^2 is a measure of the probability of finding the electron in a specified region.

The Schrödinger equation is soluble for ψ provided sufficient information is available. It is assumed that ψ varies continuously with changing values of x, y, and z but that it has a single, finite value at any one location which drops to zero at infinite distance from the nucleus. These conditions can be met only when E is given certain specific values, known as *eigenvalues*, and the corresponding solutions for ψ are known as *eigenfunctions*.

These eigenvalues are found to correspond to the 'energy levels' of the Bohr model. By using Schrödinger's equation it becomes possible to calculate the electron densities one would expect at various distances from the nucleus. For example, it can be shown that maximum probability of finding the electron in a hydrogen atom occurs at approximately 0.05 nm from the nucleus. This corresponds to the electron in the ground state, i.e. in the first shell. From calculations such as these it is possible to deduce the shapes of orbitals of various types. The *s*-orbitals, for example, can be shown to be spherically shaped around the nucleus. The shapes of *p*- and *d*-orbitals deduced in this manner are shown in Figure 2.27.

Suggestions for further reading

JAFFE, BERNARD *Moseley and the Numbering of the Elements*, Heinemann Science Study Series No. 40 (London: Heinemann Educational Books, 1974)

ROMER, ALFRED *The Restless Atom*, Heinemann Science Study Series No. 10 (London: Heinemann Educational Books, 1963)

HOLDEN, ALAN *The Nature of Atoms* (Oxford: Oxford University Press, 1971)

WRIGHT, STEPHEN (ed.) *Classical Scientific Papers – Physics* (London: Mills and Boon, 1964)

Da ANDRADE, E. N. *Rutherford and the Nature of the Atom*, Heinemann Science Study Series No. 29 (London: Heinemann Educational Books, 1965)

Problems

1. Atoms can be ionized by impact with light of suitable wavelength. Light of 242 nm wavelength will just ionize sodium atoms. Calculate the ionization energy of sodium. (The velocity of light = 3×10^8 m s^{-1}.)
2. Naturally occurring lithium contains isotopes of mass numbers 6 and 7 with percentage abundances of 7.40 and 92.60 respectively. Calculate the relative atomic mass of lithium.
3. In experiments with thyratron valves of the type illustrated in Figure 2.21, a rapid escalation of current was observed at the following potentials:

   (*a*) Neon-filled valve: 21.6 V
   (*b*) Argon-filled valve: 15.8 V

   Calculate the first ionization energies of neon and argon.
4. Give a concise account of the structure of the atom, including in your account definitions of the terms *proton, neutron, isotope, electronic energy level.*

   Explain how a knowledge of the structure of the atom has led to an understanding of (*a*) the periodicity of the elements and (*b*) atomic emission spectra. (Oxford and Cambridge 'A')
5. Explain the formation of the line spectra of hydrogen and show how the line spectra are related to the ionization energy of the atom.

   How, using X-rays, could the identity of a specimen of a metal be established? (Cambridge 'A')
6. Outline the use of the mass spectrometer in the determination of atomic masses. Why is $^{12}C = 12$ now used as a standard in these determinations?

   The mass spectrum of an element X contained three peaks or lines at m/e 24, 25, and 26 of relative intensities 7.88: 1.01: 1.11. Explain these data and calculate the atomic mass of X. (Cambridge 'A')
7. Calculate, showing your reasoning, the relative atomic mass of chlorine to three significant figures given that the element consists of two isotopes with relative atomic masses of 34.98 and 36.98 and relative abundances of 75.4% and 24.6%. (Oxford 'A')

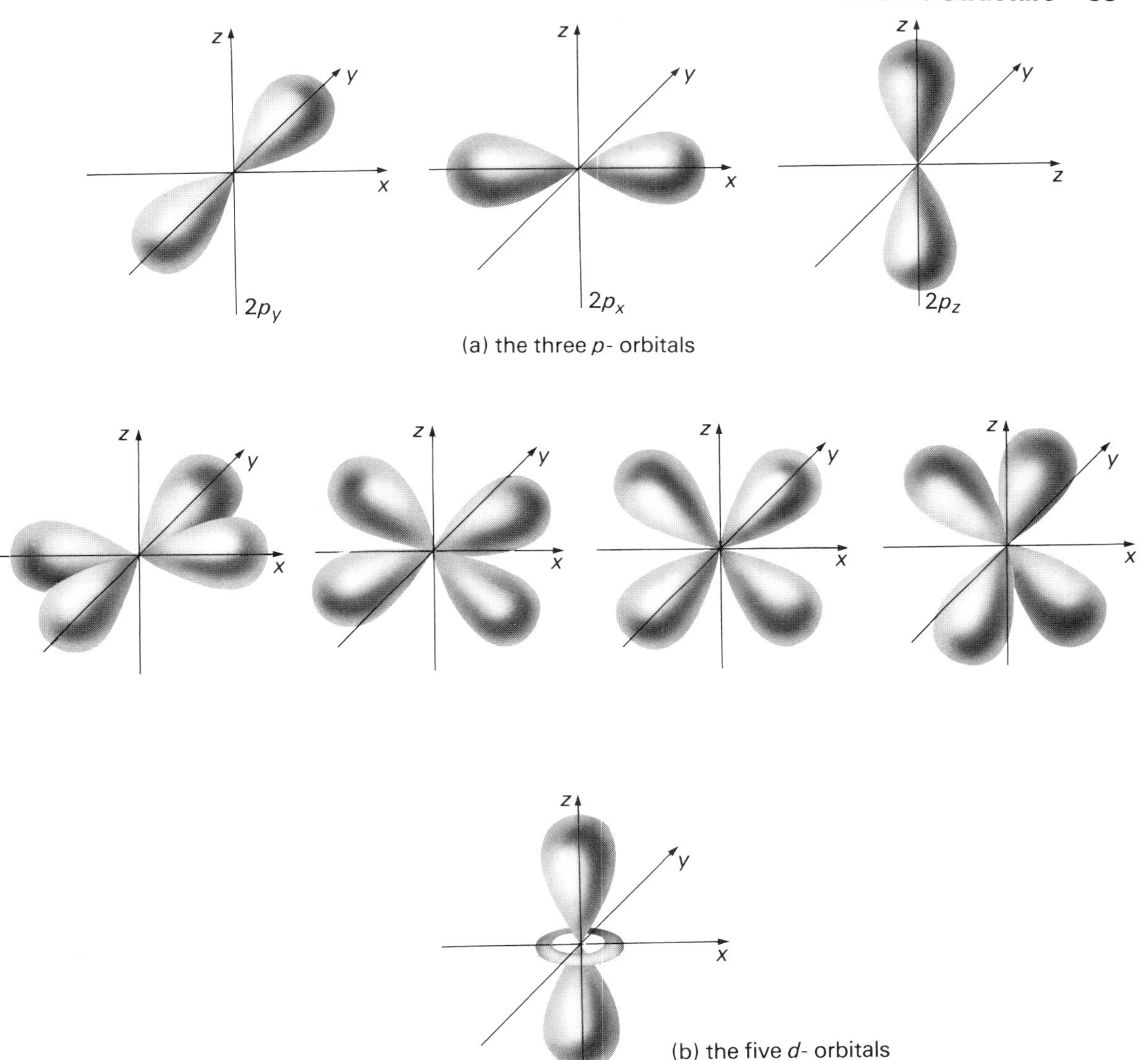

Figure 2.27 The shapes of *p* and *d*-orbitals

8. The following table lists the logarithmic values of the successive ionization energies (in kJ mol^{-1}) of the element sodium.

Number of electrons removed	1	2	3	4	5	6
Logarithm of ionization energy	2.69	3.66	3.84	3.97	4.12	4.22
Number of electrons removed	7	8	9	10	11	
Logarithm of ionization energy	4.30	4.41	4.46	5.15	5.30	

(*a*) Plot a graph of the logarithm of ionization energies against the number of electrons removed.

(*b*) Explain how from your graph information about the electronic structure of a sodium atom may be obtained.

(*c*) Discuss how a knowledge of successive ionization energies of the elements can help the chemist in explaining the chemical properties of elements. (London 'A')

3. Spectroscopy and Analysis

AROUND the turn of the eighteenth century, Sir Isaac Newton showed that white light could be dispersed into the colours of the visible spectrum by a glass prism. In 1800, Sir William Herschel discovered radiation in the portion of the spectrum of white light beyond the red by its heating action on the bulb of a thermometer. This radiation, invisible to the eye, became known as *infra-red*. Shortly afterwards further radiation was discovered before the blue portion of the visible spectrum by the fluorescence it induced in substances called 'phosphors'. This radiation was named *ultra-violet*. Several other forms of radiation were discovered and found to belong to the same 'family', the members of which are characterized by having the same velocity in a vacuum, but differing in wavelength and thus frequency. The family is known as the *electromagnetic spectrum* because it consists of the lines of force of a magnetic field perpendicular to the lines of force of an electrical field. This combination of electrical and magnetic fields can travel through space without requiring the presence of a material medium as, for example, sound waves do. The complete electromagnetic spectrum is shown in Figure 3.1.

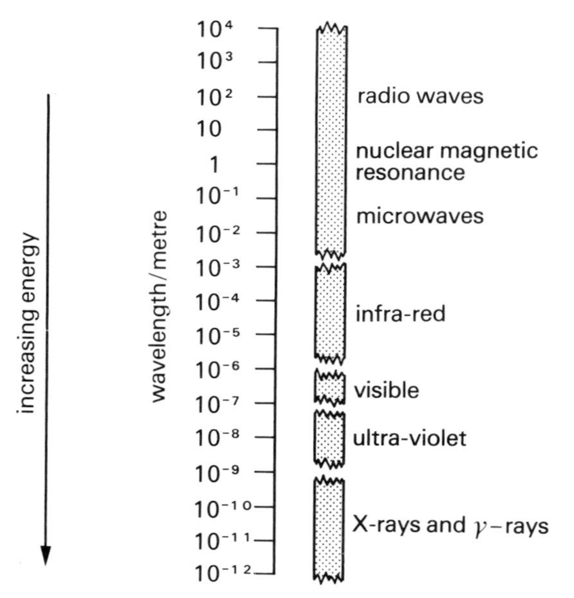

Figure 3.1 The electromagnetic spectrum

Chapter 2 discussed the importance of studies on the emission spectra of atoms in the determination of their electronic configurations. Spectroscopy has also been developed into an extremely useful technique of analysis.

Atomic emission spectroscopy

Apparatus of the type shown in Figure 2.18 (page 26) can be used to determine the concentration of certain elements in a given sample. The test material is generally excited by striking an arc between carbon electrodes, one of which holds the test sample. The intensity of a selected line in the spectrum photographed over a given time is directly related to the concentration of the element giving rise to that line. By comparing this intensity with spectra formed when standards of known composition are used, the concentration of a particular element can be deduced. In more sophisticated systems, the photographic plate is replaced by a photo-sensitive device placed at the appropriate position in the spectrum. This is connected to a meter which can be calibrated directly in concentration units for the element. Table 3.1 lists the wavelengths of lines commonly used for this purpose and the sensitivity of the technique for a representative selection of elements.

Table 3.1

Element	*Wavelength/nm*	*Sensitivity in parts per million (p.p.m.)*
Barium	393.4	0.2
Calcium	422.7	0.005
Copper	324.8	0.1
Iron	372.0	0.5
Manganese	403.3	0.02
Potassium	766.5	0.001
Sodium	589.0	0.0001
Strontium	460.7	0.01

It will be seen that the technique compares favourably, in terms of sensitivity, with conventional chemical analysis and provides results much more quickly. It is thus extremely useful, for example, to check individual batches of steel where even a small deviation from the specified composition may render the sample unsuitable for certain purposes.

This technique depends on the energy emitted by atoms in excited states, i.e. those with electrons with energy above the ground state. Even for such a readily excitable element as sodium, only about 1.5 per cent of the atoms are in an excited state at 5000 K and this must impose a limit on the sensitivity of the method. Moreover, because the percentage of excited atoms is temperature dependent, short-term variations of temperature in the discharge source can lead to poor reproducibility of results.

These problems are largely avoided in the technique of *atomic absorption spectroscopy.*

Atomic absorption spectrophotometry

An atomic absorption spectrophotometer is shown in Figure 3.2, and a schematic diagram of the apparatus in Figure 3.3. The hollow cathode source emits light from the particular element under test. The test sample is drawn up into the 'nebulizer' in which it is mixed with a fuel gas (such as propane or ethyne) and an oxidant gas (such as air or dinitrogen monoxide). The mixture from the nebulizer is burnt, and the atoms of the element under test contained in the flame can then absorb the light from the hollow cathode source thereby promoting electrons to excited states. The extent of this absorption is thus a measure of the concentration of the particular element in the test sample.

Atomic absorption spectrophotometry is rapid, sensitive, and applicable to a wide range of elements. It finds many applications in the metallurgical industries for the rapid analysis of alloy samples. Some particularly interesting applications are found in the field of clinical chemistry. Figure 3.4 shows the read-out from the analysis of the urine of three patients treated by gold injections for rheumatoid arthritis. It is essential to monitor

Figure 3.2 An atomic absorption spectrophotometer

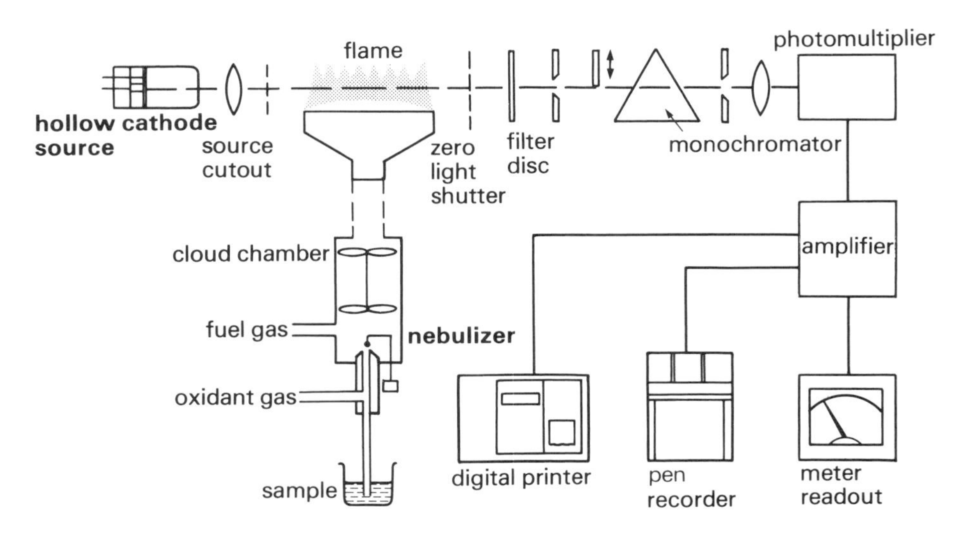

Figure 3.3 Schematic diagram of an atomic absorption spectrophotometer

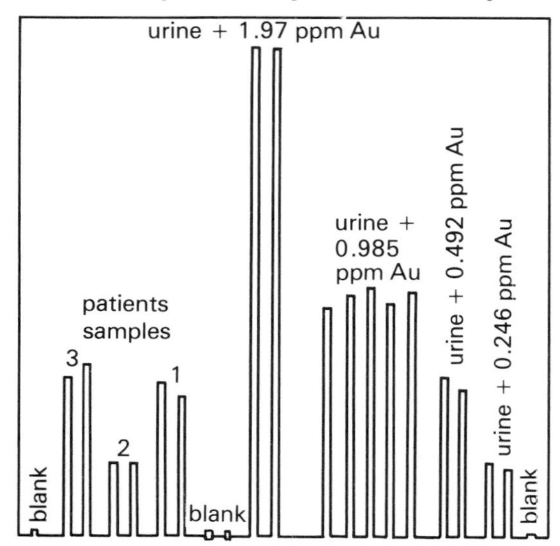

Figure 3.4 An atomic absorption spectrophotometer read-out

Figure 3.5 A colorimeter

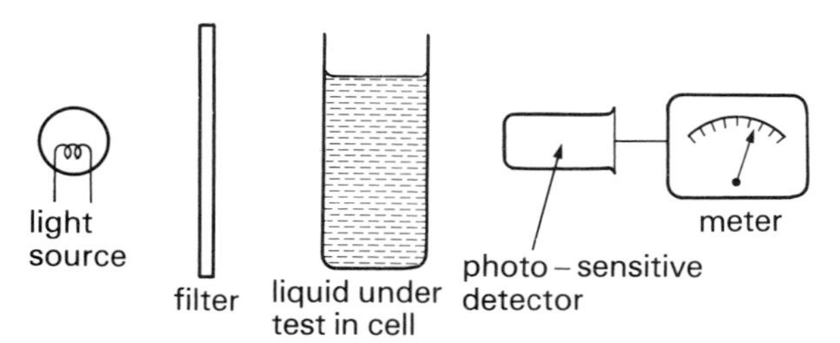

Figure 3.6 Schematic diagram of colorimeter

their urine and serum gold levels as these could rise to hazardous concentrations. The read-out also shows the readings for normal urine to which known concentrations of gold have been added to provide standards. Each reading is taken twice to avoid the intrusion of artefacts. The gold levels of patients 1, 2, and 3 can be seen to approximate to 0.5, 0.25, and 0.5 p.p.m. respectively.

Colorimetry

Colorimeters of the type shown in Figure 3.5 and represented schematically in Figure 3.6 can be used for solving a wide range of routine analytical problems. The filter is chosen to select a narrow band of wavelengths that can be absorbed by the liquid under test. The technique can be used for determining the concentrations of coloured ions or molecules in solution, or concentrations of such species that can be reacted with a suitable reagent to produce a coloured solution. It can be shown that the intensity of the light emerging from the cell containing solution (I), the intensity of the light emerging from cell containing pure solvent (I_0), and the concentration, C, of the absorbing species in the solution are related by the equation:

$$\lg \frac{I_0}{I} \propto C$$

If we assume that the meter readings are proportional to the intensity of light incident upon the detector, we can write:

$$\lg \frac{\text{Meter reading with solvent in cell } (m_0)}{\text{Meter reading with solution in cell } (m)} = kC$$

As this relationship is not obeyed strictly a calibration curve must be plotted for the accurate determination of any particular substance.

As is well known, a small quantity of chlorine is added to the water in swimming pools for disinfection purposes. Considerable discomfort would be inflicted on the bathers if the concentration of chlorine were allowed to rise beyond a certain level. The chlorine level can be regularly checked using colorimetry by adding a fixed volume of the water to a solution of potassium iodide, when a quantity of iodine proportional to the chlorine concentration will be liberated.

$$Cl_2(aq) + 2I^-(aq) \rightarrow 2Cl^-(aq) + I_2(aq)$$

The intensity of the brown colour developed in the solution is thus a direct measure of the chlorine concentration. A calibration curve for a particular

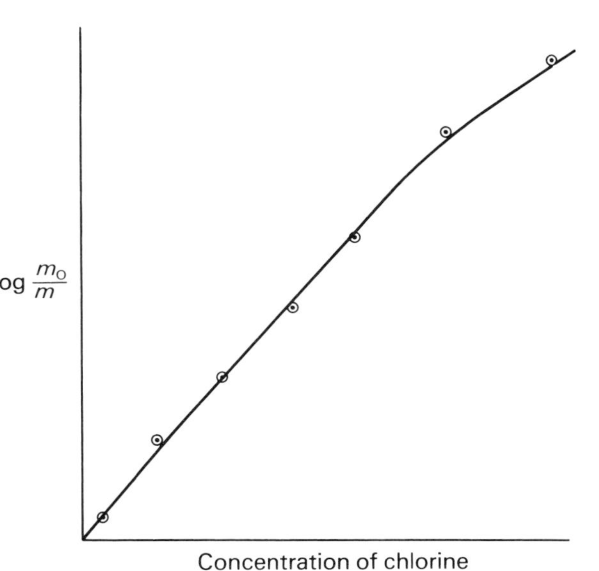

Figure 3.7 A colorimeter calibration curve for determination of chlorine

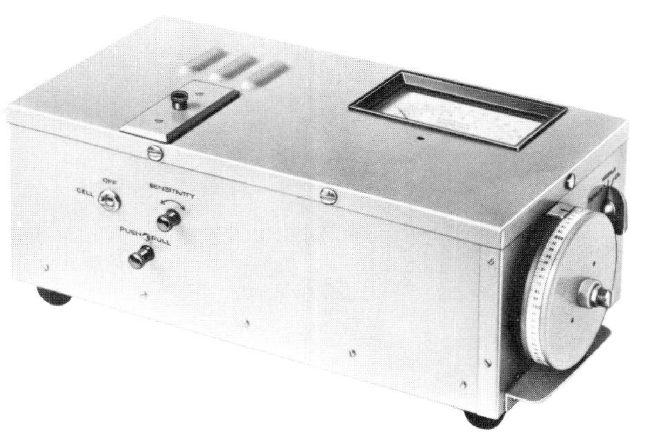
Figure 3.8 A single-beam visible spectrophotometer

Figure 3.9 The read-out meter

colorimeter and filter is shown in Figure 3.7. From this, and the colorimeter reading for the solution produced as above, the chlorine concentration can be deduced. This technique is not as sensitive as atomic absorption spectroscopy, but it is adequate for many purposes and has the advantage that it can be carried out by non-specialist personnel with a minimum of training.

Visible region and ultra-violet spectrophotometry

The simplest instrument for determining the visible region absorption spectrum of solutions operates on a similar principle to the colorimeter described above. Instead of selecting a narrow band of wavelengths with a filter, however, a diffraction grating is employed to select wavelengths at suitable intervals, at each of which the percentage transmittance is determined. The percentage transmittance is defined as:

$$\frac{\text{Intensity of light emerging from solution}}{\text{Intensity of light emerging from same thickness of solvent}} \times 100$$

In practice, the intensity of light is taken to correspond to the appropriate reading on the electrical meter. A typical 'single-beam' instrument of this type is shown in Figure 3.8. After setting the wavelength selector control (seen on the extreme right of the instrument) at the first chosen wavelength and with water in the path of the light, the sensitivity control is adjusted so that the meter, shown in close-up in Figure 3.9, reads 100 per cent transmittance. The solution is then moved into the path of light and the percentage transmittance recorded. These measurements are repeated at the other selected wavelengths. When graphs of these determinations are plotted, the results are of the type shown in Figure 3.10. A simple use of this technique is to confirm the identity of a particular coloured ion; its visible absorption spectrum acts as a 'finger-print'.

Instruments of this type are not capable of showing the fine structure of an absorption spectrum; they are said to be of 'low resolution'. High resolution instruments are of the 'double-beam' variety. The optical diagram of such an instrument is shown in Figure 3.11.

This instrument operates in both the visible and the ultra-violet regions of the spectrum, and has separate sources for use in each region. The rotating star wheel 'chops' the beam, directing alternate portions through the cuvette containing the sample under test and the reference cuvette containing

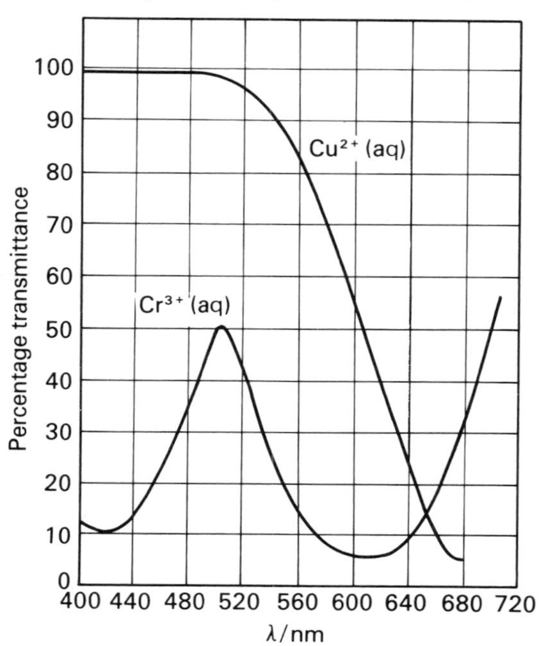

Figure 3.10 Visible absorption spectra of two solutions of coloured ions

Figure 3.11 A double-beam spectrophotometer for operation in visible and ultra-violet regions

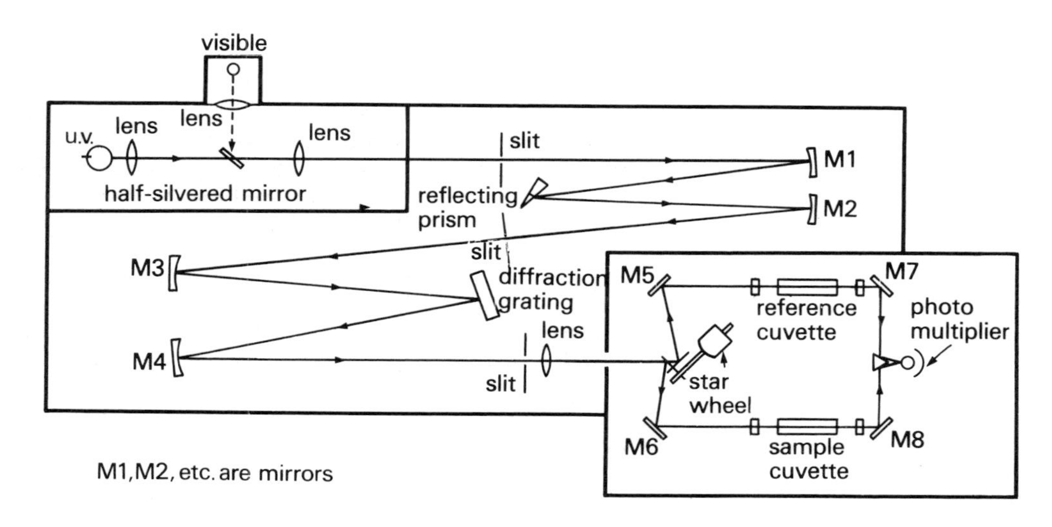

the solvent used in preparing the sample solution. The difference in intensities of the radiations reaching the photomultiplier is automatically converted into a percentage transmittance. The wavelength range required is automatically scanned by rotation of the diffraction grating and the absorption spectrum plotted directly on a pen-recorder. The spectrum of iodine vapour obtained in this way is shown in Figure 3.12. The troughs of this spectrum result from the dissociation of diatomic molecules of iodine into single atoms:

$$I_2(g) \rightarrow 2I(g)$$

Such absorption spectra are discontinuous (i.e. show a series of peaks and troughs) because during the course of dissociation there is a change in the interactions between the energy levels of each of the two atoms in the molecules, and these energy levels are quantized. When the molecule is completely dissociated, however, the separate atoms can absorb energy of any wavelength in that region as kinetic energy; in this region, therefore, the absorption becomes continuous. If we calculate the convergence limit for the absorption peaks in the spectrum the limiting value corresponds to the bond energy of the iodine molecule. (This procedure is analogous to that applied in Chapter 2 to obtain the ionization energy of hydrogen.) In the case of iodine the convergence limit of wavelength obtained in this fashion corresponds to a frequency of 3.80×10^{14} Hz. We can use Planck's relationship to determine the bond energy as follows:

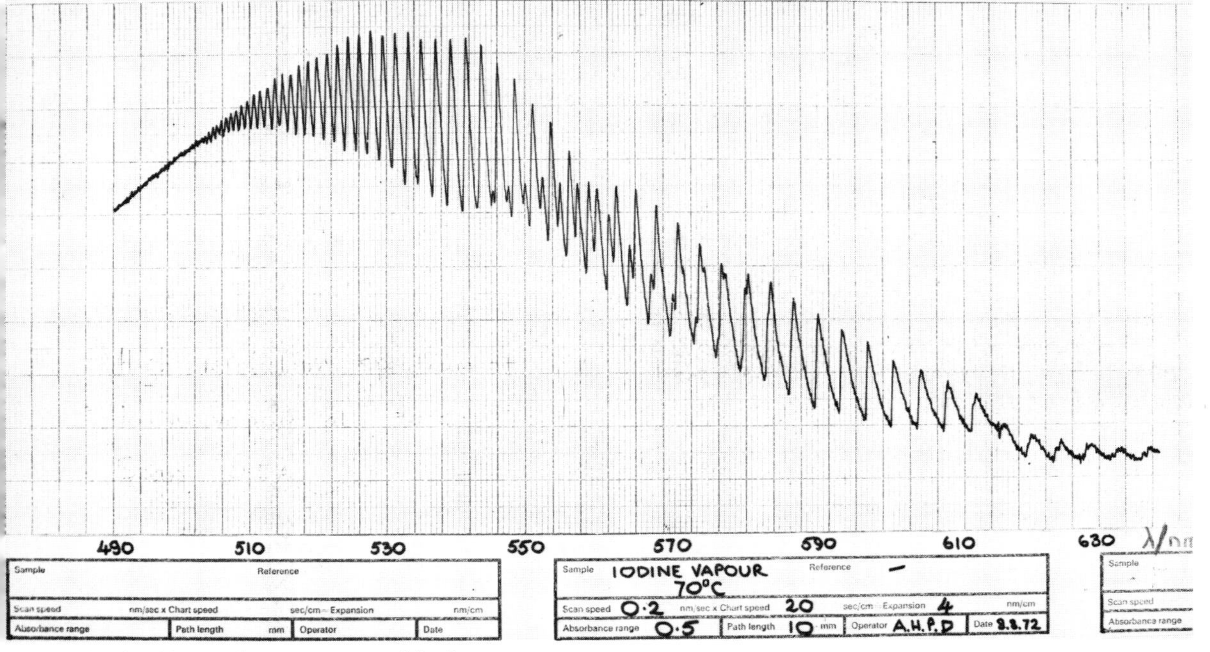

Figure 3.12 Absorption spectrum of iodine vapour

$$E = h\nu$$

$$E = (3.99 \times 10^{-13}\ \text{kJ mol}^{-1}\ \text{Hz}^{-1}) \times (3.80 \times 10^{14}\ \text{Hz})$$

$$= 152\ \text{kJ mol}^{-1}$$

N.B.—This method can only be used to determine the bond energies of diatomic molecules because of the more complicated spectra of polyatomic species resulting from the additional molecular deformations to those involved in the dissociation process. The absorption of radiation by molecular deformations will be considered in greater detail in the next section.

Infra-red spectrophotometry

Absorption in this region of the spectrum results from vibrations between groups of atoms bonded to each other. The frequency at which absorption occurs is characteristic of the bonds involved, but some variation in the precise location of the maximum absorbance is observed due to the effect of neighbouring groups of atoms. Such absorbances are observed at wavelengths of order one micrometre (μm). An alternative and very common method of expressing absorption troughs in this region of the spectrum is in terms of 'wave-number'. This is the reciprocal of the wavelength and is given the symbol $\bar{\nu}$.

The optical diagram of a typical infra-red spectrophotometer is shown in Figure 3.13. Radiation from the source is split by the reflecting mirrors into two parallel beams, one of which passes through the sample and the other (reference beam) travels directly to the detector. Each beam is aimed alternately on the detector via the rotating, or 'chopping', mirrors *A*. When each beam has the same intensity, there is zero signal from the detector and the servo-mechanism withdraws the comb completely from the reference beam. When absorption occurs at the selected frequency, however, the comb is inserted into the reference beam by the servo-mechanism until the signals again balance. The movement of the comb can thus be directly coupled to the pen recorder to register the percentage transmittance. The prism, normally of potassium bromide or calcium fluoride, is rotated to scan the frequency range required and a read-out of the type shown in Figure 3.14 is automatically traced.

It can be shown that, for a molecule composed of n atoms, there are $(3n - 5)$ possible vibrational modes if it is linear and $(3n - 6)$ modes if it is non-linear. Figure 3.15 shows the possible vibrational modes of a non-linear triatomic molecule.

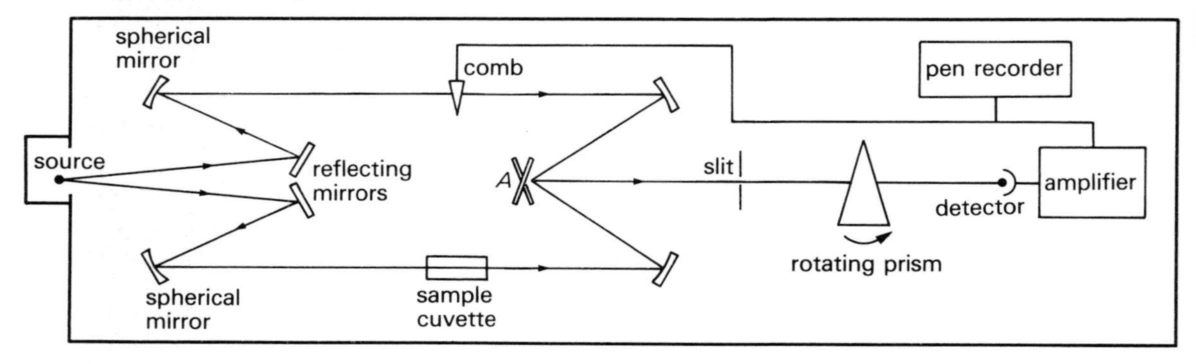

Figure 3.13 Optical diagram of an infra-red spectrophotometer

By inspection of a series of infra-red absorption spectra of related compounds, such as those shown in Figure 3.14, it is possible to associate troughs of absorption at particular wavelengths with the presence of particular groups of atoms in the material under test. For example, the presence of such a trough at about 3.5×10^{-6} metres in the spectra of hexane and trichloromethane and its absence from the spectrum of tetrachloromethane implies that it results from the vibrations associated with the group:

$$\geq C-H$$

On such an empirical basis it is possible to construct correlation tables of the type found in specialist publications on absorption spectroscopy. For work in the infra-red range instruments are often calibrated in 'wave-numbers'; the normal units of which are cm^{-1}.

Just as no two human individuals have identical fingerprints, each organic compound has a unique infra-red absorption spectrum, and thus by comparing the spectrum of an unknown substance with the spectra of known standards it is possible to make a positive identification. The infra-red absorption spectrum of acetylsalicylic acid, the active constituent of aspirin, is shown in Figure 3.16.

Only those molecular vibrations which result in an overall displacement of charge absorb infra-red radiation and thus show in the spectrum. Vibrations that do not result in overall displacement of charge may, however, be shown up by *Raman spectroscopy*.

Raman spectroscopy

When light passes through a homogeneous medium a phenomenon known as *Rayleigh scattering* can be observed, in which some photons are scattered through various angles without change in frequency. The Indian physicist Chandrosekhara Raman discovered experimentally in 1928 that the scattered light also contained weak radiation at discrete frequencies not present in the incident light. This results from inelastic collisions between photons and the molecules of the medium. Generally, the photon loses energy and promotes the molecule to an excited energy state and a line of lower frequency appears in the Raman spectrum (known as a *Stokes line*). Occasionally, a photon gains energy by collision with a molecule in the excited state and a line of greater frequency (an *anti-Stokes line*) appears in the spectrum.

Nuclear magnetic resonance spectroscopy (NMR)

Since its discovery in 1946, NMR spectroscopy has rapidly become one of the organic chemist's most powerful techniques for the elucidation of molecular structure. This brief introduction to the technique will be confined to the most useful and developed branch of the subject—proton magnetic resonance.

A proton has magnetic properties associated with it, and if placed in a uniform magnetic field it can take up one of two orientations with respect to that field—a low-energy orientation in which it is aligned with the field, and a high-energy

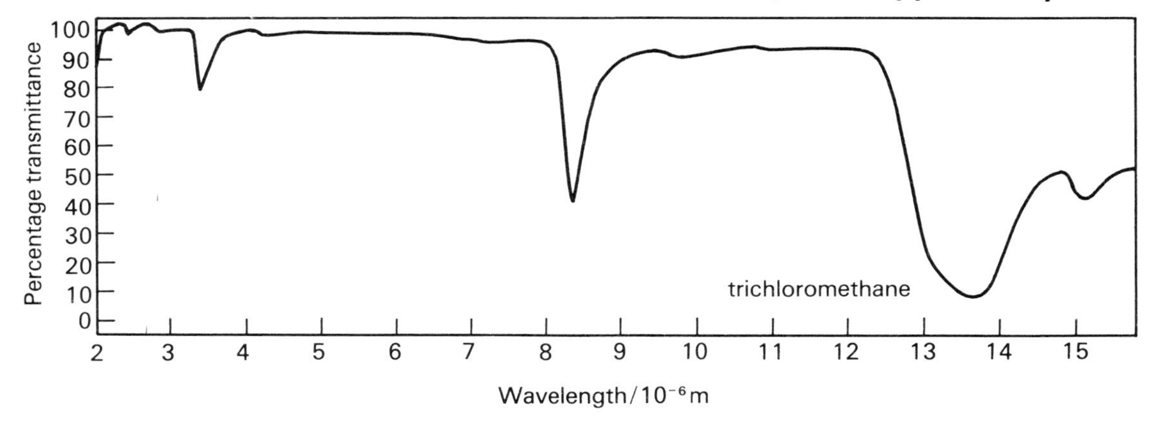

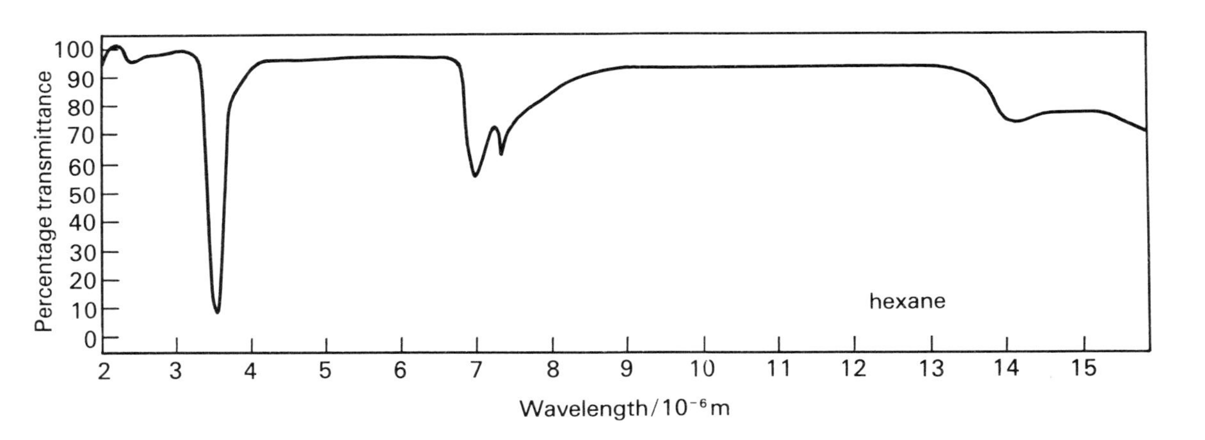

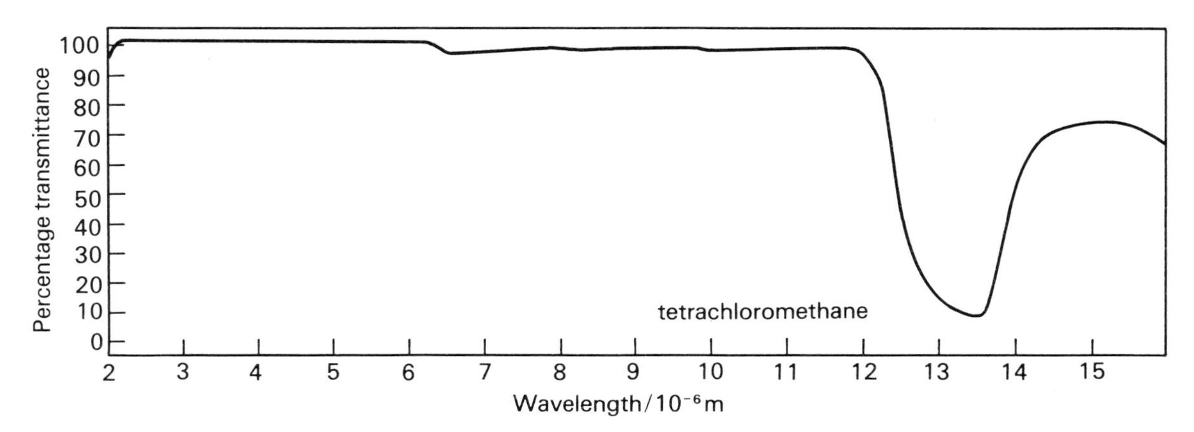

Figure 3.14 The infra-red absorption spectra of three related compounds

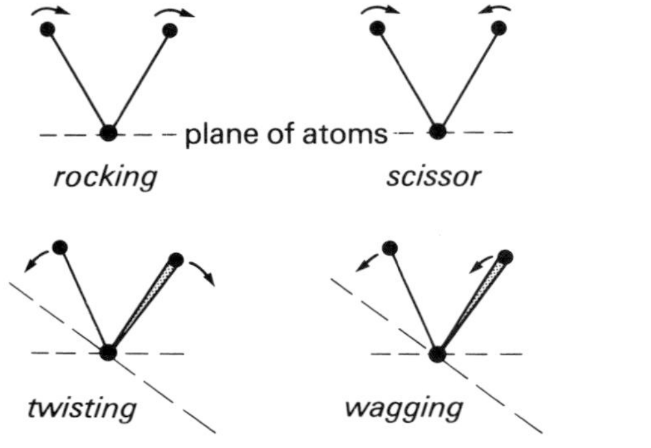

(*a*) bending, or deformation, molecular vibrations

(*b*) stretching, or valence, molecular vibrations

Figure 3.15 Vibrational modes of a non-linear triatomic molecule

orientation in which it is aligned against the field. Transition between these energy states results from the absorption of electromagnetic radiation of suitable frequency. With magnetic field strengths of the order 10^4 A m^{-1}, the energy required to 'flip' the proton lies in the radio-frequency portion of the electromagnetic spectrum. The instrumentation for NMR thus includes an homogeneous magnetic field, a radio-frequency generator, and a radio-frequency receiver. It is convenient to maintain a steady magnetic field and the standard instruments normally contain in addition a 'sweep generator' feeding 'sweep coils' which are used to modify magnetic field over a small range. A schematic diagram of a NMR spectrometer is shown in Figure 3.17.

The precise frequency at which a proton (the nucleus of a standard hydrogen atom) 'flips' is affected by the molecular environment in which it is situated. The interpretation of NMR spectra requires a high degree of skill and experience if the maximum amount of structural information is to be deduced.

Mass spectrometry

The principle of the mass spectrometer and its use in investigating the isotopic composition of elements is discussed on page 24, but the technique can also be used to provide valuable information about the structure of compounds, particularly in the field of organic chemistry. Even with

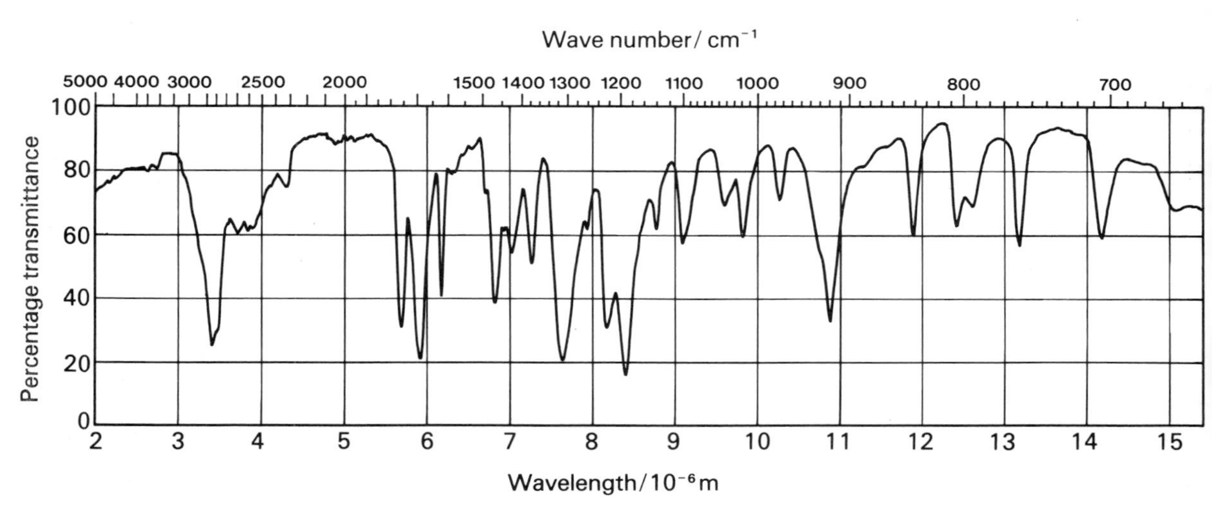

Figure 3.16 The infra-red absorption spectrum of acetylsalicylic acid

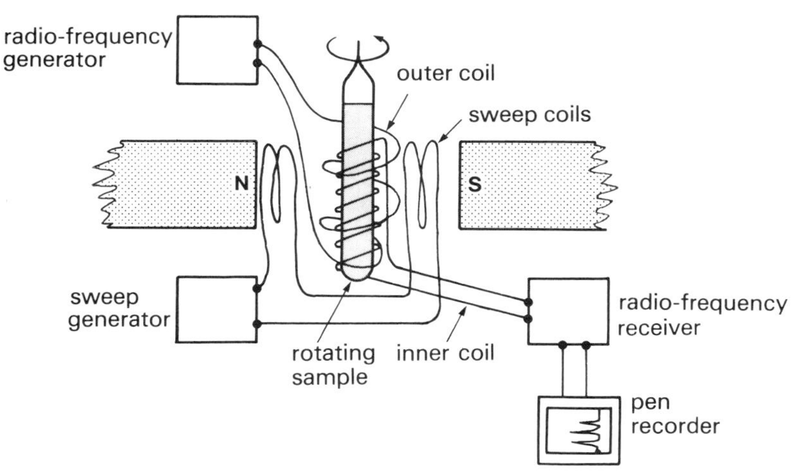

Figure 3.17 An NMR spectrometer

the high degree of vacuum encountered in the mass spectrometer, some air molecules are always present and the mass spectrum of air is therefore superimposed on that of the sample under test. Figure 3.18 shows the mass spectrum of air; the proportion of ions with a multiple charge (such as Ar^{2+}) will depend on the magnitude of the voltage applied between the filament and anode of the ionization chamber because this determines the energy of the ionizing electrons.

Figure 3.19 shows the mass spectrum of propanone. The peaks due to ionized air molecules within the mass range scanned should be noted. Of the remaining peaks in such a spectrum, that with the highest mass corresponds to the ion formed when the complete molecule is given a single positive charge. The relative molecular mass of propanone is thus identified as 58. The remaining peaks are due to fragmentation of some of the molecules, and by identifying the fragments formed structural information concerning the parent molecule may be deduced. The fragments responsible for the remaining peaks in the mass spectrum of propanone have been marked appropriately.

Some modern developments in mass spectrometry are discussed in Chapter 14, page 214.

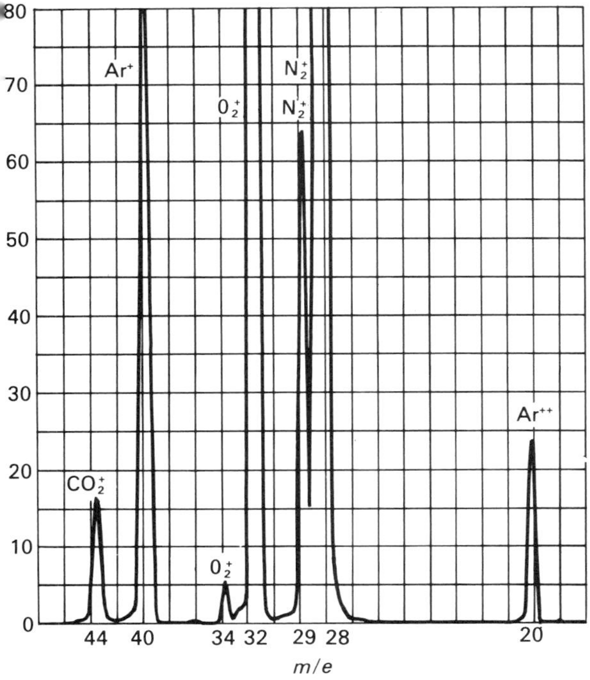

Figure 3.18 Mass spectrum of air

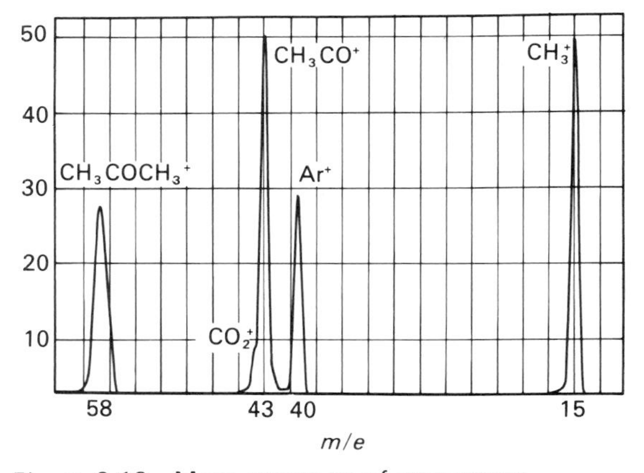

Figure 3.19 Mass spectrum of propanone

Suggestions for further reading

BROWNING, D. R. (ed.) *Spectroscopy* (London: McGraw-Hill, 1969)

WILLIAMS, D. H. and FLEMING, I. *Spectroscopic Methods in Organic Chemistry* (London: McGraw-Hill, 1966)

DENNEY, R. C. *A Dictionary of Spectroscopy* (London: Macmillan, 1973)

CROOKS, J. E. *The Spectrum in Chemistry* (London: Academic Press, 1978)

Interpretation of spectra

An introduction to the interpretation of spectra will be found in JOHNSTONE, A. H. and SHARP, D. W. A. *A Worksheet Introduction to Chemical Spectroscopy* (London: Heinemann Educational Books, 1971)

Problems

1. In the visible absorption spectrum of chlorine, convergence of the discrete lines occurs at 478.5 nm. Calculate the bond dissociation energy of chlorine in kJ mol^{-1}.

2. Solutions of nickel(II) sulphate are green in colour and solutions more concentrated than 0.1 *M* can be determined directly using a colorimeter. With more dilute solutions, however, it is preferable to add drops of dimethylglyoxime solution, which gives an intense red colouration, and then use the colorimeter. Two drops of dimethylglyoxime were added (*a*) to a 0.0001 *M* solution of nickel(II) sulphate and (*b*) to an unknown solution of the salt, and when these were placed in a colorimeter fitted with a suitable filter meter readings of 86 and 20 respectively were obtained. Assuming that the lg of the meter reading is proportional to the concentration, calculate the concentration of the unknown solution of nickel(II) sulphate.

3. Which single analytical technique discussed in this chapter would be most suitable for supplying a rapid answer to each of the following analytical problems? Give reasons for your choice.

(*a*) A passenger is apprehended at an international airport because a customs officer suspects that a medicine bottle in her handbag contains a drug which it is illegal to import into his country. She claims that the tablets are aspirin. Is she telling the truth?

(*b*) The ratio of potassium to sodium ions is found to range only between narrow limits in the urine of healthy adults; however, in certain diseases the ratio is disturbed. How would you investigate the urine of such a patient in order to diagnose disease?

4. Explain how electromagnetic radiation is absorbed or emitted by atoms. How do the spectra of diatomic molecules differ from atomic spectra? Discuss the value of the information which may be obtained from the interpretation of molecular spectra.

(Cambridge Entrance and Awards
—Physical Science)

4. Thermochemistry

CONSIDER 1 mole of magnesium metal and 1 dm^3 of a 2 M solution of hydrochloric acid separately. It is not possible to evaluate the *total* quantity of energy associated with these substances; however, when they react together according to the equation:

$$Mg(s) + 2H^+(aq) \rightarrow Mg^{2+}(aq) + H_2(g)$$

it is possible to determine the *change* in total internal energy resulting from the reaction. This change in the total internal energy associated with a chemical reaction is given the symbol ΔU, where the Greek capital delta (Δ) denotes a finite, measurable change in the quantity U, the internal energy.

$$\Delta U = U_{\text{final state}} - U_{\text{initial state}}$$

If the reaction is carried out in the apparatus shown in Figure 4.1, the change in internal energy is seen to be made up of two components; the change in heat energy, q, and the work performed by the hydrogen gas against the pressure of the atmosphere, w. Following the most recent recommendations of the International Union of Pure and Applied Chemistry (IUPAC), all energies *received* by the system are regarded as *positive* and those *evolved* by the system as *negative*. Thus

$$\Delta U = q + w$$

(The student is warned that this convention is not universally followed, particularly in older textbooks.)

The heat change may be determined using a suitable calorimeter (discussed later in this chapter); in our example the heat change -461.8 kJ mol^{-1}. (The term mol^{-1} is taken to mean the mole quantities specified in the equation and thus refers to two moles of hydrogen ion and one mole of each of the other three materials.)

The work performed can be determined in the following manner:

$$\text{work done} = \text{force} \times \text{distance moved}$$

$$\text{but pressure} = \frac{\text{force}}{\text{area}}$$

$$\text{thus work done} = \text{pressure} \times \text{area} \times \text{distance moved}$$

$$= \text{pressure} \times \text{volume change}$$

$$= p\Delta V$$

Assuming that our reaction is conducted at s.t.p.:

$$p = 1 \text{ atmosphere} = 1.01 \times 10^5 \text{ Pa}$$

$$\Delta V = 22.4 \text{ dm}^3 = 2.24 \times 10^{-2} \text{ m}^3$$

$$p\Delta V = (1.01 \times 10^5 \text{ Pa}) \times (2.24 \times 10^{-2} \text{ m}^3)$$

$$= 2.26 \times 10^3 \text{ J}$$

$$= 2.26 \text{ kJ}$$

An alternative method of calculating the work done involves the general gas equation in the modified form, $p\Delta V = \Delta nRT$.

$$\text{work done} = \Delta nRT$$

$$= (1 \text{ mol}) \times (8.31 \text{ J K}^{-1} \text{ mol}^{-1}) \times (273 \text{ K})$$

$$= 2.27 \times 10^3 \text{ J}$$

$$= 2.27 \text{ kJ}$$

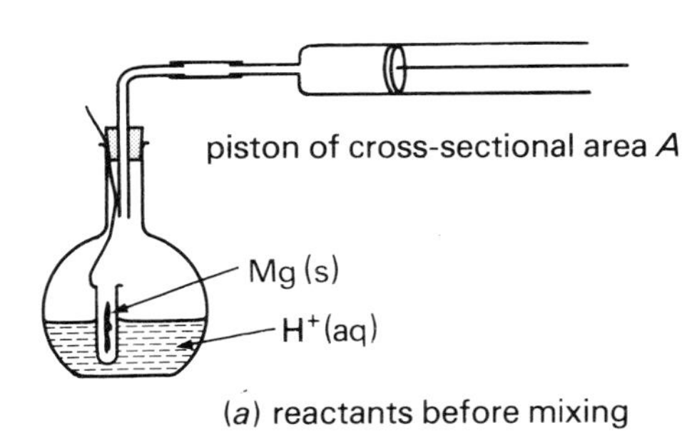

(*a*) reactants before mixing

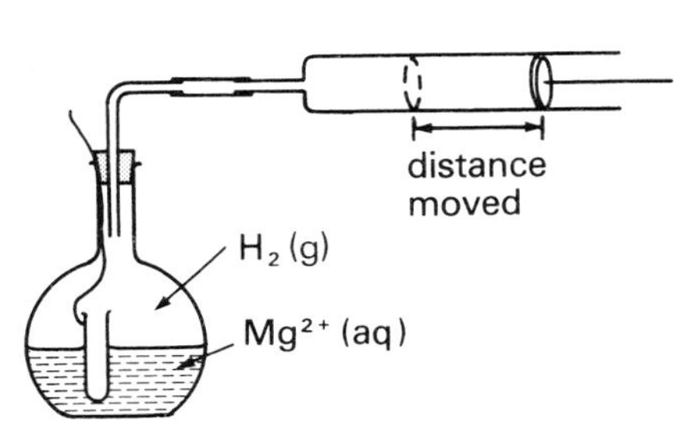

(*b*) products after mixing

Figure 4.1 Energy changes in the reaction between magnesium and hydrochloric acid

These energy quantities can be combined to give the total internal energy change:

$$\begin{aligned}\Delta U &= q + w \\ &= -461.8 + (-2.3)\ \text{kJ mol}^{-1} \\ &= -464.1\ \text{kJ mol}^{-1}\end{aligned}$$

It should be noted that when no change of volume occurs the change in internal energy depends solely on the heat exchanged. Consideration of this example shows that the contribution to the internal energy change of work done is often small compared with the contribution of heat exchanged.

Enthalpy, or heat content

In general, chemical reactions are carried out in the laboratory in apparatus open to the atmosphere, i.e. under conditions of constant pressure. The heat exchanged under these conditions is known as the *enthalpy change* and is given the symbol ΔH.

$$\Delta H = q_{\text{(at constant pressure)}} = \Delta U - p\Delta V$$

Calorimetric techniques

Simple calorimeters

Some calorimeters are illustrated in Figure 4.2. In calorimeter (*a*) the apparatus should be inverted after the reactants have been added to ensure that the thermometer can be read without the mercury thread being hidden by the bung. Care must be taken with the other calorimeters to ensure that this masking of the working range of the thermometer does not occur.

When calculating the enthalpy change from the results obtained using apparatus of this type it is usual to assume that aqueous solutions have the same specific heat capacity as water. This quantity is 4.18 J g^{-1} K^{-1}. An example of this type of calculation is given below.

To determine the enthalpy change of the reaction:

$$NH_4NO_3(s) + aq \rightarrow NH_4NO_3(aq\ 1M)$$

2 grams of ammonium nitrate (0.025 mole) were added to 25 cm^3 of water in a simple calorimeter. A temperature *fall* of 5.2 °C was observed.

Energy *absorbed* in reaction =

$$(4.18\ \text{J g}^{-1}\ \text{K}^{-1}) \times (25\ \text{g}) \times (5.2\ \text{K})$$
$$= 543\ \text{J}$$

This energy resulted from 0.025 mole of ammonium nitrate dissolving to form a 1 M solution; thus for 1 mole of reaction occurring:

Energy absorbed in reaction =

$$543 \times \frac{1}{0.025}\ \text{J mol}^{-1}$$
$$= 217\ \text{kJ mol}^{-1}$$

As the reactants have less energy than the products in this example, ΔH for the reaction is positive. The result can be expressed by the statement:

$$NH_4NO_3(s) + aq \rightarrow NH_4NO_3(aq\ 1M);\quad \Delta H = +217\ \text{kJ mol}^{-1}$$

Figure 4.2 Some simple calorimeters

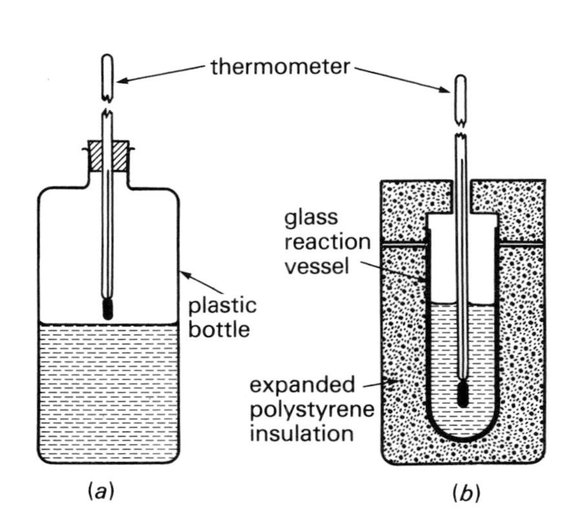

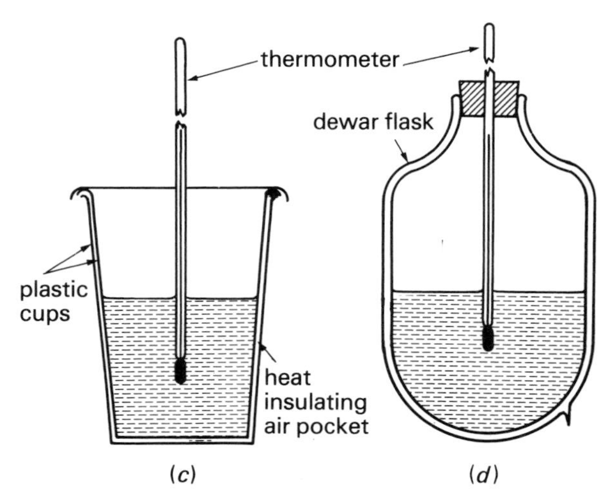

The electrical compensation calorimeter

The simple calorimeters in Figure 4.2 suffer in accuracy from the difficulty in estimating heat losses generally, and in particular that proportion of the heat that is used to change the temperature of the apparatus and not the solution. Such difficulties are largely overcome in the *electrical compensation calorimeter* (Figure 4.3). The example below will make the principle behind the method clear.

To investigate the enthalpy change of the reaction:

$$Cu^{2+}(aq) + Zn(s) \rightarrow Cu(s) + Zn^{2+}(aq)$$

100 cm^3 of 0.2 M copper(II) sulphate (containing 0.02 moles of Cu^{2+}(aq)) were placed in the calorimeter and a small excess of zinc metal was added. The maximum temperature rise was noted while the flask was swirled to give good mixing. The electrical supply was then switched on, and it was found that 4320 joules had to be supplied to produce the same temperature rise as that recorded during the reaction.

Energy evolved during reaction

$$= 4.32 \text{ kJ}$$

$$\Delta H = -4.32 \times \frac{1}{0.02} \text{ kJ mol}^{-1}$$

$$= -216 \text{ kJ mol}^{-1}$$

The enthalpy changes of some other displacement reactions are given in Table 4.1.

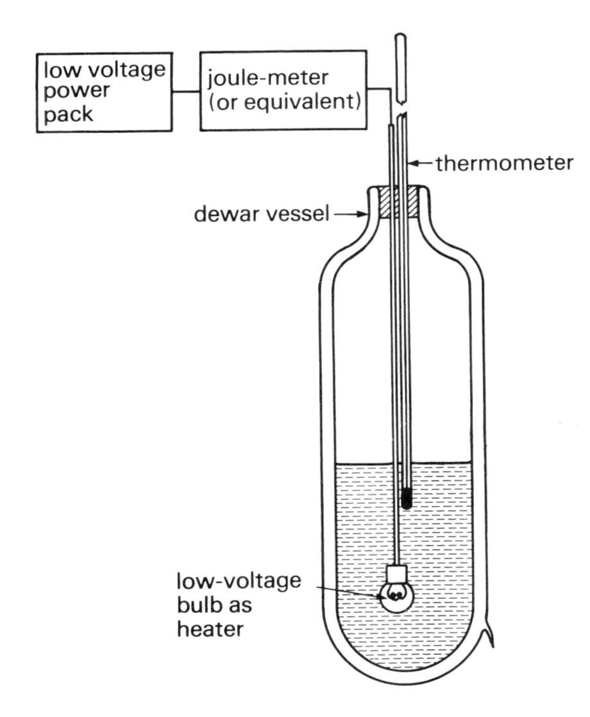

Figure 4.3 An electrical compensation calorimeter

Table 4.1 Enthalpies of some displacement reactions

Reaction	ΔH/kJ mol^{-1} *
$Cu(s) + 2Ag^{+}(aq) \rightarrow Cu^{2+}(aq) + 2Ag(s)$	−147
$Zn(s) + 2Ag^{+}(aq) \rightarrow Zn^{2+}(aq) + 2Ag(s)$	−363
$Mg(s) + Cu^{2+}(aq) \rightarrow Mg^{2+}(aq) + Cu(s)$	−526
$Mg(s) + 2Ag^{+}(aq) \rightarrow Mg^{2+}(aq) + 2Ag(s)$	−673

* The term mol^{-1} here refers to the quantities specified in the equation.

The Thiemann or fuel calorimeter

A spirit-burner containing the liquid under test is weighed and ignited inside the apparatus, as shown in Figure 4.4. The apparatus is constantly stirred to ensure uniformity of temperature and burning is continued until a convenient temperature rise is obtained. The spirit-burner is extinguished and quickly reweighed. The electrical supply is switched on and the energy required to give a temperature rise identical with that given by the reaction is determined.

The specimen results given below show the principle of the calculation; they correspond to the reaction in which ethanol burns in air according to the equation:

$$C_5H_5OH(l) + 3O_2(g) \rightarrow 2CO_2(g) + 3H_2O(l)$$

Initial mass of spirit burner + ethanol = 56.384 g
Final mass of spirit burner + ethanol = 52.661 g

Mass of ethanol = 3.723 g

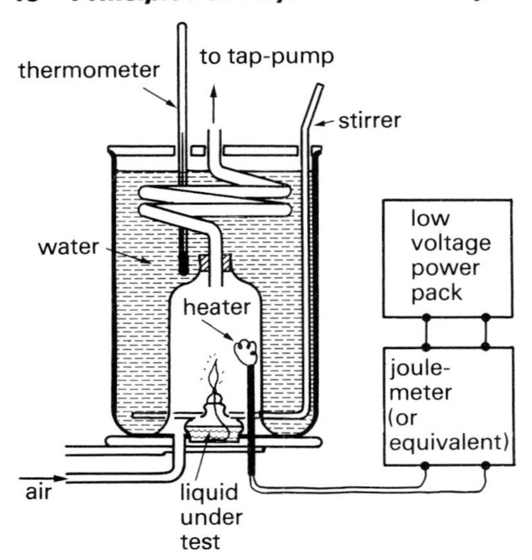

Figure 4.4 The Thiemann or fuel calorimeter

Electrical energy to give same rise in temperature as the burning of 3.723 g ethanol = 110 kJ.
1 mole of ethanol is 46 g.
Enthalpy of combustion of ethanol,

$$\Delta H_{comb} = -110 \times \frac{46}{3.723} \text{ kJ mol}^{-1}$$
$$= -1360 \text{ kJ mol}^{-1}$$

Results obtained using this apparatus can also be processed by estimating a thermal capacity for the apparatus from the specific heat capacities and masses of its components. The electrical method described above, however, is both more convenient and more accurate.

With suitable modification, the Thiemann apparatus can be used to determine the enthalpies of combustion of gases and some solids.

The bomb calorimeter

The apparatus is shown diagrammatically in Figure 4.5 prior to use with a solid test material. With suitable modification the technique can be adapted for use with liquids and gases. The sample is ignited and the energy liberated is distributed evenly by the stirrer. The maximum temperature rise is noted and when the thermometer records a constant value, the immersion heater is switched on and the energy required to give an identical temperature rise electrically is determined.

In an experiment in which graphite was burned according to the equation:

$$C(s) + O_2(g) \rightarrow CO_2(g)$$

2.361 g of graphite on burning gave the same temperature rise as 77.2 kJ of electrical energy. 1 mole of carbon atoms is 12 g.

$$\Delta H_{comb} = -77.2 \times \frac{12}{2.361} \text{ kJ mol}^{-1}$$
$$= -392 \text{ kJ mol}^{-1}$$

It should be noted that the bomb calorimeter makes determinations under constant *volume* conditions and the energy change thus corresponds to ΔU. When, however, as in our example, the number of moles of gas, and hence the volume, does not change between reactants and products this value also corresponds to ΔH. For reactions in which the number of moles of gas changes between reactants and products, the work done when constant pressure conditions are restored must be calculated and the appropriate correction applied.

In making accurate determinations with the bomb calorimeter it is necessary to make an allowance for the energy supplied in igniting the sample and the energy of combustion of the ignition filament. The calculation can also be carried out by estimating a value for the thermal capacity of the apparatus or by carrying out a calibration experi-

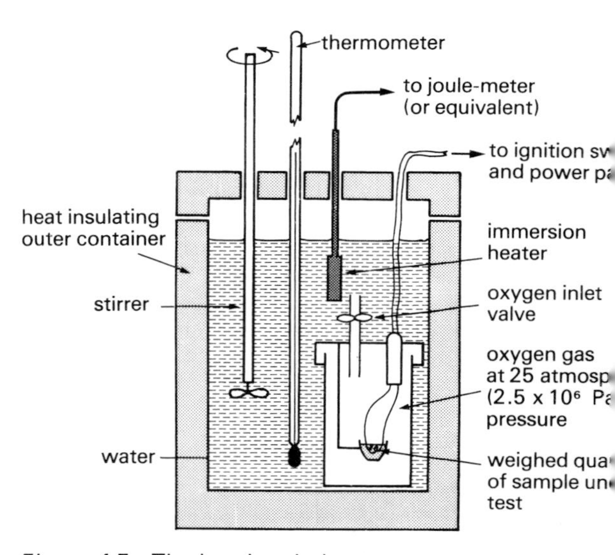

Figure 4.5 The bomb calorimeter

ment with a substance of known enthalpy of combustion. Benzenecarboxylic acid (benzoic acid), $\Delta H_{comb} = -3228 \text{ kJ mol}^{-1}$, is commonly used for this purpose.

Measuring energy electrically

In using the electrical compensation calorimeter, Thiemann apparatus, and the bomb calorimeter it is necessary to measure energy injected into the system electrically. One of the three equivalent circuits shown in Figure 4.6 can be used for this purpose.

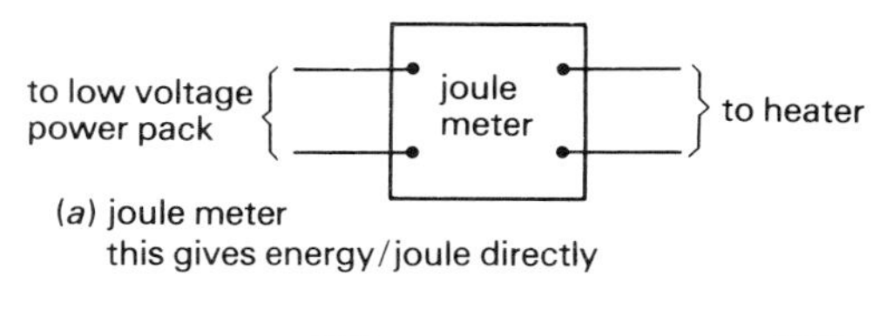

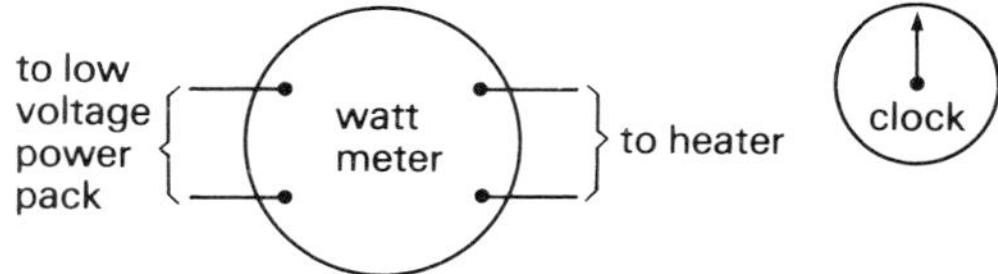

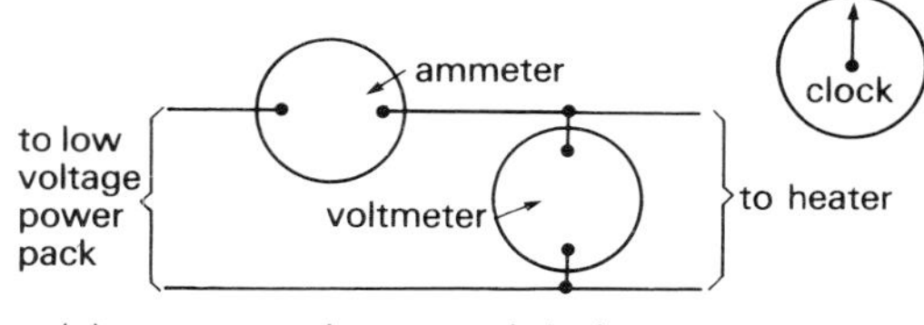

Figure 4.6 Equivalent circuits to measure energy electrically

The standard state

The numerical values for enthalpy changes can be affected by:

(*a*) The temperature at which the determination is carried out.
(*b*) The physical state (solid, liquid, or gas) of reactants and products.
(*c*) The pressure of gaseous reactants and products.
(*d*) The concentrations of solutions involved.

For tabulation purposes, therefore, it is necessary to refer enthalpy changes to a standard state. The enthalpy change under standard conditions is given the symbol $\Delta H^{\ominus}$. The standard conditions chosen for this purpose are as follows. Gases must be at a pressure of one atmosphere (760 mmHg or 101 kPa), solutions must have unit concentration. The reaction must be carried out at a stated temperature (unless otherwise specified this is taken as 298 K). The substances must be in their normal physical states at the stated temperature.

$$C(s) + O_2(g) \rightarrow CO_2(g); \quad \Delta H^{\ominus}_{298} = -393.5 \text{ kJ mol}^{-1}$$

N.B.—When the reaction involves substances such as carbon that can exist in different allotropic forms, the stable form at the stated temperature is taken as standard. Thus the equation above refers to carbon in the form of graphite.

Two alternative thermochemical equations are possible in discussing the combustion enthalpy of hydrogen:

$$H_2(g) + \tfrac{1}{2}O_2(g) \rightarrow H_2O(l); \quad \Delta H^{\ominus} = -286 \text{ kJ mol}^{-1}$$

$$2H_2(g) + O_2(g) \rightarrow 2H_2O(l); \quad \Delta H^{\ominus} = -572 \text{ kJ mol}^{-1}$$

When such ambiguity is possible it is essential that the enthalpy change is referred to the appropriate equation. A useful additional definition that allows us to avoid the need to write the equation for certain reactions without ambiguity is the *standard enthalpy of formation* of a compound.

The *standard enthalpy of formation* of a compound is the standard enthalpy change that occurs when one mole of the substance is made from its constituent elements under standard conditions. The symbol given to this quantity is $\Delta H_f^{\ominus}$. Thus

$$\Delta H_f^{\ominus}[CO_2(g)] = -393.5 \text{ kJ mol}^{-1}$$
$$\Delta H_f^{\ominus}[H_2O(l)] = -286 \text{ kJ mol}^{-1}$$

N.B.—When the temperature subscript is suppressed, 298 K is assumed. Similarly the standard enthalpy of combustion of a substance is the standard enthalpy change that occurs on the complete combustion of one mole of the substance under standard conditions. Thus

$$\Delta H^{\ominus}_{comb}[C(s)] = -393.5 \text{ kJ mol}^{-1}$$
$$\Delta H^{\ominus}_{comb}[H_2(g)] = -286 \text{ kJ mol}^{-1}$$

By convention the standard enthalpies of formation of elements is taken as zero. Thus

$$\Delta H_f^\ominus [C(s)] = 0 \text{ kJ mol}^{-1}$$
$$\Delta H_f^\ominus [H_2(g)] = 0 \text{ kJ mol}^{-1}$$

Exothermic and endothermic reactions

Reactions in which heat is evolved (i.e. for which *ΔH is negative*) are said to be *exothermic*, while those in which heat is absorbed (i.e. for which *ΔH is positive*) are said to be endothermic. Substances with negative standard enthalpies of formation are sometimes called *exothermic compounds* and those with positive standard enthalpies of formation may be described as *endothermic compounds.*

The law of constant heat summation

This law was first enunciated by G. H. Hess in 1840 and, in modern wording, states: *the total enthalpy change involved in a chemical reaction is constant and is unaffected by the route along which the reaction proceeds.*

Hess's Law is a direct consequence of the Law of Conservation of Energy, also known as the First Law of Thermodynamics. This law states: *energy can neither be created nor destroyed.* It is one of the best established laws in science.

The results in Table 4.1 provide a neat experimental confirmation of Hess's Law. When zinc metal displaces silver ions:

$$(1)\ Zn(s) + 2Ag^+(aq) \rightarrow Zn^{2+}(aq) + 2Ag(s); \quad \Delta H^\ominus = -363 \text{ kJ mol}^{-1}$$

If, however, the zinc is first used to displace copper(II) ions:

$$(2)\ Zn(s) + Cu^{2+}(aq) \rightarrow Zn^{2+}(aq) + Cu(s); \quad \Delta H^\ominus = -216 \text{ kJ mol}^{-1}$$

and the copper displaced is then used to displace silver ions:

$$(3)\ Cu(s) + 2Ag^+(aq) \rightarrow Cu^{2+}(aq) + 2Ag(s); \quad \Delta H^\ominus = -147 \text{ kJ mol}^{-1}$$

If Hess's Law holds, $\Delta H^\ominus$ for reaction (1) should equal the sum of the $\Delta H^\ominus$ values for reactions (2) and (3); this is seen to be the case.

The tabulation of energy data

An enormous quantity of data concerning enthalpy changes of reaction have accumulated since the classical work in this field of D. Berthelot in the last century. One convenient form in which this energy can be tabulated is as standard enthalpies of formation. Although these quantities can be determined directly for some compounds, notably oxides, by the calorimetric techniques discussed previously, it is not always possible to do this. The standard enthalpy of formation of ethane, for example, cannot be directly determined:

$$2C(s) + 3H_2(g) \rightarrow C_2H_6(g)$$

There are two main reasons why this is so; firstly the direct reaction of carbon and hydrogen is exceedingly slow and secondly such direct combination would allow the possibility of forming all the other hydrocarbons (an indefinitely large number), it being impossible to demand the formation of one particular product. Such enthalpies may, however, be determined using a suitable *energy cycle.* The construction of such energy cycles assumes the validity of Hess's Law of Constant Heat Summation.

Simple energy cycles

This device makes use of the fact that the direct determination of standard enthalpies of formation of oxides is normally possible. The cycle is shown in Figure 4.7(*a*) in the general case, and in Figure 4.7(*b*) specifically applied to ethane.

$$\Delta H_1^\ominus = \Delta H_f^\ominus [C_2H_6(g)] = \Delta H_2^\ominus - \Delta H_3^\ominus$$

$$\Delta H_2^\ominus = 2 \times \Delta H_f^\ominus [CO_2(g)] + 3 \times \Delta H_f^\ominus [H_2O(1)]$$
$$= (-787) + (-588) \text{ kJ mol}^{-1}$$
$$= -1645 \text{ kJ mol}^{-1} \text{ (from tables)}$$

$$\Delta H_3^\ominus = \Delta H_{comb}^\ominus [C_2H_6(g)]$$
$$= -1560 \text{ kJ mol}^{-1} \text{ (from tables)}$$

thus, $\Delta H_f^\ominus [C_2H_6(g)]$
$$= -1645 - (-1560) \text{ kJ mol}^{-1}$$
$$= -85 \text{ kJ mol}^{-1}.$$

Another example of the use of energy cycles of this type is the oxidation of methanol (methyl alcohol) to methanal (formaldehyde) according to the equation:

$$CH_3OH(l) + \tfrac{1}{2}O_2(g) \rightarrow HCHO(g) + H_2O(l)$$

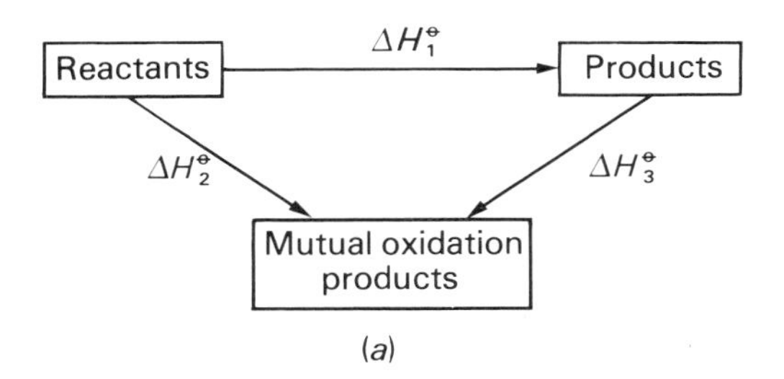

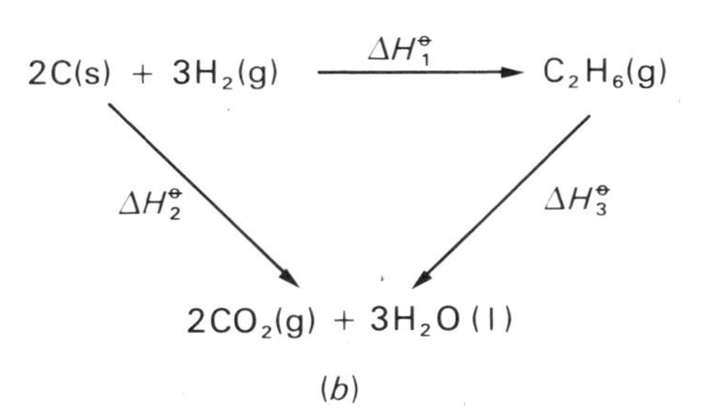

Figure 4.7 Simple energy cycle for the indirect determination of standard enthalpies of formation applied to ethane

As the reaction involves the incomplete oxidation of methanol it is not possible to determine the standard enthalpy change directly. The appropriate energy cycle is shown in Figure 4.8.

$$\Delta H_1^\ominus = \Delta H_2^\ominus - \Delta H_3^\ominus$$

$$\Delta H_2^\ominus = \Delta H_{comb}^\ominus\,[CH_3OH(l)]$$
$$= -726 \text{ kJ mol}^{-1} \text{ (from tables)}$$

$$\Delta H_3^\ominus = \Delta H_{comb}^\ominus\,[HCHO(g)]$$
$$= -550 \text{ kJ mol}^{-1} \text{ (from tables)}$$

$$\Delta H_1^\ominus = -725 - (550) \text{ kJ mol}^{-1}$$
$$= -175 \text{ kJ mol}^{-1}.$$

$CH_3OH(l) + \frac{1}{2}O_2(g)$ $\xrightarrow{\Delta H_1^\ominus}$ $HCHO(g) + H_2O(l)$

$\Delta H_2^\ominus$ $\Delta H_3^\ominus$

$CO_2(g) + 2H_2O(l)$

Figure 4.8 Using an energy cycle to determine the standard enthalpy change for the oxidation of methanol to methanal

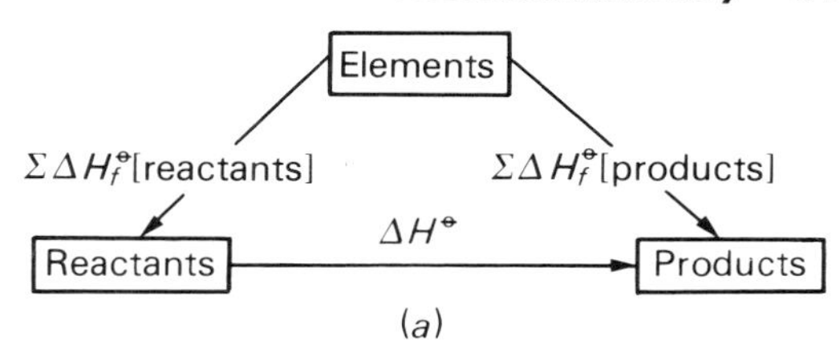

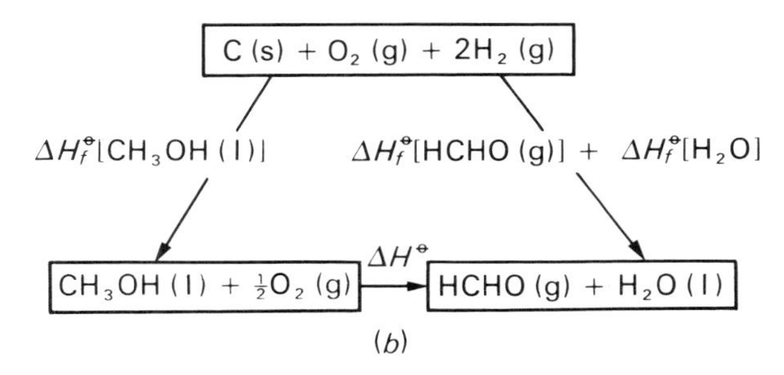

Figure 4.9 An energy cycle using standard enthalpies of formation to give a standard enthalpy of reaction indirectly

An alternative energy cycle for the determination of this standard enthalpy of reaction is illustrated in general in Figure 4.9(*a*), and specifically applied to this reaction in Figure 4.9(*b*). It will be observed that the standard enthalpy of reaction is the sum of the standard enthalpies of formation of the products minus the sum of the standard enthalpies of formation of the reactants.

$$\Delta H^\ominus = \Sigma\Delta H_f^\ominus\,[\text{products}] - \Sigma\Delta H_f^\ominus\,[\text{reactants}]$$

Substituting the appropriate table values for the reaction we are considering:

$$\Delta H^\ominus = \{-117 + (-286)\} - (-238) \text{ kJ mol}^{-1}$$
$$= -165 \text{ kJ mol}^{-1}.$$

The poor agreement between the values calculated from the different cycles illustrates the difficulty in obtaining data which are self-consistent, even from the same set of published tables.

Standard enthalpy of atomization of an element, $\Delta H_{atom}^\ominus$

This is the enthalpy change involved when one mole of gaseous atoms is formed from the element in the defined physical state under standard conditions. Two methods by which this important quantity can be determined are as follows.

(1) This method is applicable to gaseous diatomic elements such as chlorine. Examination of the appropriate region of its absorption spectrum reveals a series of convergent lines. As discussed in Chapter 3, the convergent limit of these lines corresponds to the bond dissociation energy of the molecule.

$$Cl_2(g) \rightarrow 2Cl(g); \Delta H^{\ominus} = +242 \text{ kJ mol}^{-1}$$

thus, $\Delta H^{\ominus}_{atom}[Cl_2(g)] = +121 \text{ kJ mol}^{-1}$.

(2) For an element such as tin, the standard enthalpy of atomization refers to the process:

$$Sn(s) \rightarrow Sn(g); \Delta H^{\ominus}_{atom} = +301 \text{ kJ mol}^{-1}$$

N.B.—The data refers to the 'white' allotrope of tin, which is the stable form at 298 K. This quantity is obtained by summing the following calorimetrically obtained data:

(*a*) Energy required to raise the temperature of solid tin from 298 K to its melting point, 505 K.
(*b*) Its molar latent heat of fusion, $\Delta H^{\ominus}_{fus}$
(*c*) Energy required to raise the temperature of tin from 505 K to its boiling point, 2960 K.
(*d*) Its molar latent heat of evaporation, $\Delta H^{\ominus}_{evap}$

In deducing a value for this quantity in the case of carbon, a modification of method (2) has to be employed. Several complications arise including the fact that carbon does not melt but sublimes around 4000 K. The value quoted for this important quantity, $+715 \text{ kJ mol}^{-1}$, is thus one of the most unreliable in the whole of thermochemistry.

Bond energies

These are the energies required to break one mole of bonds of the specified type. As we have seen (page 39), this quantity can be determined for the bonds between the two atoms in a diatomic molecule directly by spectroscopy. For example, the hydrogen–chlorine bond strength can be determined from examination of the absorption spectrum of hydrogen chloride gas:

$$HCl(g) \rightarrow H(g) + Cl(g); \Delta H^{\ominus} = +431 \text{ kJ mol}^{-1}$$

The bond energy of hydrogen–chlorine, given the symbol *E*(H–Cl), is thus 431 kJ mol^{-1}.

It is not possible to determine the bond energies involved in polyatomic molecules in this way because of the more complex nature of their spectra.

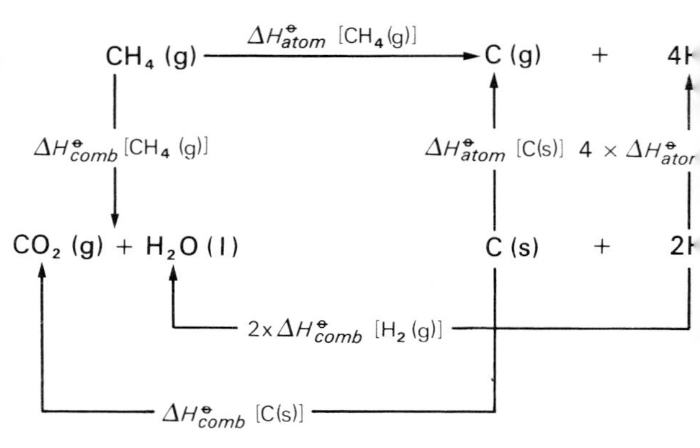

Figure 4.10 Energy cycle to deduce the standard enthalpy of atomization of methane

Such information can, however, be obtained from the standard enthalpy of atomization of the compound. In the case of compounds this quantity is defined as the standard entropy change that occurs when one mole of the compound is converted into gaseous atoms. (Compare this definition carefully with that given previously for the standard enthalpy of atomization of an element.) Using methane as an example, this quantity refers to the process:

$$CH_4(g) \rightarrow C(g) + 4H(g); \quad \Delta H^{\ominus}_{atom} = +1662 \text{ kJ mol}^{-1}$$

The value of standard enthalpy of atomization of the compound must be obtained from an energy cycle of the type illustrated in Figure 4.10. In the case of methane, atomization of one mole involves breaking four moles of carbon–hydrogen bonds, therefore

$$4 \times E(\text{C–H}) = 1662 \text{ kJ mol}^{-1}$$
$$E(\text{C–H}) = 415 \text{ kJ mol}^{-1}$$

The corresponding atomization equation for ethane is:

$$C_2H_6(g) \rightarrow 2C(g) + 6H(g); \quad \Delta H^{\ominus}_{atom} = +2924 \text{ kJ mol}^{-1}$$

thus, $[6 \times E(\text{C–H})] + E(\text{C–C}) = 2924 \text{ kJ mol}^{-1}$

Substituting the previously deduced value for *E*(C–H) and rearranging yields:

$$E(\text{C–C}) = 2924 - (6 \times 415) \text{ kJ mol}^{-1}$$
$$= 434 \text{ kJ mol}^{-1}.$$

Values for bond dissociation energies deduced from

standard enthalpies of atomization vary quite considerably when deduced using different compounds because of the variable effect of neighbouring atoms on the bond. The data in Table 4.2 illustrate this.

Table 4.2 Bond strengths of carbon–chlorine in some compounds

Compound	*Formula*	E(C–Cl)*kJ mol*$^{-1}$
Chloromethane	CH_3Cl	335
Tetrachloromethane	CCl_4	327
Chloroethane	C_2H_5Cl	342

The values quoted in tables of data are thus averages and should not be regarded as accurately stating the magnitude in any particular molecular environment; these values are normally derived from considering larger molecules than those containing one or two carbon atoms. Values of E(C–H) and E(C–C) more generally applicable than those calculated above can be deduced from data for the atomization of butane and pentane:

$$C_4H_{10}(g) \rightarrow 4C(g) + 10H(g); \quad \Delta H^{\ominus}_{atom} = +5165 \text{ kJ mol}^{-1}$$

$$C_5H_{12}(g) \rightarrow 5C(g) + 12H(g); \quad \Delta H^{\ominus}_{atom} = +6337 \text{ kJ mol}^{-1}$$

Thus $3E(\text{C–C}) + 10E(\text{C–H}) = 5165 \text{ kJ mol}^{-1}$

and $4E(\text{C–C}) + 12E(\text{C–H}) = 6337 \text{ kJ mol}^{-1}$

Treating these as simultaneous equations and solving yields:

$$E(\text{C–H}) = 412 \text{ kJ mol}^{-1}$$

and $$E(\text{C–C}) = 347 \text{ kJ mol}^{-1}$$

Trends in combustion enthalpies of related compounds

In Figure 4.11 the enthalpies of combustion of (*a*) the first seven straight-chain alkanes and (*b*) the first seven straight-chain alcohols are plotted against the number of carbon atoms in each compound. The linearity of each plot provides additional evidence that each bond has a particular amount of energy associated with it. The linearity occurs because each compound in an homologous series has one more carbon–carbon bond and two more carbon–hydrogen bonds than its immediate predecessor. This would lead one to expect that the slopes of both graphs would be parallel and this, to a good approximation, is found to be the case.

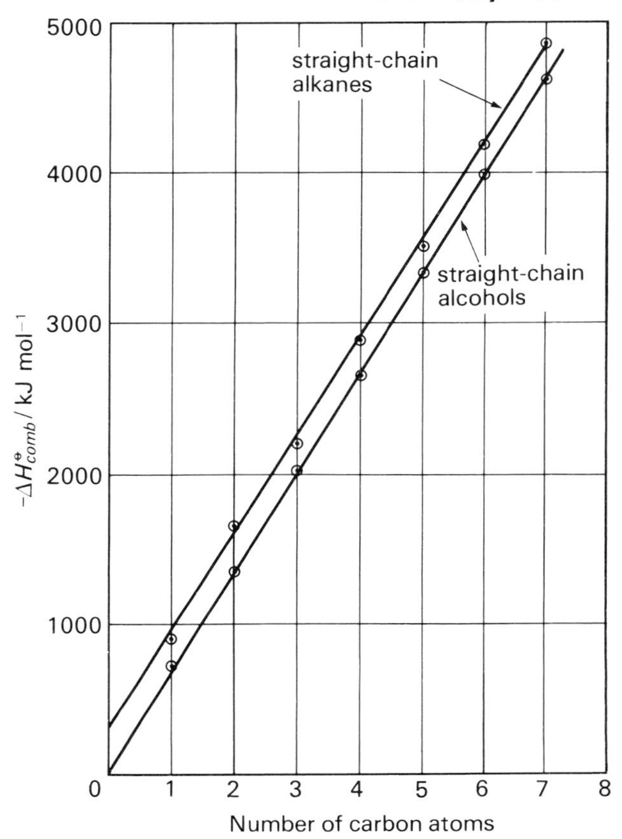

Figure 4.11 Enthalpies of combustion of alkanes and alcohols

The alcohols have the general formula $C_nH_{2n+1}OH$. When $n = 0$, this gives the formula for water; because $\Delta H^{\ominus}_{comb}[H_2O(l)] = 0$, the plot for the alcohols passes through the origin. The alkanes have the general formula C_nH_{2n+2}. When $n = 0$ in this case we obtain the formula of hydrogen; $\Delta H^{\ominus}_{comb}[H_2(g)] = -286 \text{ kJ mol}^{-1}$ and this agrees reasonably with the intercept of the graph in Figure 4.11.

Estimating enthalpy changes using bond energies

Bond energies can be used to estimate enthalpy changes that cannot be directly determined. However, as bond energies are variable the results so obtained must of necessity be approximate. The following example demonstrates the method.

To estimate ΔH for the reaction

$$C_2H_6(g) + Br_2(l) \rightarrow C_2H_5Br(l) + HBr(g)$$

Energy to break bonds:

$$1 \times E(\text{C–H}) = 412 \text{ kJ mol}^{-1}$$
$$1 \times E(\text{Br–Br}) = 193 \text{ kJ mol}^{-1}$$
$$\text{Total} = 605 \text{ kJ mol}^{-1}$$

Energy from making bonds:

$$1 \times E(\text{C–Br}) = 209 \text{ kJ mol}^{-1}$$
$$1 \times E(\text{H–Br}) = 366 \text{ kJ mol}^{-1}$$
$$\text{Total} = 575 \text{ kJ mol}^{-1}$$

ΔH = [total energy from making bonds] – [total energy to break bonds]
= 605 – 575 kJ mol^{-1}
= 30 kJ mol^{-1}

Electron affinity

This is the enthalpy change that occurs when a mole of isolated gaseous atoms gain an electron each to become an anion. In the case of chlorine atoms, for example:

$$Cl(g) + e^- \rightarrow Cl^-(g);\ \Delta H = -364 \text{ kJ mol}^{-1}$$

It will be seen that this enthalpy change is numerically equal, but opposite in sign, to the first ionization energy of the chloride ion.

The second electron affinity, the enthalpy change occurring when a second electron is added to a mole of isolated gaseous singly charged anions, is generally positive because of the repulsion effect of the electron already present:

$$O(g) + e^- \rightarrow O^-(g);\ \Delta H = -141 \text{ kJ mol}^{-1}$$
$$O^-(g) + e^- \rightarrow O^{2-}(g);\ \Delta H = +791 \text{ kJ mol}^{-1}$$

Lattice energy (L.E.)

This is the standard enthalpy change for the formation of the crystal lattice from its constituent gaseous ions. In the case of sodium chloride, for example, this corresponds to the standard enthalpy change for the process:

$$Na^+(g) + Cl^-(g) \rightarrow NaCl(s)$$

According to electrostatic theory, the force F between two charged particles in a vacuum is given by

$$F = \frac{q_1 \times q_2}{4\pi\varepsilon_0 r^2}$$

In which q_1 and q_2 are the magnitudes of the charges, r is the distance between the centres of the particles, and ε_0 is the permittivity of a vacuum. This expression can be integrated to yield the work done in bringing together such a pair of charged particles from infinity to a separation of r.

$$\text{Work done} = -\int_{\infty}^{r} \frac{q_1 \times q_2}{4\pi\varepsilon_0 r^2} dr = -\frac{q_1 \times q_2}{4\pi\varepsilon_0 r}$$

The distance separating the centres of ions in a crystal lattice, r, can be determined using X-ray diffraction as discussed in Chapter 5. When the calculation is extended to take account of the interactions involved when Avogadro's number of ions form into rows, then into two-dimensional networks, and finally into a three-dimensional lattice, an expression for the lattice energy is obtained.

$$\text{L.E.} = -\frac{LMz^+z^-e^2}{4\pi\varepsilon_0 r}$$

in which L is Avogadro's constant, z^+ and z^- are the number of charges carried by the cation and anion respectively, e is the charge carried by an electron, and M is the Madelung constant (this takes account of the successive additions of ions to form a lattice and its magnitude depends on the particular arrangement of the ions within the crystal under consideration). This equation allows the calculation of theoretical values for lattice energies.

Lattice energies can also be deduced using a special form of energy cycle, known as a *Born–Haber cycle*, in which experimentally determined quantities are inserted. Lattice energies determined from a Born–Haber cycle are thus *experimental* values.

The Born–Haber cycle for sodium chloride

This is shown in Figure 4.12. It may be regarded as an energy 'balance sheet' for the standard enthalpy of formation of sodium chloride. Starting from a datum line with sodium metal and chlorine gas, energy expenditure is indicated by the upward arrows and energy gains by the downwards arrows. All the quantities required to complete the cycle, except the lattice energy, can be determined *experimentally* by methods discussed previously. Inserting these values, therefore, allows an experimental value for the lattice energy of sodium chloride to be deduced.

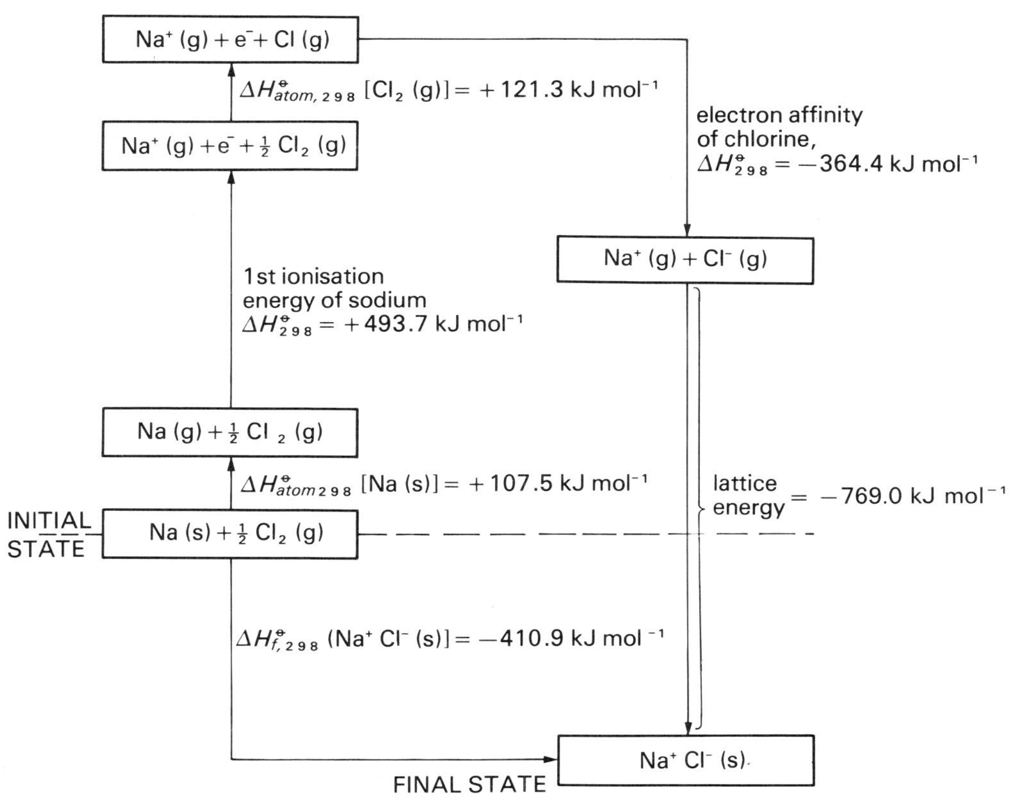

Figure 4.12 Born–Haber cycle for the formation of sodium chloride

Table 4.3 compares some experiment values for lattice energies determined using a Born–Haber cycle with the theoretical values. Examination of this table shows that strikingly good agreement occurs (within one per cent) between theoretical and experimental values for lattices energies of the sodium and potassium halides. This suggests that the electrostatic model adopted for the calculation

Table 4.3 Comparison of theoretical and experimental values of lattice energy for some compounds

		Lattice energy/kJ mol⁻¹		
Compound	*Formula*	*Theoretical value*	*Experimental value*	*% difference**
Sodium chloride	NaCl	−766	−769	0.39
Sodium bromide	NaBr	−731	−736	0.68
Sodium iodide	NaI	−686	−688	0.29
Potassium chloride	KCl	−692	−698	0.86
Potassium bromide	KBr	−667	−672	0.74
Potassium iodide	KI	−631	−632	0.16
Silver chloride	AgCl	−769	−916	16
Silver bromide	AgBr	−759	−908	16
Silver iodide	AgI	−736	−865	15

* $\frac{\text{Experimental value} - \text{Theoretical value}}{\text{Experimental value}} \times 100$

of lattice energies is a good one in these cases. The silver halides show considerable discrepancy (around 15 per cent) between theoretical and experimental values of lattice energy. This suggests that the purely electrostatic model is not adequate to explain the structure of silver halides. Moreover, because the experimental values are *consistently greater* than the theoretical ones for these compounds, it seems likely that the discrepancy is due to the electron clouds around the ions interacting instead of remaining discrete as postulated by the electrostatic model. The types of bonding possible in crystals are discussed more fully in Chapter 5.

Stability

Claude Berthelot suggested that reactions between materials will occur when the products of reaction have a lower enthalpy than the reactants. According to this theory, compounds with negative values of enthalpy of formation will be stable with respect to their constituent elements and, in general, reactions for which the standard enthalpy change is negative will be those that occur. It is now known that this is not a completely reliable rule; several stable endothermically formed substances are known, and many endothermic reactions are known. The rule is, however, a reasonable general guide, particularly when the value of $\Delta H^\ominus$ is numerically large. The problem is discussed further in Chapter 11.

A further warning is essential in the application of this rule. Consider the thermochemical data for the combustion of sucrose (cane-sugar) in oxygen:

$$C_{12}H_{22}O_{11}(s) + 12O_2(g) \rightarrow 12CO_2(g) + 11H_2O(l); \quad \Delta H^\ominus = -5650 \text{ kJ mol}^{-1}.$$

The large negative value of the standard enthalpy change for this reaction suggests that cane-sugar on exposure to air will spontaneously burn to form carbon dioxide and water; this appears to be contrary to everyday experience. The dilemma is resolved when we realize that enthalpy changes tell us nothing about the *rate* at which chemical reactions occur and in our example the combustion is occurring at so slow a rate at room temperature that in terms of everyday experience cane-sugar appears to be stable in air. Cane-sugar is an example of a substance that is energetically unstable but kinetically stable in air at room temperature. Rates of reaction (chemical kinetics) are discussed in Chapter 10.

Enthalpy changes and stoichiometry

Only one stable anhydrous chloride of magnesium is known and this corresponds to the formula $MgCl_2$. We can use energetics data and suitable Born–Haber cycles to show why this is so. As we have seen previously, there is good agreement between the values obtained for the lattice energy of *ionic* compounds when calculated from electrostatic theory and when obtained from experimental results using a Born–Haber cycle. We can be confident, therefore, that the values we can calculate for such ionic compounds are valid even if the compound is non-existent. A chloride of magnesium of formula MgCl would, for example, have a lattice energy of -753 kJ mol^{-1} and one of formula $MgCl_3$ a lattice energy of -5440 kJ mol^{-1}. In constructing Born–Haber cycles it is necessary to have all the necessary data for the energy changes involved except for one; we can thus construct the Born–Haber cycles for MgCl and $MgCl_3$ using our theoretical values of lattice energy and in which the standard enthalpy of formation is the only unknown quantity.

Figure 4.13 shows such cycles for MgCl, $MgCl_2$, and $MgCl_3$. This shows that MgCl, with $\Delta H_f^\ominus$ of -113 kJ mol^{-1}, and $MgCl_3$, with $\Delta H_f^\ominus$ of $+3904$ kJ mol^{-1}, are both energetically unstable with respect to $MgCl_2$, with $\Delta H_f^\ominus$ of -653 kJ mol^{-1}. Careful analysis of these cycles shows how the energy changes involved are balanced in determining the sign and magnitude of the standard enthalpy change. A large negative value of standard enthalpy change is favoured by a large value of lattice energy; this in turn will be favoured by a greater charge on the magnesium ion. As the second ionization energy of magnesium is only a little greater than the first ionization energy, this extra energy expenditure is more than compensated for by the increased lattice energy of $MgCl_2$ compared with MgCl. There is a large increment, however, between the second and third ionization energies of magnesium which is not compensated for by the difference between the lattice energies of $MgCl_2$ and $MgCl_3$.

Algebraic treatment of thermochemical equations

As an alternative to using energy cycles to combine thermochemical data, it is sometimes more convenient to treat the thermochemical statements algebraically. The example below should make this method clear.

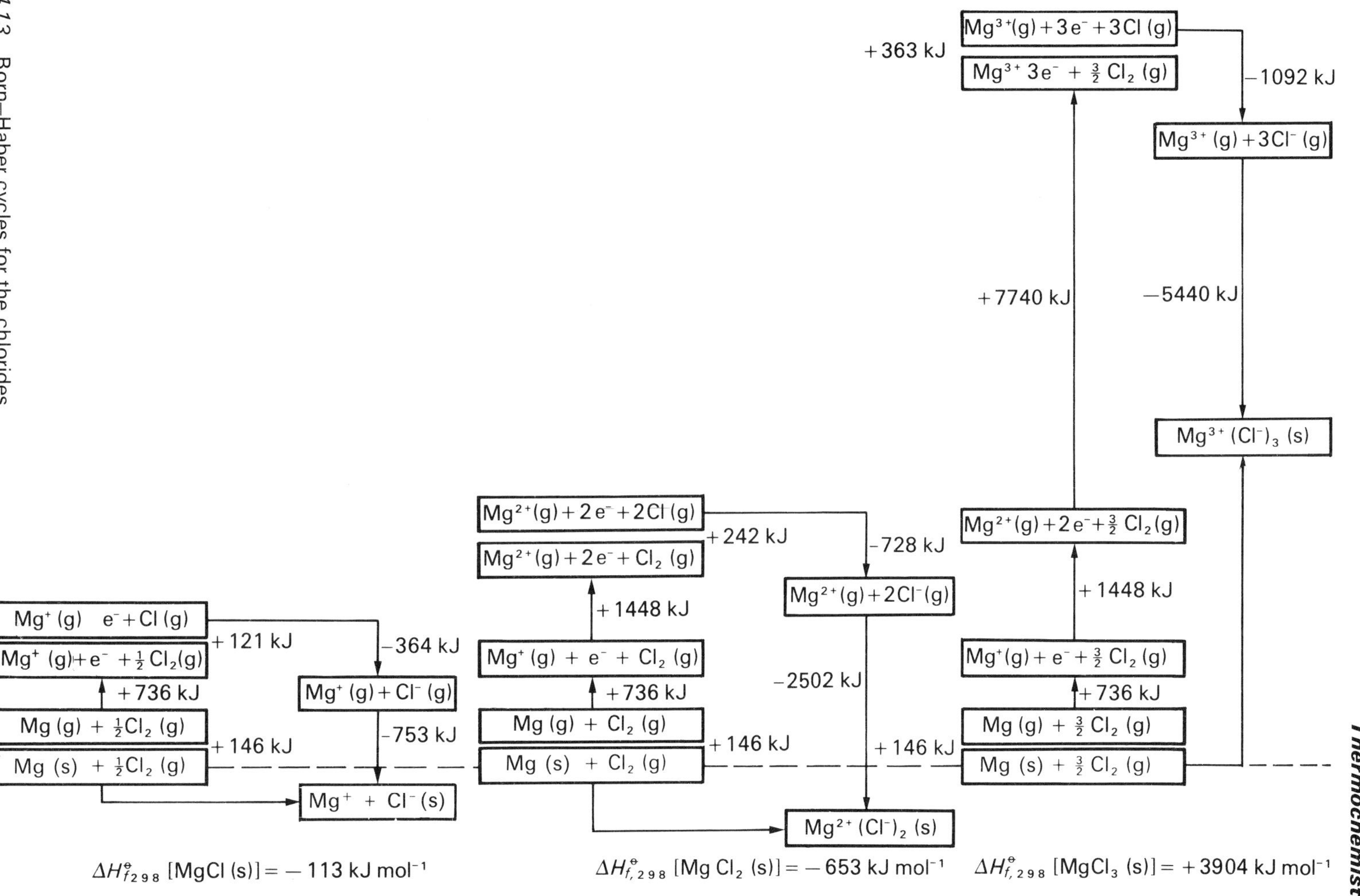

Figure 4.13 Born–Haber cycles for the chlorides of magnesium

Given the data:

$$N_2(g) + 2O_2(g) \rightarrow 2NO_2(g); \quad \Delta H^{\ominus} = +88 \text{ kJ mol}^{-1} \quad (1)$$

$$N_2(g) + 2O_2(g) \rightarrow N_2O_4(g); \quad \Delta H^{\ominus} = +10 \text{ kJ mol}^{-1} \quad (2)$$

if it is required to calculate the standard enthalpy change for:

$$2NO_2(g) \rightarrow N_2O_4(g)$$

this can be accomplished by subtracting equation (1) from equation (2) and rearranging:

$$2NO_2(g) \rightarrow N_2O_4(g);$$

$$\Delta H^{\ominus} = 10 - (+88) \text{ kJ mol}^{-1} = -78 \text{ kJ mol}^{-1}$$

Relativity and thermochemistry

In 1905 Albert Einstein published his *Special theory of relativity* which showed that matter and energy can be interconverted according to the equation

$$E = mc^2$$

where E is the energy, m is the mass, and c is the velocity of light. To accommodate this new knowledge, it became obvious that the separate laws of conservation of matter and energy must be combined into a joint law of conservation of mass energy. Because of the large value of the velocity of light, only a small mass change is associated with very large quantities of energy. Consider the complete conversion of 1 kg of matter to energy:

$$E = (1 \text{ kg}) \times (3 \times 10^8 \text{ m s}^{-1})^2 = 9 \times 10^{16} \text{ kg m}^2 \text{ s}^{-2} = 9 \times 10^{16} \text{ J}$$

By comparison, the mass change associated with the energies involved in thermochemical processes would be far too small for detection by any chemical balance. The much greater energies involved in nuclear processes, however, have correspondingly greater mass changes associated with them; this is discussed in Chapter 14.

Problems

1. Calculate the work done by expansion when 1 mole of aluminium reacts with dilute sulphuric acid according to the equation:

$$2Al(s) + 6H^+(aq) \rightarrow 2Al^{3+}(aq) + 3H_2(g)$$

2. 100 cm^3 of copper(II) sulphate solution were placed in an electrical compensation calorimeter and an excess of iron metal added. The reaction occurring was:

$$Fe(s) + Cu^{2+}(aq) \rightarrow Fe^{2+}(aq) + Cu(s)$$

3040 joules had to be supplied electrically to give the same rise in temperature that resulted from the reaction. Calculate $\Delta H^{\ominus}$ for the reaction in kJ mol^{-1}.

3. 3.715 g of propanone was burned in a Thiemann calorimeter. 116.4 kJ of energy had to be supplied electrically to raise the temperature by the same amount as the reaction. Calculate the standard enthalpy of combustion of propanone.
4. Using information taken from a suitable data book, (*a*) use bond energies to estimate the enthalpy of hydrogenation of ethene; (*b*) calculate the lattice energy of magnesium oxide.
5. State Hess's Law and explain the principles on which it is based.

   The complete combustion of 0.030 g of ethane (C_2H_6) and of 0.044 g of propane (C_3H_8) in a calorimeter with a heat capacity of 418 J K^{-1} causes temperature rises of 3.68 and 5.26 °C respectively. Calculate the mean energies required to dissociate the C–C and C–H bonds, given the following data:

$$O_2(g) \rightarrow 2O(g); \quad \Delta H^{\ominus} = 498 \text{ kJ mol}^{-1}$$

$$CO_2(g) \rightarrow C(g) + 2O(g); \quad \Delta H^{\ominus} = 1607 \text{ kJ mol}^{-1}$$

$$H_2O(g) \rightarrow 2H(g) + O(g); \quad \Delta H^{\ominus} = 971 \text{ kJ mol}^{-1}$$

(Oxford Entrance and Awards)

6. Describe briefly one method of determining enthalpies of reaction, pointing out the principal uses and limitations of the method.

   The standard enthalpy of formation of $NO_2(g)$ is 33.0 kJ mol^{-1} and that of $N_2O_4(g)$ 9.0 kJ mol^{-1}.

$$N_2(g) \rightarrow 2N(g); \quad \Delta H^{\ominus} = 944 \text{ kJ mol}^{-1}$$

$$O_2(g) \rightarrow 2O(g); \quad \Delta H^{\ominus} = 498 \text{ kJ mol}^{-1}$$

   Using these data, estimate the N–O bond energy in NO_2 and the N–N bond energy in N_2O_4. Indicate any assumptions you make, and discuss the extent to which you think these are valid. (Cambridge Entrance and Awards)

7. Explain what is meant by the standard enthalpy change for a reaction.

   With the aid of a suitable example in each

case, show how thermochemical measurements have been used:

(*a*) to determine indirectly the enthalpy change of a reaction

(*b*) to obtain information about bonding in metal chlorides.

The equation: $\Delta H^{\ominus} = -(660n + 250)$ kJ mol^{-1}, where n is an integer that is greater than 3, gives the enthalpy change of combustion at 25 °C of straight chain hydrocarbons (C_nH_{2n+2}) in the gaseous state. Interpret the quantities 660 and 250 respectively in relation to the structures of the hydrocarbons. (WJEC 'A')

8. Use your Book of Data to determine the lattice energy of strontium chloride. Show how you arrive at your answer.

A substance of formula $SrCl_3$ does not exist. What further energetics data would you need to explain this fact? Show how you would use the data. (Nuffield 'A')

9. The Born–Haber cycle below represents the enthalpy changes in the formation of an alkali metal halide MX from an alkali metal (Li, Na, K, Rb, Cs) and a halogen (F_2, Cl_2, Br_2, or I_2).

(*a*) Name the halogen for which the enthalpy change **2** has the *largest value.*

(*b*) Name the alkali metal for which the enthalpy change **3** has the *largest* value.

(*c*) Name the halogen for which the enthalpy change **4** has the *smallest* value.

(*d*) Name the alkali metal halide for which the enthalpy change **5** has the *smallest* value.

(*e*) The following is a list of the enthalpy changes for potassium bromide, in kJ mol^{-1}:

$K(s) \rightarrow K(g)$; $\Delta H^{\ominus} = +92$

$K(g) \rightarrow K^+(g) + e^-$; $\Delta H^{\ominus} = +418$

$\frac{1}{2}Br_2(g) \rightarrow Br(g)$; $\Delta H^{\ominus} = +96$

$Br(g) + e^- \rightarrow Br^-(g)$; $\Delta H^{\ominus} = -326$

$K^+(g) + Br^-(g) \rightarrow KBr(s)$; $\Delta H^{\ominus} = -677$

Calculate the standard enthalpy of formation, $\Delta H_f^{\ominus}$, of potassium bromide. (JMB 'A')

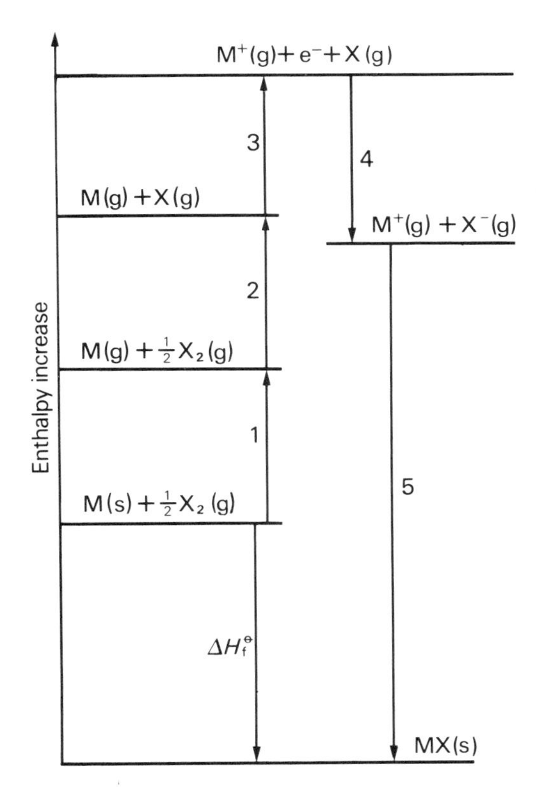

5. Structure and Bonding

Determination of the arrangement of ions in crystals

IN Chapter 4 the formation of ionic compounds, such as sodium chloride, from their elements was discussed. We shall now consider how the ions in such ionic materials are arranged. It has long been observed that all perfect crystals of sodium chloride, of whatever size, are cubic in shape. When a knife is placed on such a crystal parallel to one side and struck sharply with a hammer, the crystal breaks (or *cleaves*) regularly along the length of the knife. Moreover, if the knife is held on the crystal at certain other angles regular cleavage can also be produced. These observations suggest that the regular shapes of such crystalline materials are a reflection of the inner regularity of the ions of which they are composed.

More definite information about the arrangement of the ions in crystals developed out of work directed by Max von Laue in 1912. Using the apparatus illustrated diagrammatically in Figure 5.1, Laue obtained diffraction photographs of the type illustrated in Figure 5.2. Von Laue explained his observations by postulating that the regular arrangement of the ions in the lattice caused the crystal to act like a diffraction grating. Straight waves travelling parallel to the face of the crystal

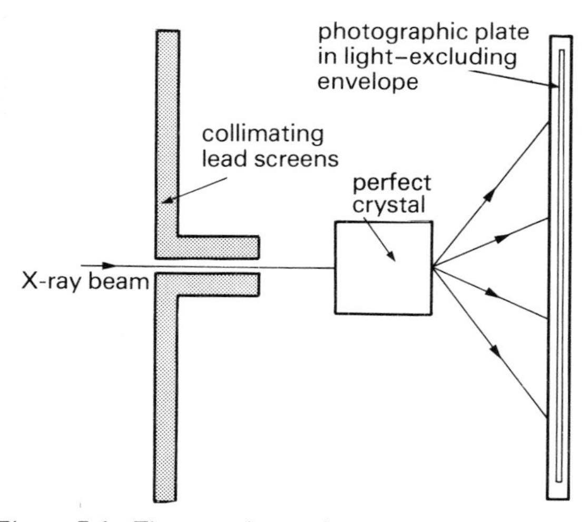

Figure 5.1 The experimental arrangement used to obtain diffraction patterns by von Laue

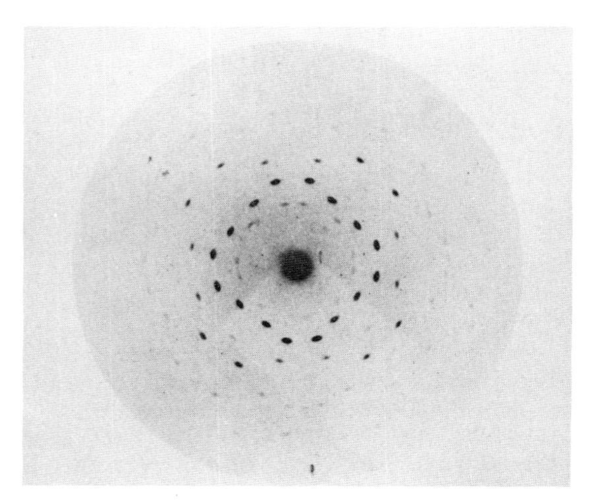

Figure 5.2 An X-ray diffraction pattern obtained by von Laue using a crystal of zinc sulphide

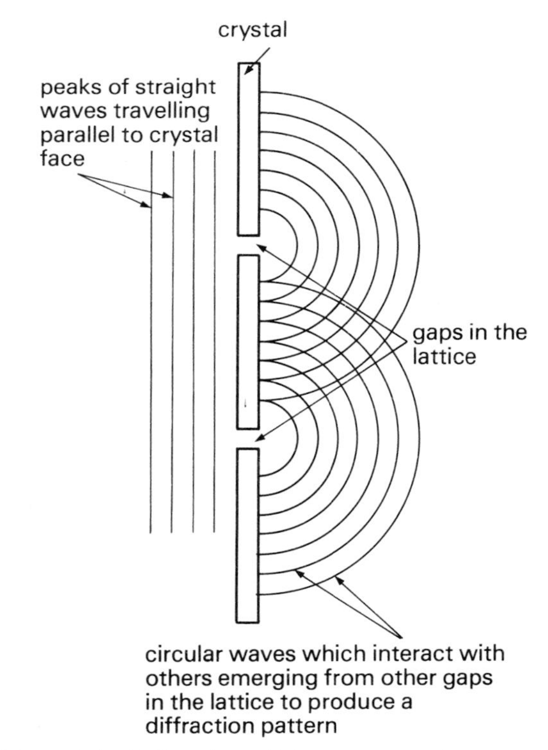

Figure 5.3 Diffraction of X-rays on passing through a crystal lattice

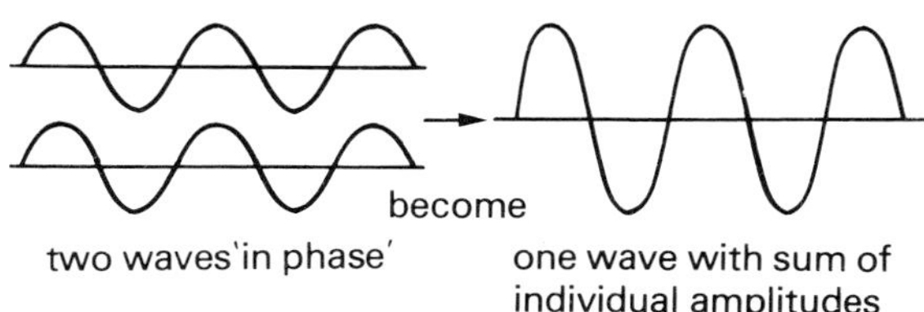

(*a*) reinforcement by constructive interference

become
two waves 'out of phase'
one wave with zero amplitude

(*b*) neutralization by destructive interference

Figure 5.4 Interference between waves

emerge as circular waves through the 'gaps' in the lattice, and these circular waves interact with waves emerging from neighbouring 'gaps' to produce a diffraction pattern. The interaction between neighbouring circular waves results from their mutual interference, as shown in Figure 5.3. When two waves meet which are 'in phase', i.e. with their peaks and troughs exactly coincident, they become one wave whose amplitude is the sum of the individual amplitudes and are thus strengthened or reinforced; when the two waves are 'out of phase', i.e. when the peak of one coincides with the trough of the other, their amplitudes cancel each other (*see* Figure 5.4). Interaction of this type between waves is known as *interference.* The interference between the waves passing through neighbouring gaps in the lattice causes the spots of high and low intensity observed on the plate in Figure 5.2. By a geometrical analysis of the position of the spots of high intensity von Laue was able to deduce the distances between 'gaps' in the crystal lattice, and hence the distances between ions in the face presented to the X-ray source. However, this analysis was extremely cumbersome and has now been superseded by the much simpler approach developed by Sir Lawrence Bragg in 1913.

The Bragg method

Bragg's approach involved the study of diffraction patterns that resulted when X-rays were reflected from crystal faces and not, as in the technique of von Laue, from their transmission through crystals.

The principle of Bragg's method of investigating crystal structures is illustrated in Figure 5.5. The apparatus (consisting of an X-ray source, X-ray detector, and crystal holder, together with turntables so that the angles between incident rays, crystal face, and detector can be varied) is called a *goniometer*. The apparatus is adjusted to find values of angle θ for which strong signals are recorded by the detector. Under these conditions reinforcement interference must be occurring. For waves Y and Z to reinforce each other in this way, the additional path travelled by Z (BC + CD) must be a whole number multiple of the wavelength, $n\lambda$, of the incident radiation so that the two waves emerge from the crystal in phase:

$$n\lambda = \mathrm{BC} + \mathrm{CD}$$
$$= 2\mathrm{BC}$$

but, $\mathrm{BC} = d \sin\theta$

thus, $n\lambda = 2d\sin\theta$

This relationship, known as the *Bragg equation*, enables the distances between the parallel layers of ions in the lattice, d, to be calculated. For values of $n = 1$ the diffraction is said to be *first-order*; a similar nomenclature is used for higher values of n.

For cubic lattices, such as those of sodium chloride and potassium chloride, it is possible to orientate the crystal in three different positions so that the radiation is incident upon parallel layers

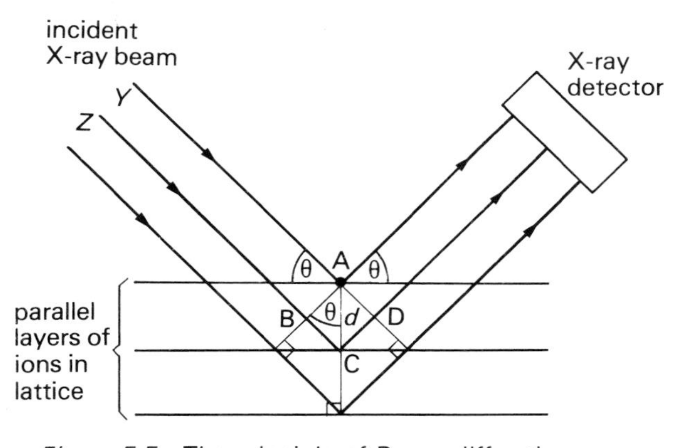

Figure 5.5 The principle of Bragg diffraction

of ions. These planes of ions are illustrated in Figure 5.6, and they correspond to the cleavage planes discussed in the opening paragraph on page 60. The student is advised to examine a model of the sodium chloride structure and to identify these planes for himself.

The X-ray detector readings obtained for sodium chloride when each of these planes is presented in turn is shown in Figure 5.7. The spacings calculated by applying the Bragg equation to these results, together with the appropriate geometry. allows complete elucidation of the crystal structure.

X-rays are diffracted from regions of high electron density; by the application of a technique known as *Fourier analysis* to the results of diffraction experiments an electron density map of the substance can be deduced. On such maps the contours are lines joining points at which the electron density is identical. The electron density map for sodium chloride obtained from X-ray diffraction data is shown in Figure 5.8. It can be seen that the spherical symmetry of electron density distribution around the ions is slightly distorted in such a crystal lattice. This is due to mutual repulsion between the electrons of neighbouring ions.

Sodium ions are smaller than chloride ions. Cations are invariably smaller than the parent atom because (*a*) a complete outer shell of electrons may be lost during ion formation, as in the case of sodium, and (*b*) the loss of one or more electrons allows the unbalanced positive charge or charges in the nucleus to attract all the electron shells more closely. Anions are invariably larger than the parent atom for converse reasons. Table 5.1 shows the comparison of atomic and ionic radii for some typical elements.

Table 5.1 Atomic and ionic radii of selected elements

Element	*Atomic radius/ nm*	*Ion*	*Ionic radius/ nm*
Na	0.186	Na^+	0.098
Mg	0.160	Mg^{2+}	0.065
Al	0.143	Al^{3+}	0.045
Cl	0.099	Cl^-	0.181
O	0.066	O^{2-}	0.146

The separation distance of ions in a lattice

The size of the ions is a factor that obviously affects the distances between neighbouring ions in a lattice. Examination of electron density maps of ionic compounds shows that there are regions between neighbouring ions where the electron density is extremely low. The ions in such a lattice should therefore not be regarded as touching. Figure 5.9 plots the variation in potential energy for a sodium and chloride ion-pair with separation distance; the

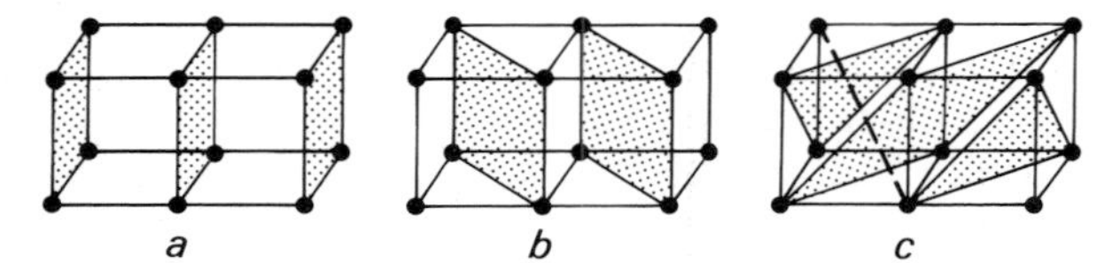

Figure 5.6 The three planes of ions in simple cubic crystals

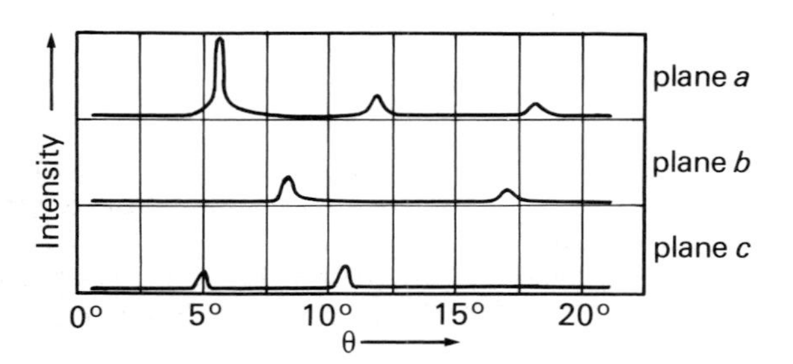

Figure 5.7 Goniometer detector readings from the three planes in sodium chloride

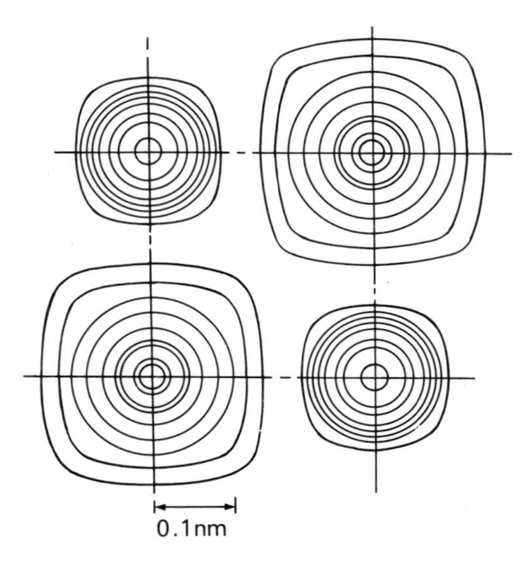

Figure 5.8 The electron density map of sodium chloride

values have been calculated from electrostatic theory. As ion-pairs approach, the force of attraction due to their opposite charges is increasingly opposed by the mutual repulsion of their similarly charged electron shells. The inter-ionic distance observed by X-ray studies on such compounds, known as the *equilibrium distance*, represents a balance between these two forces.

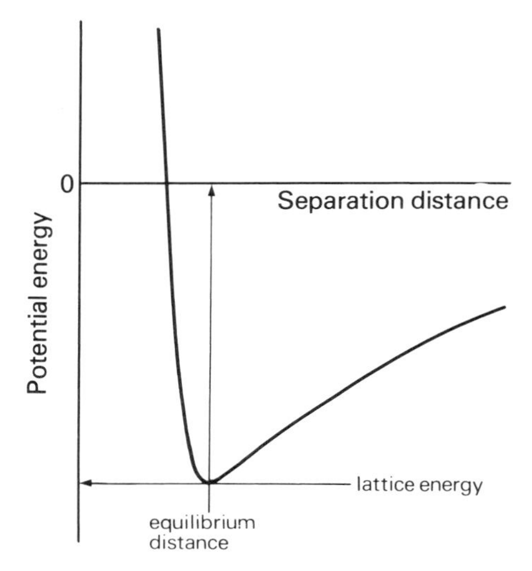

Figure 5.9 Potential energy/separation distance curve for sodium chloride

Figure 5.10 Space-filling model of sodium chloride

The sodium chloride system

Using the methods discussed above, the structure of sodium chloride has been determined and the result can be presented in the form of a space-filling model as shown in Figure 5.10. Each chloride ion has six sodium ions as its nearest neighbours, as shown in Figure 5.11, and each sodium ion has six chloride ions as nearest neighbours. The number of nearest neighbours of each ion in a structure is known as its *coordination number*. Sodium chloride is an example of a cubic system with 6:6 coordination. Crystal structures of this type are common.

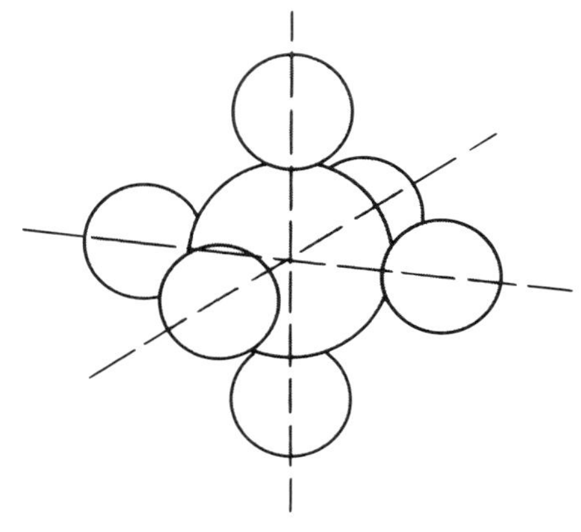

Figure 5.11 The coordination of chloride ions in sodium chloride

The crystal systems

The variety of crystalline forms exhibited by substances can be classified into six *crystal systems* according to their degree of symmetry. These six crystal systems are shown in Figure 5.12.

The unit cell

As a crystal system consists of a regularly repeating arrangement of the component particles, there must be in each case a smallest unit which retains all the symmetry observed for the macroscopic crystal. This smallest unit is known as the *unit cell*.

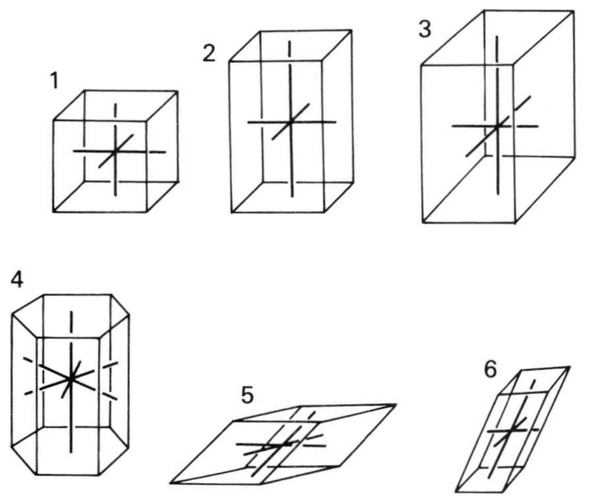

Figure 5.12 The six crystal systems

Avogadro's number from diffraction measurements

The unit cell of sodium chloride is shown in Figure 5.13. It can be seen that most of the constituent ions are shared between more than one unit cell. Including the internally situated chloride ion, however, the unit cell contains the equivalent of four pairs of sodium and chloride ions.

The molar mass of sodium chloride is 58.44 g, and the density of sodium chloride is 2.163 g cm^{-3}. Thus the total volume, including ions and space, of 1 mole of NaCl is

$$\frac{58.44}{2.163}\ \mathrm{cm^3} = 27.01\ \mathrm{cm^3}$$

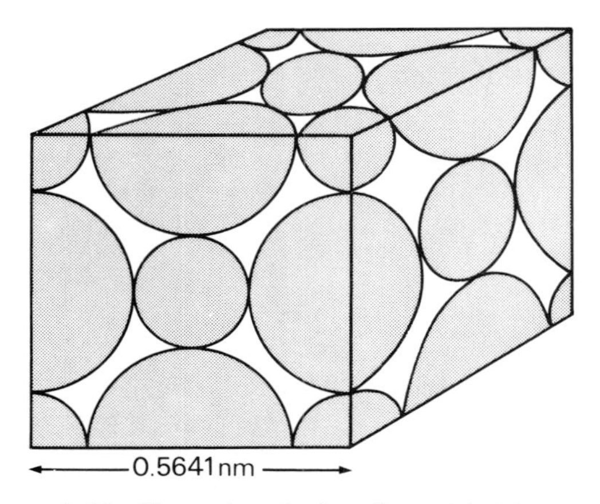

Figure 5.13 The unit cell of sodium chloride

From the dimensions of the unit cell determined by diffraction, the total volume occupied by four ion-pairs $= (0.5641 \times 10^{-7})^3\ \mathrm{cm^3}$

$$= 0.1788 \times 10^{-21}\ \mathrm{cm^3}.$$

Thus the volume occupied by one ion-pair

$$= 0.04500 \times 10^{-21}\ \mathrm{cm^3}$$
$$= 4.50 \times 10^{-23}\ \mathrm{cm^3}$$

Because 1 mole of NaCl contains Avogadro's number, L, of ion-pairs,

$$L = \frac{27.01\ \mathrm{cm^3}}{4.50 \times 10^{-23}\ \mathrm{cm^3}} = 6.03 \times 10^{23}$$

Isomorphism

In 1819, while engaged in the study of the crystalline phosphates and arsenates of sodium, potassium, and ammonium, Mitscherlich observed that salts of similar composition (e.g. $KH_2PO_4.H_2O$ and $KH_2AsO_4.H_2O$) formed crystals of the same shape. This led him to propound his *Law of Isomorphism* which, in modern wording, states:

Substances which are similar in crystalline form can generally be represented by similar formulae. Application of this law proved to be a useful method for determining the atomic masses of some elements; the example below illustrates the principle involved.

Potassium chlorate(VII) and potassium manganate(VII) are isomorphous. The formula of potassium chlorate(VII) is known to be $KClO_4$. Potassium manganate(VII) is found on analysis to contain 34.8 per cent of manganese. By application of Mitscherlich's Law, the formula of potassium manganate(VII) can be deduced as $KMnO_4$. Using values for the relative atomic masses of potassium and oxygen of 39 and 16 respectively and calling the relative atomic mass of manganese x:

$$\frac{x}{39 + x + (4 \times 16)} = \frac{34.8}{100}$$
$$\frac{x}{103 + x} = \frac{34.8}{100}$$
$$100x = 34.8(103 + x)$$
$$= 3584 + 34.8x$$
$$65.2x = 3584$$
$$x = 54.8$$

It is necessary, however, to use caution when applying this law as several exceptions are known. Lead sulphide and silver sulphide, for example, are isomorphous but their formulae are PbS and

Ag_2S. When a solution of two isomorphous substances is allowed to crystallize, the solid is homogenous but contains both substances in proportions depending on the relative concentrations in the original solution. Such solids are known as *mixed crystals*. If a crystal is suspended in a saturated solution of a compound with which it is isomorphous, the crystal continues to grow. Crystals formed in this way are known as *overgrowths*.

Use of crystal models

Students are advised to devote some time to the study of crystal models (of both the space-filling and ball-and-spoke varieties) and, if possible, to construct some of these for themselves. They should determine the coordination numbers of the constituent ions and attempt to classify each substance into the appropriate crystal system.

Covalent bonding

In ionic bonding considered above there is transfer of electrons between the reactive species and attraction between the ions formed is due to interaction between electrostatic fields radially situated around neighbouring ions. The forces involved are thus essentially non-directional. In covalent bonding, however, there is a directional force between the two atoms involved in the bond. As we have seen in the examples discussed previously, ionic compounds are formed when the atoms involved lose or gain sufficient electrons to attain the electron configuration of the nearest noble gas. In 1916 G. N. Lewis suggested that covalent bonding could be explained by the sharing of electrons in pairs so that each atom attains a noble gas configuration. In the simplest case of covalency, that of the hydrogen molecule, each atom is regarded as donating its electron into a charge cloud between the atoms which is then shared between the two nuclei. The positive nuclei exert a repulsive force on each other, but each is attracted by the negative charge cloud of electrons. The internuclear distance, also known as the *bond length*, is thus determined by the point where these opposing forces just balance. In the hydrogen molecule the bond length is 0.075 nm.

Figure 5.14 shows the electron density map of the hydrogen molecule. Comparison of this with the electron density map of sodium chloride in Figure 5.8 shows that in the case of hydrogen there is substantial electron density *between* the atoms, thus accounting for the directional nature of the bond. Figure 5.15 shows three models which are commonly used to represent the hydrogen molecule. None of these, of course, is a completely faithful representation of an actual hydrogen molecule and chemists freely interchange between models depending upon which proves the most useful in a particular context.

The electron density map of a more complex covalent substance, *cis*-butenedioic (maleic) acid is shown in Figure 5.16. Comparison of this with the structural formula shows that the position of the hydrogen atoms is not shown. This is because X-rays are diffracted from regions of high electron density, and electron density is low around hydrogen atoms.

An alternative diffraction technique, that of electrons, can be used to determine with considerable accuracy the position of hydrogen atoms in molecules and the angles between bonds.

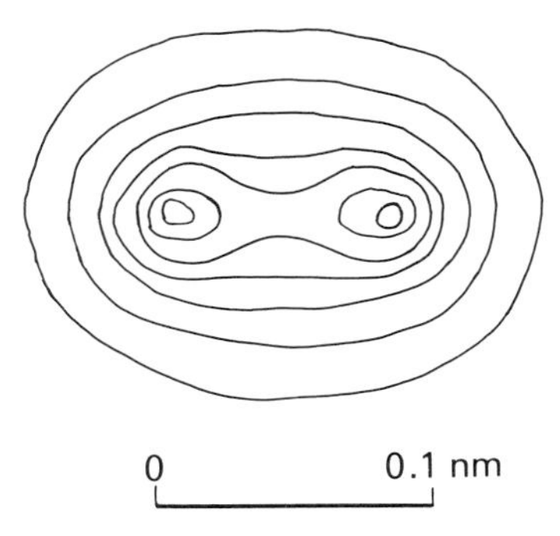

Figure 5.14 The electron density map of the hydrogen molecule

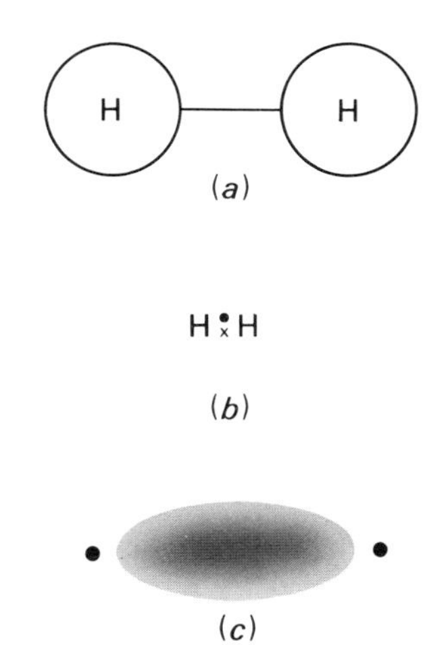

Figure 5.15 Three representations of the hydrogen molecule (*a*) valence bonds, (*b*) dot-and-cross, and (*c*) charge cloud

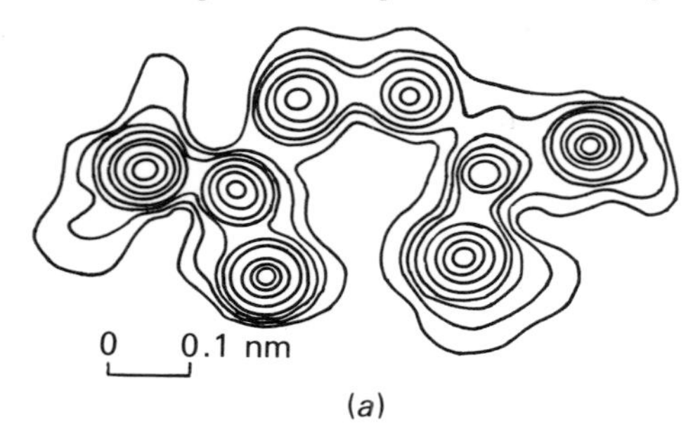

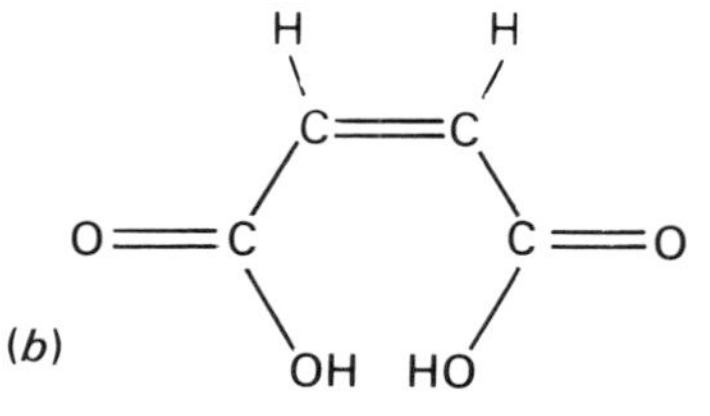

Figure 5.16 Electron density map and structural formula of *cis*-butenedioic (maleic) acid

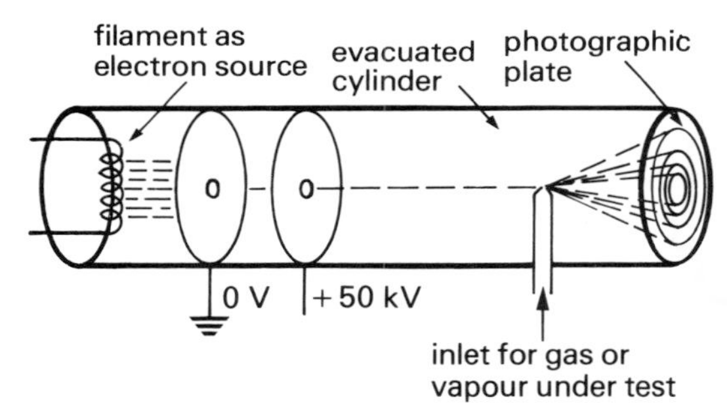

Figure 5.17 Apparatus for electron diffraction studies

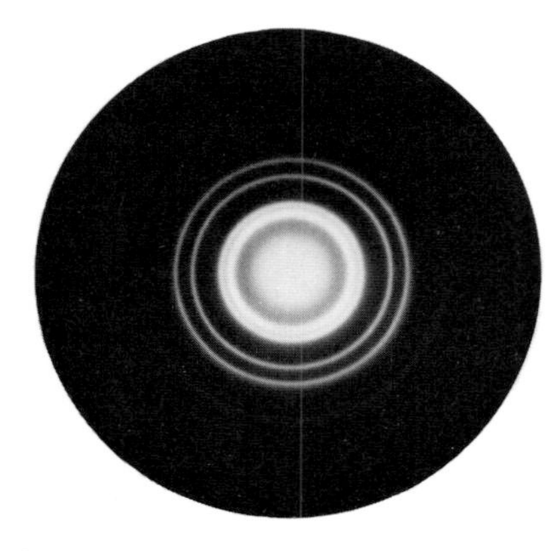

Figure 5.18 An electron diffraction pattern

Electron diffraction

The apparatus used for electron diffraction is shown in Figure 5.17. Gas or vapour of the test material enters the cylinder, which is maintained under a good degree of vacuum, and causes diffraction of the electron beam passing through it. The diffraction pattern is recorded photographically at the far end of the cylinder. A typical electron diffraction pattern is shown in Figure 5.18. The diameters of the concentric rings so formed are related to the bond lengths and angles in the molecule. A possible structure for the molecule is proposed and the electron diffraction pattern that this would produce is determined. This is compared with the actual pattern obtained and the postulated structure is refined until the electron diffraction pattern expected from that structure agrees with the one obtained in practice.

The structures of some related hydrides

Carbon has four electrons in its outer shell. It can attain the electron configuration of the noble gas neon if each of these electrons enters into covalent bonding with a hydrogen atom. Figure 5.19(*a*) shows the ground state electron configuration of carbon. It is first necessary to supply enough energy to promote one electron from the 2*s* to the unoccupied 2*p* orbital, as shown in Figure 5.19(*b*). The electrons are then shared with 1*s* electrons from four hydrogen atoms by orbital overlapping. The occupancy of the appropriate orbitals around the carbon atom in methane is shown in Figure 5.19(*c*) where the electrons from hydrogen are shown as dotted arrows. The four bonds around the carbon atom in methane, therefore, consists of electron charge clouds. These will exert mutually repulsive forces on each other, and the shape of the methane molecule will thus be dictated by the four bonds assuming positions in which these repulsive forces will be minimized. The structure corresponding to these conditions is that of a regular tetrahedron, with the carbon atom situated in the centre and with the four hydrogen atoms at its apices. The valence bond and charge cloud representations of methane, as determined by electron diffraction, are shown in Figure 5.20(*a*). It will be observed that the bond angle corresponds to that of a regular tetrahedron.

Nitrogen has five electrons in its outermost shell and thus needs to form covalent bonds with only three hydrogen atoms to attain the electron structure of neon in the formation of its hydride, am-

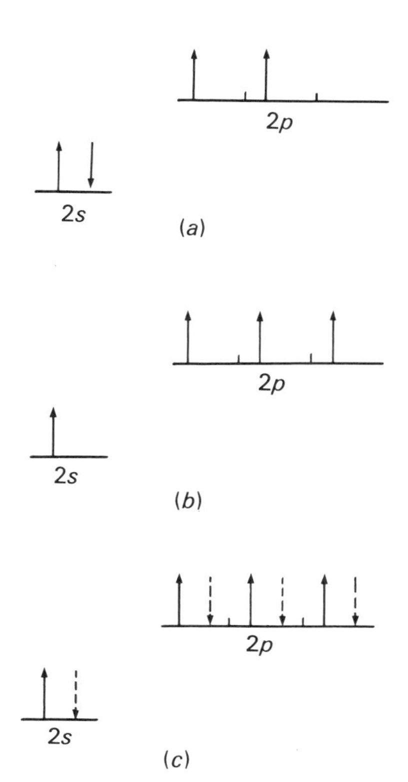

Figure 5.19 Electron configuration of the carbon atom (*a*) in the ground state, (*b*) in the promoted state, and (*c*) after covalent bond formation in methane

monia. Around the central atom in ammonia, therefore, there are three nitrogen-to-hydrogen bonds and one orbital containing a pair of electrons not involved in bonding to other atoms. Such a non-bonding pair of electrons is known as a *lone pair*. As the lone pair has no positively charged nucleus to restrain it at the opposite end to the nitrogen atom, it grows fatter thus causing greater repulsion on the bonding electron pairs. As a consequence of this the bond angle in ammonia is smaller than that of methane (*see* Figure 20(*b*)). The shape of the ammonia molecule is described as pyramidal.

Oxygen has six electrons in its outermost shell and thus bonds with two hydrogen atoms to attain the electron structure of neon in the formation of water. Water has two lone pairs of electrons and the bond angle is even smaller than that of ammonia (Figure 20(*c*)). The water molecule is described as *V*-shaped.

Fluorine has seven electrons in its outermost shell and bonds with one hydrogen atom in the formation of hydrogen fluoride. The molecule is thus linear in shape.

Methane, ammonia, water, and hydrogen fluoride together with neon all possess the same total number of electrons and are known as an *isoelectronic series*.

Boron has three electrons in its outermost shell and by covalent sharing with three hydrogen atoms it forms its hydride, BH_3. There is in this compound, however, one completely unoccupied

Figure 5.20 Valence bond and charge cloud representations of (*a*) methane, (*b*) ammonia, (*c*) water, and (*d*) hydrogen fluoride

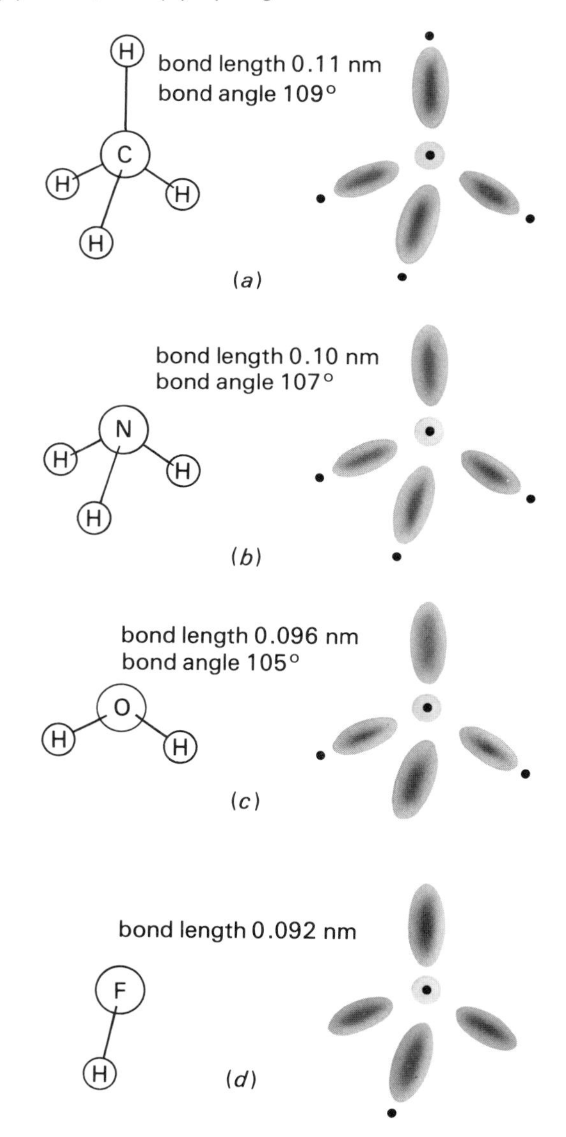

orbital. The shape of the boron trihydride molecule is therefore planar-triangular, the shape best accommodating repulsion between the three negative charge clouds.

Dative covalency

Boron trifluoride has an unoccupied orbital around the boron atom for reasons discussed above. It is a gas and reacts with gaseous ammonia readily to give a white solid of formula $NH_3.BF_3$. This can be explained in terms of the lone pair on the nitrogen atom of the ammonia molecule being donated to the unoccupied orbital of the boron atom. This is represented using the dot-and-cross notation and valence-bond notation in Figure 5.21. When such a covalent bond is formed by *both electrons being donated from the same atom*, the bonding is described as *dative covalency*. It should be stressed that a dative bond, once formed, is indistinguishable from other covalent bonds. Compounds formed in this manner are known as 'adduct' or 'addition' compounds.

lone pair from the nitrogen donated to empty boron orbital

becomes

$$H_3N + BF_3 \longrightarrow H_3N{\rightarrow}BF_3$$

Figure 5.21 Formation of an adduct compound between ammonia and boron trifluoride by dative covalency (*above*)

Figure 5.22 Dot-and-cross, valence bond, and charge cloud representations of (*a*) ethane, (*b*) ethene, and (*c*) ethyne (*right*)

Multiple bonds

Figure 5.22 shows the dot-and-cross, valence bond, and charge cloud representations of ethane, ethene (ethylene), and ethyne (acetylene). It shows how the sharing of more than one electron and hence the formation of multiple bonds leads to attainment of the neon configuration by carbon in these cases. There is an interesting relationship between the bond length, as determined by electron diffraction, and the bond energy, as determined thermochemically, in the carbon–carbon bonds of these related compounds (Table 5.2).

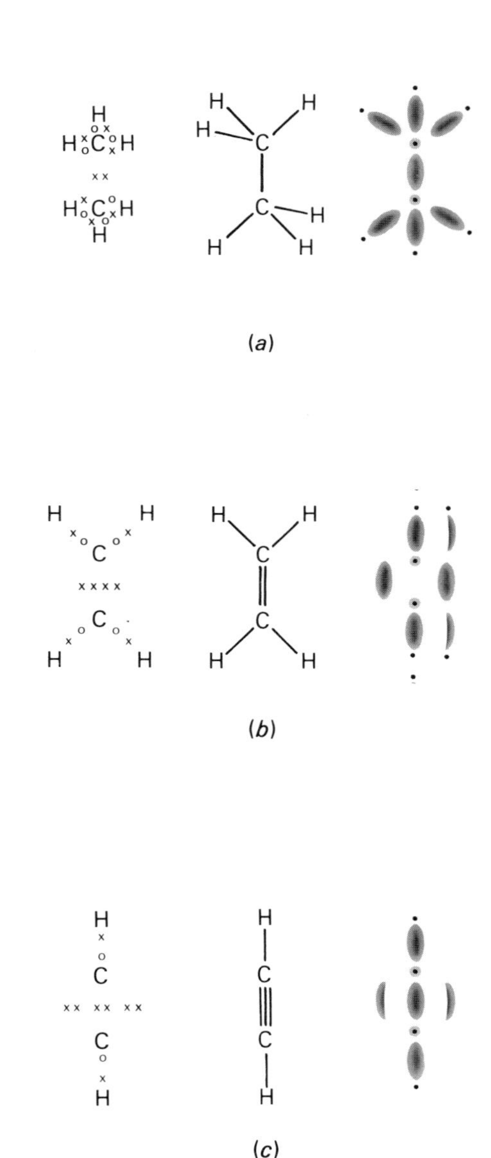

Table 5.2 Some carbon–carbon bond lengths and bond energies

Compound	*Bond length/ nm*	*Bond energy/ kJ mol^{-1}*
Ethane	0.154	346
Ethene	0.134	611
Ethyne	0.120	837

The bond length in ethene is less than that in ethane because the two negative charge clouds between the carbon atoms repel one another, causing the 'bowing out' shown in the charge cloud representation with a consequent reduction in bond length. In ethyne, with three such charge clouds between the carbon atoms, the effect is correspondingly greater. The bond energy of ethene might be expected to be double that in ethane involving as it does two charge clouds between the carbon atoms; that the observed value is slightly less than double is due to the energy required to overcome the mutual repulsion. That the bond energy in ethyne is rather less than three times that in ethane can be similarly explained.

Effect of lone pairs of electrons on strengths of adjacent bonds

Examination of the data in Table 5.3 shows that the bond energies of the hydrogen halides decrease regularly with increasing atomic number of the halogen. On the other hand, the bond energy of chlorine is greater than that of fluorine but less than that of bromine, which in turn is greater than that of iodine. The apparently anomolous value for the bond energy in the fluorine molecule can be explained in terms of repulsion between lone pairs of electrons on adjacent atoms. As the fluorine atom is the smallest of the halogen atoms, the fluorine–fluorine bond length is the shortest for all the halogen molecules and the repulsion effect is the greatest in this case. The effect does not occur in the case of the hydrogen halides as hydrogen has no lone pairs of electrons. The low value of bond energy in the fluorine molecule is a major factor contributing to the extremely high reactivity of this element; consider the effect on the enthalpy change of substituting this value for fluorine for that of chlorine in Born–Haber cycles such as those shown in Figure 4.12.

The nitrogen molecule, on the other hand, has a triple bond between the atoms in its molecules. As can be seen in the charge cloud representation of this molecule in Figure 5.25, the sole lone pairs associated with each atom are positioned so that

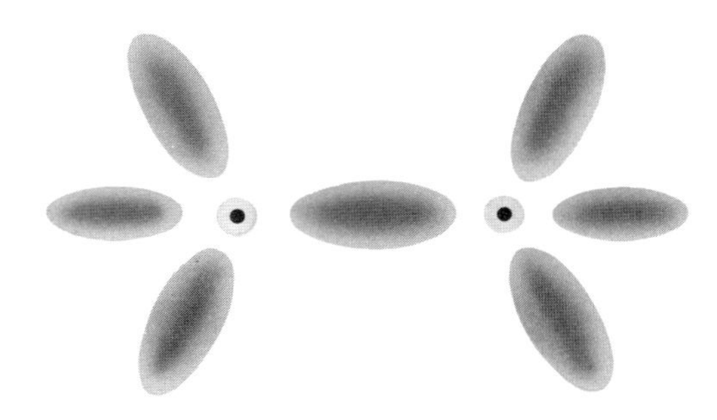

Figure 5.23 Charge cloud representation of the fluorine molecule (F–F)

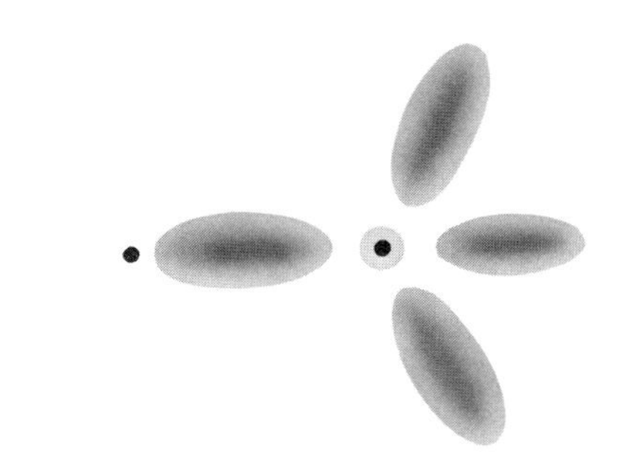

Figure 5.24 Charge cloud representation of the hydrogen fluoride molecule (H–F)

Table 5.3 Bond energies of the halogens and hydrogen halides

Halogen	*Bond energy/kJ mol^{-1}*	*Hydrogen halide*	*Bond energy/kJ mol^{-1}*
Fluorine	151	Hydrogen fluoride	561
Chlorine	239	Hydrogen chloride	428
Bromine	190	Hydrogen bromide	362
Iodine	149	Hydrogen iodide	295

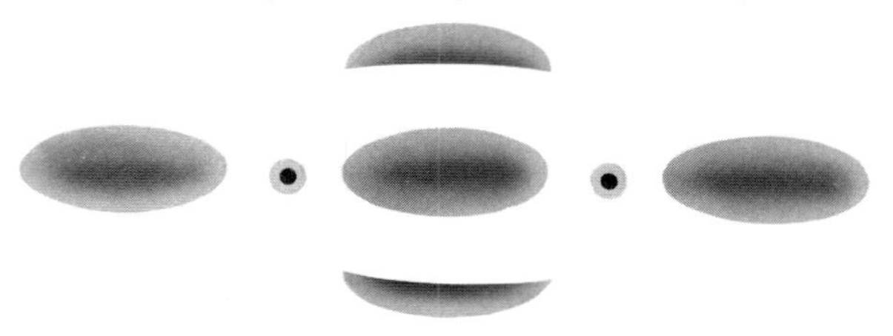

Figure 5.25 Charge cloud representation of the nitrogen molecule

there is minimum mutual repulsion between them; these factors combine to give the nitrogen molecule a bond energy of 945 kJ mol^{-1}. This extremely high value, together with the necessity of adding three moles of electrons to each mole of nitrogen atoms to give the stable ion, N^{3-}, accounts for the extremely unreactive nature of nitrogen.

Bond polarization

In the covalent bond formed between two hydrogen atoms it is reasonable to assume that the electron pair is evenly shared between the two nuclei; similarly equal sharing seems probable between the atoms in the chlorine molecule. In hydrogen chloride, however, because the electron pair is shared between two different nuclei the sharing is not necessarily equal. Such uneven distribution of the electrons in a bond would result in distortion of the electron charge cloud towards the atom gaining the greater share. This would result in the molecules having a permanent electrostatic dipole moment, which is the product of the charge difference and the distance separating the charges. Figure 5.26 shows two notations by which molecular dipoles can be represented.

Evidence for the existence of such molecular dipoles and a measurement of the dipole moment can be obtained by measuring the relative permittivity (dielectric constant) observed when such substances are placed between the parallel plates of a suitable condenser. Figure 5.27 shows the experimental arrangement for this investigation.

Figure 5.26 Representation of the dipole moment in the hydrogen chloride molecule

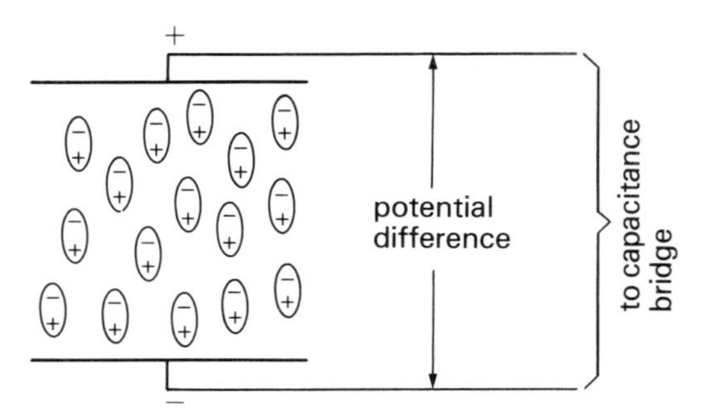

Figure 5.27 Use of capacitance measurements to determine dipole moments of molecules

Polar molecules become orientated as shown in the electrostatic field between the plates and the magnitude of their dipole affects the value of the capacitance of the system observed. The relative permittivity of the substance, ε_r, is the ratio of the capacitance measured with the substance between the plates, C, to that measured with a vacuum between them, C_0,

$$\frac{C}{C_0} = \varepsilon_r$$

All substances have relative permittivities greater than unity because the electrostatic field induces a temporary dipole even in substances which have no permanent dipole moment. From the value of the relative permittivity the dipole moment of the substance can be calculated.

Table 5.4 Dipole moments of the hydrogen halides

Substance	*Dipole moment/10^{-30} C m*
Hydrogen fluoride	6.36
Hydrogen chloride	3.43
Hydrogen bromide	2.60
Hydrogen iodide	1.27

The results in Table 5.4 show that, of the halogens considered, fluorine has the greatest tendency to draw electrons towards itself during covalent bonding and iodine has the least. Several attempts have been made to give a quantitative measure to this tendency, known as the *electronegativity* of the element, of which the most generally accepted is that of the American chemist Linus Pauling. His values of electronegativity for selected elements are given in Table 5.5.

Table 5.5 Pauling's electronegativity values for certain elements

Element	*Electronegativity value*
Fluorine	4.0
Oxygen	3.5
Chlorine	3.0
Carbon	2.5
Hydrogen	2.1
Copper	1.9
Magnesium	1.2
Lithium	1.0
Sodium	0.9
Potassium	0.8
Francium	0.7

Binary compounds formed between two elements with little difference in electronegativity values thus have small molecular dipoles and approach the purely covalent state. Binary compounds of elements having large differences in electronegativity values are those normally regarded as ionic. Pauling introduced the term 'percentage ionic character' to interpret this spectrum of bond types between pure covalent and pure ionic bonding, and the relationship between this quantity and electronegativity differences is shown in Figure 5.28.

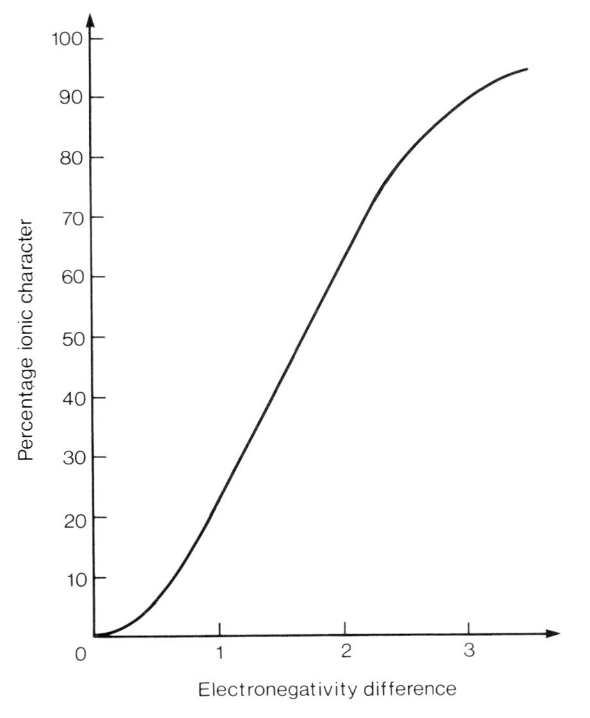

Figure 5.28 Relationship between percentage ionic character and electronegativity difference

Dipole moments are vector quantities and the values obtained in polyatomic molecules are therefore the resultant of those for the individual bonds involved. This resultant can often be zero as in the linear carbon dioxide molecule and the tetrahedral molecule of tetrachloromethane.

Hydrogen bonding

When hydrogen forms bonds with elements having high electronegativity values, such as fluorine, oxygen, and chlorine, the resulting distortion of the electron charge cloud away from the hydrogen nucleus causes a uniquely high degree of exposure of excess positive charge at the hydrogen end of the molecule because of weak shielding from a single electron system. This positive region can therefore form particularly strong dipole–dipole interactions with regions of high electron density in neighbouring molecules, or similar sites on adjacent portions of the same molecule. These particularly powerful interactions are known as *hydrogen bonds* and are generally represented by dotted lines in diagrams. Water, with a dipole moment of 9.13×10^{-30} C m, is an important example of a hydrogen bonded compound.

The structure of ice is shown in Figure 5.29. It will be observed that hydrogen bonding occurs between the hydrogen atoms and the lone pairs of electrons on adjacent molecules. The crystal structure of ice is thus essentially tetrahedral. The bonding *between* molecules of water in ice (known as intermolecular bonding), although weak compared with bonding between atoms in any one molecule (known as intramolecular bonding), is stronger than the van der Waals' forces between the molecules of many covalent solids and therefore water has relatively high values for molar latent heat of fusion and melting point. When water melts these hydrogen bonds are progressively broken, as shown in Figure 5.30. This facilitates closer packing of the molecules, and thus an initial reduction in the volume of the liquid occurs before the usual expansion effect from raising the temperature is observed. It is for this reason that water has its maximum density at 4 °C and not, as in the case of most liquids, at its melting point. The importance of the polarity of liquids such as water when acting as solvents is discussed in Chapter 11.

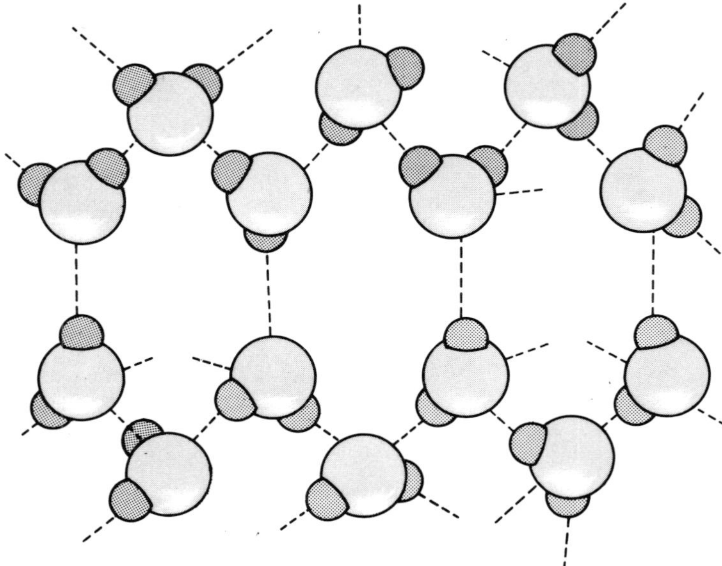

Figure 5.29 The structure of ice

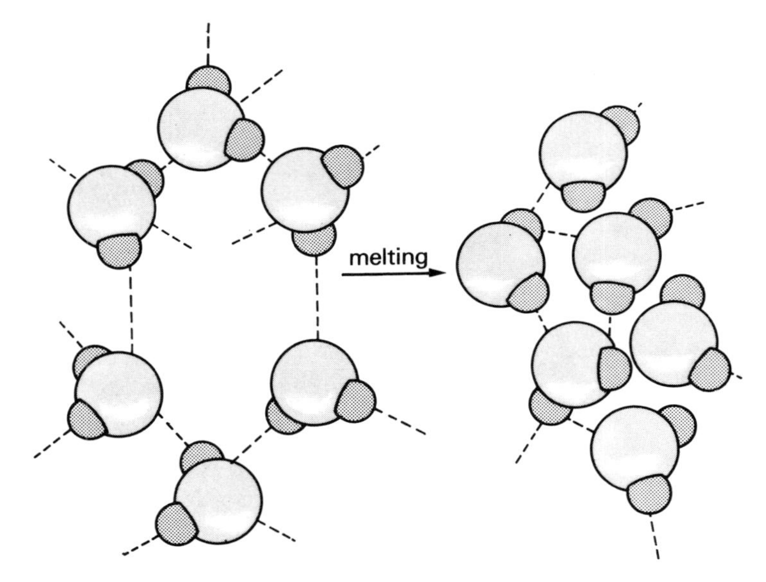

Figure 5.30 Change in molecular packing when ice melts

Bond delocalization

At the end of the eighteenth century some streets in London were illuminated by burning a gas obtained from the destructive distillation of certain natural animal oils such as whale oil. After some time, the cylinders in which this gas was stored were found to contain a colourless flammable liquid. In 1820 this liquid was analysed by Michael Faraday at the Royal Institution who declared its formula to be C_6H_6. An identical liquid was later obtained by heating benzoic acid, obtained from *gum benzoin*, and given the name benzene. Although it is possible to write valence bond structures involving multiple bonds, for a hexane derivative corresponding to this formula, a major difficulty arose in explaining why only one monosubstituted derivative, such as chlorobenzene, was found to exist. To explain this, any proposed structure would have to show all six hydrogen atoms occupying equivalent positions. The problem occupied the attention of all chemists throughout the mid-nineteenth century. In 1865 Friedrich Kekulé, Professor of Chemistry in the University of Ghent, developed a solution fulfilling the necessary conditions while dozing before his fire. In his dream he saw snake-like chains of atoms twining and twisting when one snake suddenly seized its tail in its own mouth. Kekulé woke with a start and realized that a cyclical structure for benzene would provide the solution for which he and his colleagues were searching. Kekulé's proposed structure for benzene, shown in Figure 5.31, was widely accepted for almost a hundred years. Recent discoveries, however, have necessitated that his concept be modified.

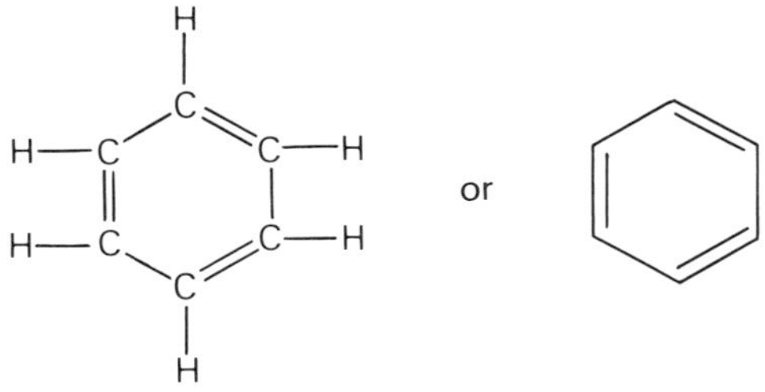

Figure 5.31 The Kekulé structure for benzene

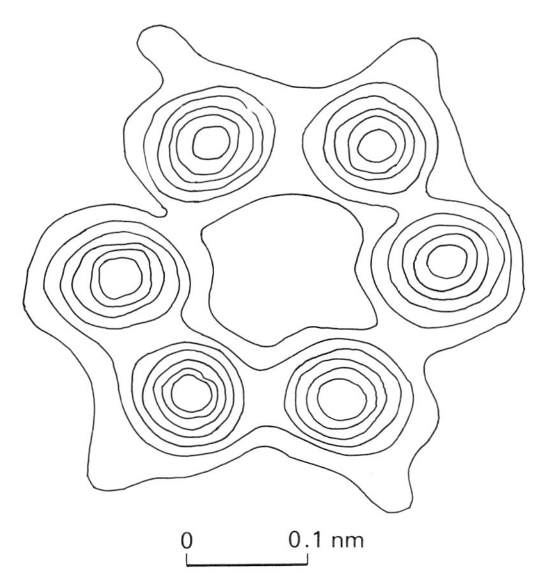

Figure 5.32 Electron density map of benzene

(*a*) *Structural considerations* As shown in Table 5.2 (page 69), carbon-to-carbon single and double bond lengths are not the same. The Kekulé structure for benzene would thus assume the shape of a puckered ring. The electron density map of benzene deduced from diffraction measurements (Figure 5.32) shows that the actual structure of benzene is that of a planar regular hexagon. Moreover, the bond length observed, 0.139 nm, is intermediate between that observed for single and double carbon-to-carbon bonds.

(*b*) *Thermochemical considerations* The hydrogenation enthalpies obtained from thermochemical cycles as discussed in Chapter 4, for cyclohexene and Kekulé benzene are shown in Figure 5.33. As the enthalpy of hydrogenation of cyclohexene involves the addition of one mole of hydrogen across one mole of carbon-to-carbon double bonds, and that for Kekulé benzene the addition of three moles of hydrogen in this manner, the latter quantity would be expected to be three times the former. It can be seen that *real* benzene is thermochemically more stable than Kekulé benzene by 152 kJ mol^{-1}. When other cycles are constructed, such as the atomization enthalpy of Kekulé benzene, and these are compared with figures obtained from cycles using data for real benzene, a discrepancy of almost identical magnitude is obtained.

(*c*) *Chemical considerations* Compounds which contain carbon-to-carbon double bonds, such as ethene and cyclohexene, characteristically undergo *addition reactions* in which a chemical species, such as bromine or hydrogen bromide, adds across the double bond, for example,

$$CH_2 = CH_2(g) + Br_2(l) \rightarrow CH_2Br - CH_2Br(l)$$

From the Kekulé structure, benzene might be expected to undergo an analogous reaction, but real benzene is found to be most reluctant to do this.

From these considerations it is obvious that a more accurate formulation for benzene would show all the carbon–carbon bonds identical and intermediate between single and double bonds. This is represented as in Figure 5.34; the 'extra bonds' are shown as shared equally between the six carbon atoms and not specifically located between any particular pairs of adjacent atoms as in the Kekulé formula. This situation is described as *bond delocalization* or *resonance*. Real benzene is said to be energetically more stable than Kekulé benzene by 152 kJ mol^{-1}, this energy term being referred to as its *delocalization* or *resonance energy*.

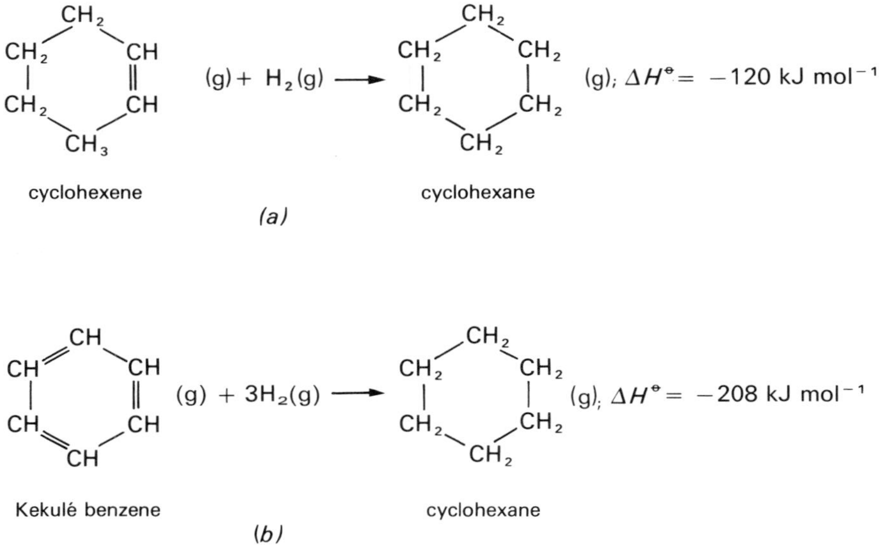

CH$_2$, CH$_2$, CH, CH$_2$, CH, CH$_3$ (cyclohexene) (g) + H_2(g) ⟶ CH$_2$, CH$_2$, CH$_2$, CH$_2$, CH$_2$, CH$_2$ (cyclohexane) (g); $\Delta H^{\ominus} = -120$ kJ mol^{-1}

cyclohexene　　　　cyclohexane

(a)

CH, CH, CH, CH, CH, CH (Kekulé benzene) (g) + 3H_2(g) ⟶ CH$_2$, CH$_2$, CH$_2$, CH$_2$, CH$_2$, CH$_2$ (cyclohexane) (g); $\Delta H^{\ominus} = -208$ kJ mol^{-1}

Kekulé benzene　　　　cyclohexane

(b)

Figure 5.33 Hydrogenation enthalpies of (*a*) cyclohexene and (*b*) Kekulé benzene

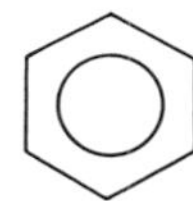

Figure 5.34 Structure of benzene showing bond delocalization, or resonance

Evidence for delocalization is found in compounds containing alternate single and double bonds in their conventional representation; use of a ruler on Figure 5.16, for example, will show this to be the bonding situation in *cis*-butenedioic (maleic) acid.

Delocalization of bonds in this manner is not restricted to molecules containing carbon. Figure 5.35 shows the structure of nitric acid as determined by electron diffraction. Although it seems from the dot-and-cross representation that of the two oxygen atoms bonded to the nitrogen but *not* also to hydrogen, one is a double and the other a single bond, they are found to have identical bond lengths. This bond length is, moreover, intermediate between that found for nitrogen-to-oxygen single (0.136 nm) and double (0.114 nm) bonds in compounds that contain only one nitrogen-to-oxygen bond. From this evidence it can be concluded that delocalization of these bonds occurs. In the nitrate ion, formed from nitric acid by loss of a proton, all the nitrogen-to-oxygen bonds are identical and the electron diffraction evidence shows that delocalization occurs between all three bonds.

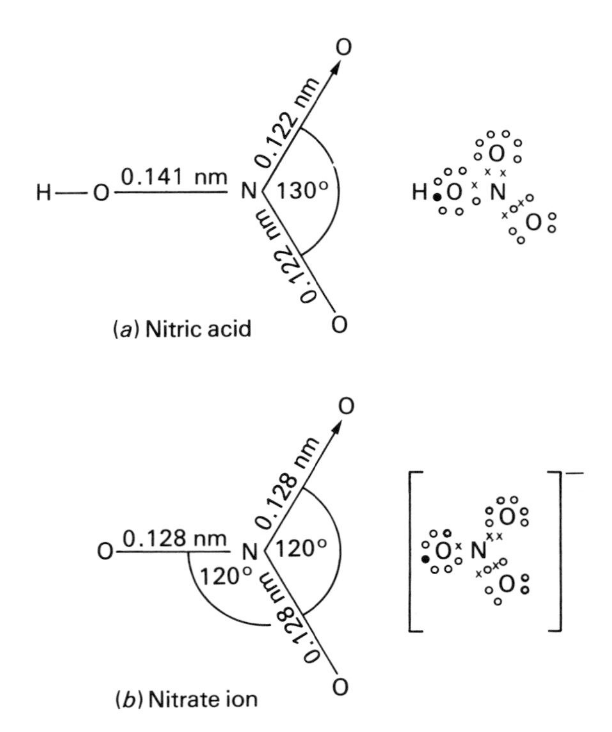

Figure 5.35 Valence bond and dot-and-cross representations of (*a*) nitric acid and (*b*) nitrate ion

Isomerism

When more than one compound is known to have the same molecular formula, but different physical and/or chemical properties, the phenomenon is known as *isomerism*. The existence of such isomers is possible because of the different structures possible for molecules composed of certain combinations of atoms. Isomerism can arise in a variety of ways.

Structural isomerism

Figure 5.36 shows the valence bond structures of the two isomers having the formula $C_2H_4Br_2$. It should be appreciated that because there is complete freedom of rotation around the carbon-to-carbon single bond, and these formulae represent 'squashed out' tetrahedral structures, these are the only two possible molecular structures corresponding to this formula.

Geometrical isomerism

Figure 5.37 shows the valence bond structures of the three isomers having the formula $C_2H_2Br_2$. These are planar structures and there is no permitted rotation about the carbon-to-carbon double bond. Two isomers corresponding to 1,2-dibromoethene are therefore possible. These are distinguished by the latin prefixes *cis* (meaning 'on this side of') and *trans* (meaning 'across from').

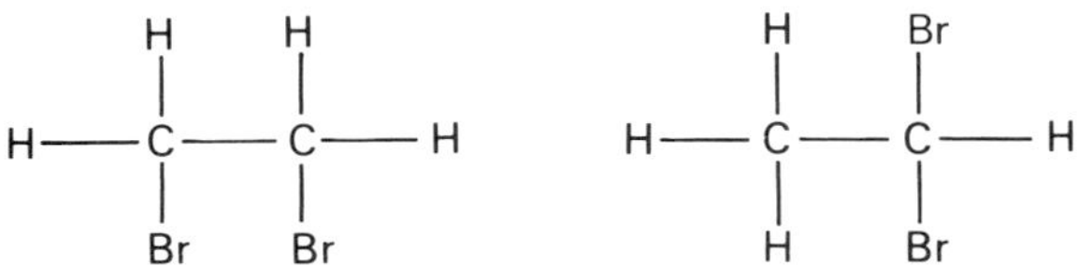

1,2-dibromoethane 1,1-dibromoethane

Figure 5.36 Isomers of formula $C_2H_4Br_2$

Stereo-isomerism

When four different atoms, or groups of atoms, are attached to the same carbon atom two different arrangements are possible. These isomers are not superimposable upon each other, each being the mirror image of the other. Such isomers are known as *enantiomers* or *enantiomorphs*. The central carbon atom is said to be an 'asymmetric carbon atom'. Important examples of stereo-isomerism are the aminoacids, the basic units from which proteins are built. The structure of the aminoacid alanine is given in Figure 5.38. The two enantiomorphs of such substances are shown in Figure 5.39.

An important property of stereo-isomeric materials is the effect they have on plane-polarized light. Light can be regarded as vibrating at right angles to its plane of propagation in an infinite number of planes. Certain materials have the property of absorbing these vibrations, except for those in one particular plane. The light emerging from such materials is thus said to be *plane-polarized*.

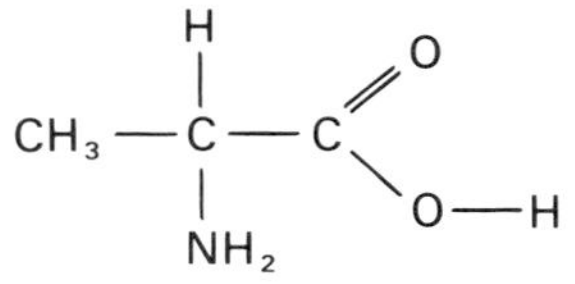

Figure 5.38 The structure of the amino acid alanine

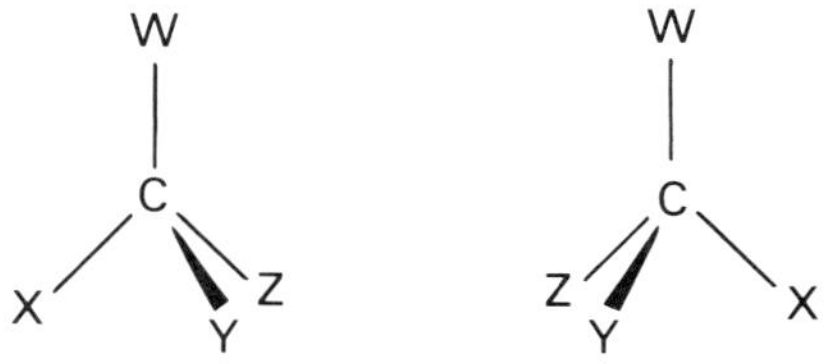

Figure 5.39 Arrangement of four different atoms of groups around an asymmetric carbon atom

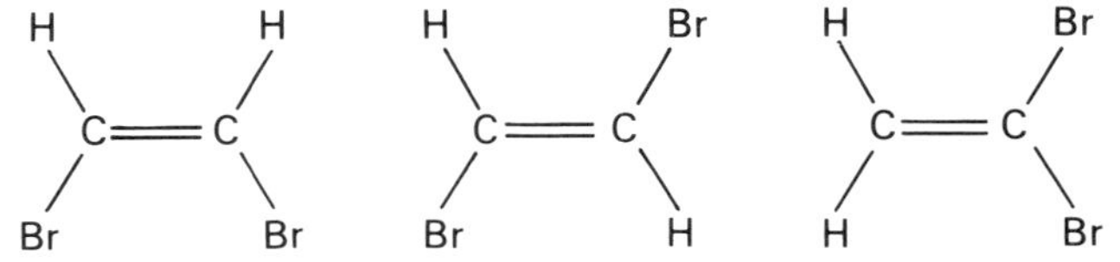

cis — 1,2-dibromoethene *trans* — 1,2-dibromoethene 1,1-dibromoethene

Figure 5.37 Isomers of formula $C_2H_2Br_2$

(Polaroid sunglasses contain such material whose function is to absorb the stray reflected light from the scene while transmitting that striking the lenses normally; in this manner glare is controlled.) The optical activity of stereoisomers can be investigated using a *polarimeter*, the basic construction of which is shown in Figure 5.40. In simple instruments the polarizing material is usually polaroid plastic; in more sophisticated apparatus special prisms, known as Nicol prisms, are employed. The light is plane-polarized by the polarizer, passes through the test material (usually in aqueous solution), and then through an analyser also of polarizing material. When the polarizer and analyser are set with no solution between them so that light passing through the former is plane-polarized at right angles to the direction of absorption by the analyser, the eye observes minimum light intensity emerging from the instrument. If the optically active solution is then placed in position, the observer sees a restoration of light intensity. The analyser is then rotated through an angle α to find the new position of minimum intensity. The angle α is the angle of rotation of the beam caused by the solution. The angle of rotation is found to depend on the following factors:

1. The material under test.
2. The concentration of the solution.
3. The length of solution through which the light beam passes.
4. The wavelength of the light.
5. The temperature of the experiment.

The rotating powers of different substances can be compared in terms of their molar rotations, $[\alpha]$, defined as the rotation produced by a 1 M solution with a path length of 1 metre at a stated temperature and with monochromatic light of a specified wavelength.

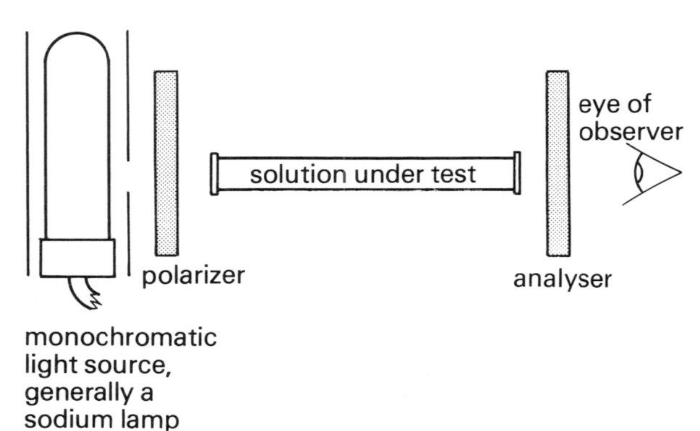

Figure 5.40 A polarimeter

It has been known for some time that the molar rotation of a substance varies with the wavelength of light employed. When graphs of molar rotation against wavelength are plotted many empirical relationships are found between the form of the graph and the structure of the molecules, particularly in the case of organic carbonyl compounds. This technique, called *optical rotatary dispersion*, can thus yield structural information.

The structure of metals

Metals generally crystallize in one of two crystal habits, *close-packed* or *body-centred cubic*.

There are two variations on the close-packed structure; in each of these the atoms are packed in layers in which each atom is surrounded by six neighbours in the plane of the layer. In simple *hexagonal close packing* such layers are built on top of each other in such a manner that every alternate layer lies directly on top of each other. Thus if the lowest layer is designated an *a* layer, the next is designated a *b* layer, and so on, the structure that results can be described as *abab* close-packing. Alternatively, the atoms can be arranged so that every third layer lies directly on top of each other, a situation described as *abcabc* close-packing. From this latter structure it is possible to deduce a face-centred unit cell, and the structure is thus known as *face-centred cubic packing*. Copper is a common example of a metal crystallizing in this habit and its structure is illustrated in Figure 5.41. Both close-packed structures contain about one-quarter of empty space.

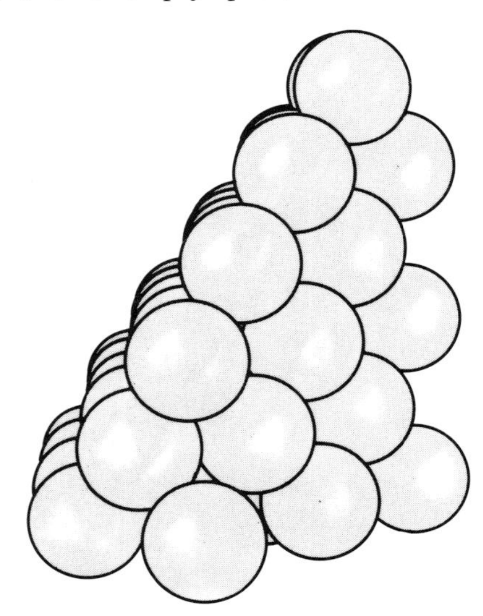

Figure 5.41 The structure of metallic copper

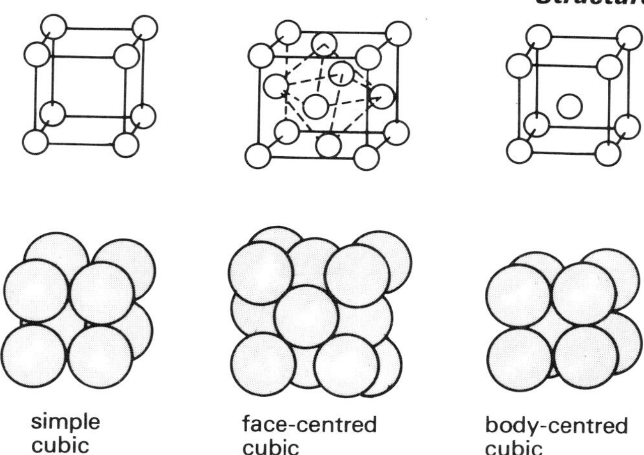

Figure 5.42 The unit cells of the cubic crystal systems

The other common packing arrangement, generally observed in the most electropositive metals such as sodium and potassium, is known as body-centred cubic. This is not a close-packed structure and it contains about one-third of empty space.

The structure of the cubic crystal systems is shown in Figure 5.42.

The 'plasma' model of metal structure

According to this theory of the bonding between atoms in metallic elements, which successfully explains many of their observed properties, each atom in the crystal is regarded as donating a proportion of its electrons into a widely delocalized electron cloud. The ions that remain are regarded as being embedded in this charge cloud and held in position by their electrostatic attraction to it. The situation is illustrated in Figure 5.43.

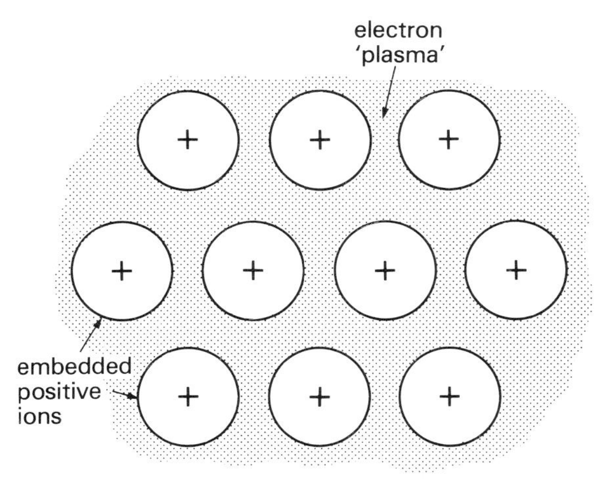

Figure 5.43 The 'plasma' model of metal structure

The good electrical conductivity of metals can be explained in terms of the capacity of the plasma to flow through the structure under the influence of an applied electrical potential difference. The good thermal conductivity can be explained if it is postulated that the metal ions oscillate about a mean position within such a structure. When heat is applied to one end of the metal, the amplitude of this oscillation is increased at that point and passed along the metal by the resulting perturbations in the plasma, thus transmitting the heat.

Diffraction studies on complex materials

In recent times the use of X-ray diffraction has been extended to elucidate the structures of complex materials such as proteins and dioxyribosenucleic acid (DNA). DNA is now accepted as the chemical agency whereby inherited characteristics are transmitted from one generation to the next in man and other living organisms. Such discoveries have therefore given rise to a new science at the interface between genetics and chemistry known as molecular biology. The X-ray diffraction patterns of materials with a high degree of crystallinity show a series of well-defined concentric rings; materials with low degrees of crystallinity show a much more diffuse diffraction pattern. The results of such studies on a sample of rubber is illustrated in Figure 5.44.

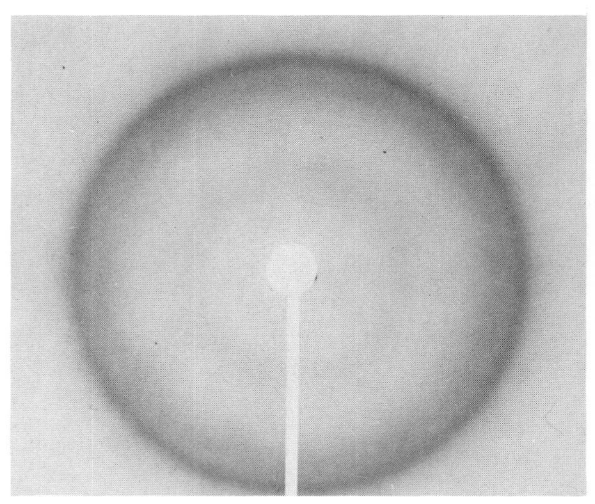

Figure 5.44 X-ray diffraction study of a sample of rubber

Bond Hybridization

As discussed in Chapter 2, *s* orbitals are spherical in shape and *p* orbitals have the 'dumbell' shape shown in Figure 2.27. As the four covalent bonds formed between carbon and other atoms in compounds such as methane are directed tetrahedrally and all four bonds are equivalent and identical, there must be some adjustment of these orbitals consequent to bond formation. In Figure 5.45 the way in which one *s* and three *p* orbitals hybridize to form four identical hybrid orbitals is illustrated. The four hybrid orbitals in this example are known as sp^3 orbitals, and it is these hybrid orbitals that overlap with the orbitals of the other atoms to form the tetrahedrally arranged bonds in such compounds. The bonds formed in this manner are known as sigma (σ) bonds. The formation of a sigma bond between the two carbon atoms in ethane is shown in Figure 5.46.

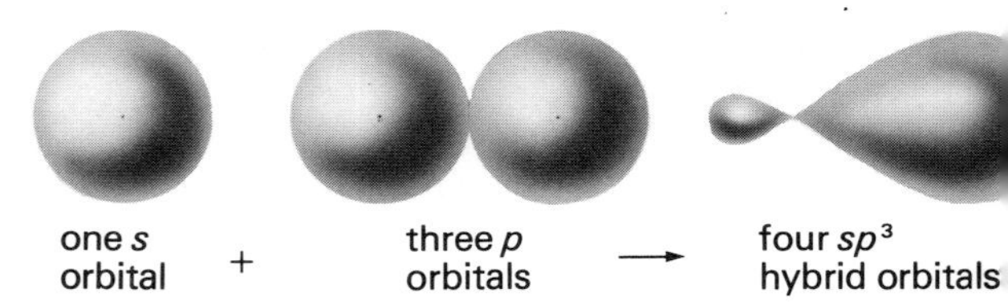

Figure 5.45 sp^3 hybridization

Figure 5.46 Sigma bond formation in ethane

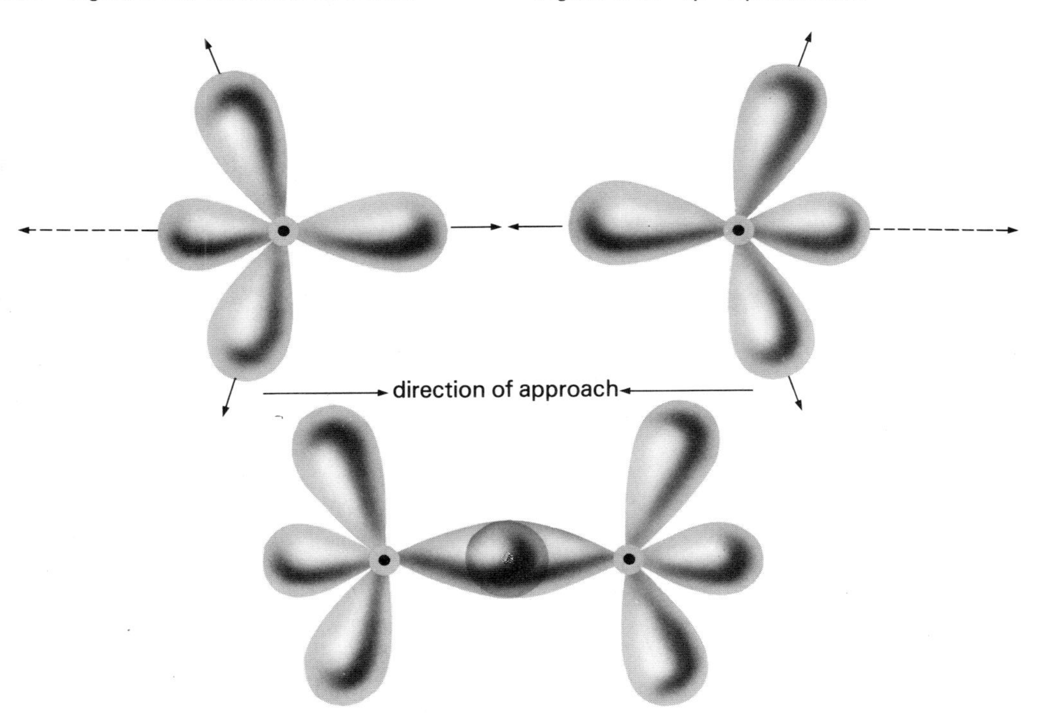

In compounds containing carbon-to-carbon double bonds, such as ethene, hybridization between one *s* and two *p* orbitals occurs, giving rise to three sp^2 hybrid orbitals. These orbitals lie in the same plane with angles of 120° between them, as shown in Figure 5.47. These hybrid orbitals overlap with the orbitals of the atoms to which they are bonded to form sigma bonds. The remaining *p* orbital interacts with that of the neighbouring atom to form a pi (π) bond. This is shown in Figure 5.48.

In benzene, in which the carbon atoms are sp^2 hybridized, the π bonds are delocalized as discussed previously. The structure of benzene is shown in Figure 5.49.

In carbon-to-carbon triple bond compounds, such as ethyne, one *s* and one *p* orbital hybridize to form two *sp* hybrid orbitals, as shown in Figure 5.50. These hybrid orbitals overlap with orbitals of neighbouring atoms to form σ bonds, and the unhybridized *p* orbitals form two π bonds mutually at right angles. This situation is illustrated for ethyne in Figure 5.51.

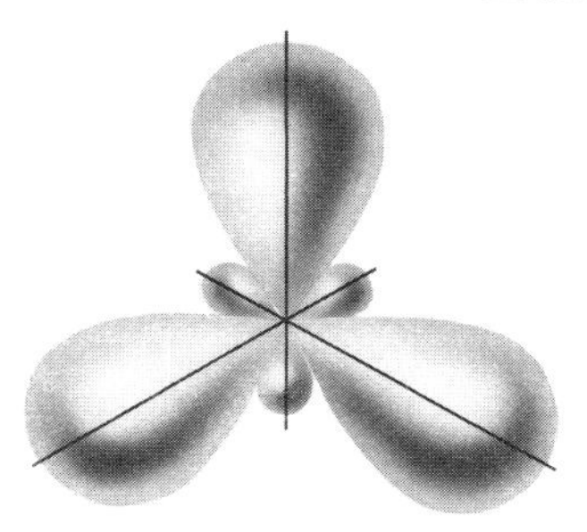

Figure 5.47 Arrangement of the sp^2 hybrid orbitals

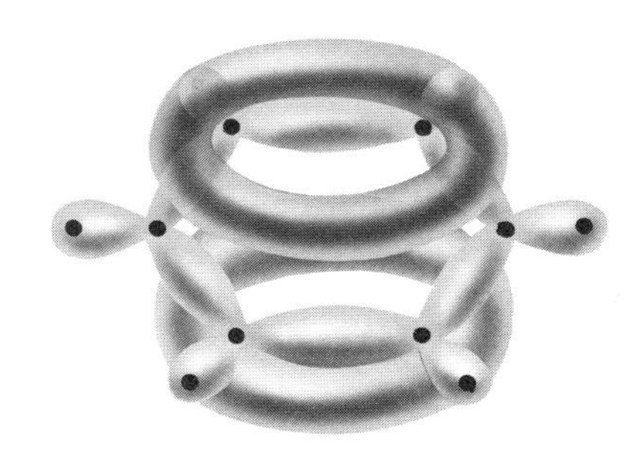

Figure 5.49 The bonding in benzene

Figure 5.48 Structure of ethene in terms of σ and π bonding

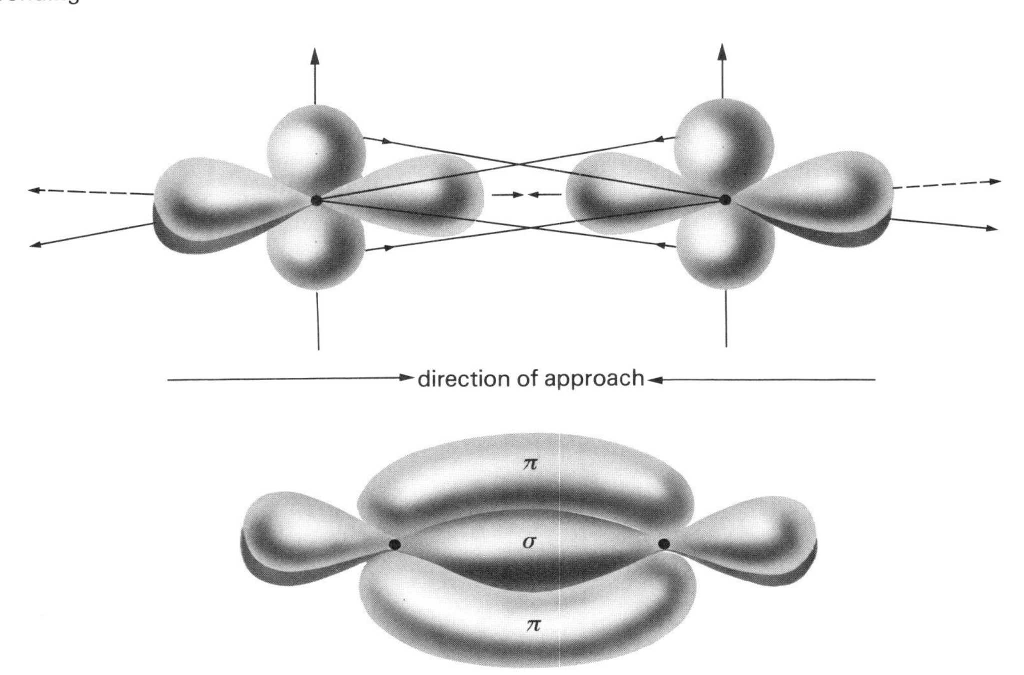

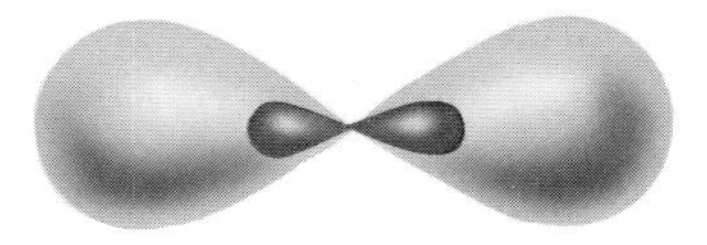

Figure 5.50 The colinear *sp* hybrid orbitals in ethyne

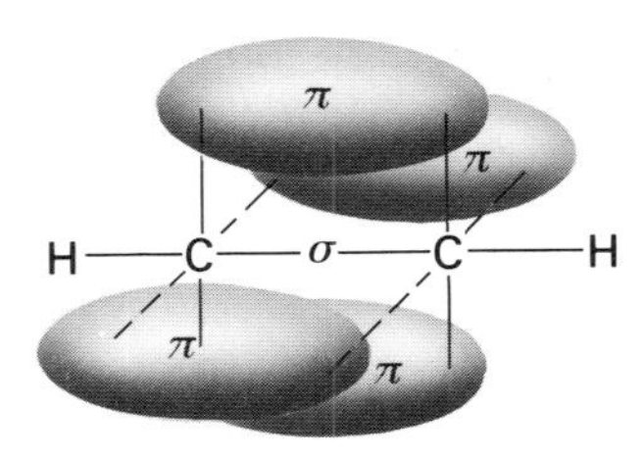

Figure 5.51 σ and π bonding in ethyne

Suggestions for further reading

HOLDEN, ALAN *Bonds between Atoms* (Oxford: Oxford University Press, 1971)

BROWN, P. J. and FORSYTH, J. B. *The Crystal Structure of Solids* (London: Edward Arnold, 1973)

Problems

1. What are *covalency* and *electrovalency*? Illustrate your answer with reference to chlorine and potassium chloride.

   Give the shapes of *s* and *p* orbitals and, in terms of orbitals, account for the shape of the tetrachloromethane (carbon tetrachloride) molecule.

   Explain why hydrogen chloride, which is covalent in the gaseous state, is ionic in water. (Cambridge 'A')

2. How can the shapes of simple molecules be explained in terms of electron pair repulsions? Your answer should include at least *one* example from each of *four* different shapes.

   What effect does the presence of a lone pair of electrons on the nitrogen atom have on
   (*a*) the H–N–H angle in ammonia
   (*b*) the properties of the ammonia molecule? (JMB 'A')

3. (*a*) A molecule of hydrogen chloride is sometimes written $\overset{\delta+}{H} - \overset{\delta-}{Cl}$. Discuss the nature of the bonding in this molecule and indicate

   (i) why there is a charge separation
   (ii) why the physical properties of the molecule demonstrate this
   (iii) why the above formula is not relevant to the situation of a dilute solution of hydrochloric acid.

   (*b*) State briefly what is understood by 'van der Waals forces'. Show, by choosing one suitable example in each case, how these forces are of importance when describing the melting of a solid and the vaporization of a liquid.

   (*c*) Describe the nature of the bonding in the ammonium ion and show how it leads to an explanation of the following facts. A sample of gaseous ammonia, labelled with radioactive hydrogen atoms, was dissolved in water. When a sample of ammonia gas was regenerated from this solution and its radioactivity measured at s.t.p. it was found to be less than that of the original gas measured under identical conditions. (JMB 'A')

4. Give one example of an ionic lattice, a molecular lattice, and a giant molecular crystal structure. Indicate how the physical properties of a crystalline substance are dependent on the nature of its lattice.

   How may X-rays be used to investigate the structure of crystals? (Oxford 'S')

5. What factors determine whether the bonding in a binary compound (one that contains two elements only) is predominantly covalent or ionic? Quote examples to illustrate your answer, giving the experimental evidence for the presence of the bonds that you describe. (Nuffield 'S')

6. Information concerning the structure of a substance can be obtained when:
   (*a*) electromagnetic radiation is emitted or absorbed by the substance;
   (*b*) electromagnetic radiation or a stream of particles such as electrons interacts with the substance to give diffraction patterns;
   (*c*) the substance interacts with an electric or magnetic field.

   Show how **one** of these methods can yield information about the structure of a substance. You should give an account of the theoretical basis which relates observations to structure, the technique itself, and an illustration of the interpretation of results for **one** actual substance. (Nuffield 'A')

7. What factors do you regard as having an

important influence on the shapes which molecules and ions adopt? Illustrate your discussion by reference to at least some of the following:

$AlCl_3$, H_2CO, NH_3, NO_3^-, PBr_5, SO_2, ClO_4^-.

(Cambridge Entrance and Awards)

8. X-ray diffraction is one of the most powerful techniques for the elucidation of the structure of crystalline solids. Give a brief outline of the principles on which this technique is based.

Crystalline solids may be classified, according to the nature of their structural units, as ionic, molecular, or atomic crystals. For each of these classes give an appropriate example, state how the particles of the substances chosen are held together in the solid state and discuss briefly the relationship between the structure and properties of these materials.

(London 'A')

9. Aluminium crystallizes with a face-centred cubic structure. The side-length of the unit cell is 0.4041 nm. Taking the density of aluminium as 2.702 g cm^{-3} and its relative atomic mass as 27.01, calculate a value for Avogadro's number.

10. What do you understand by the terms *ionic bond*, *covalent bond*, and *hydrogen bond*? Illustrate your answer with one example of each type of bond.

What factors determine the arrangement of atoms in (*a*) a molecule held together by covalent bonding and (*b*) an ionic crystal?

(Oxford and Cambridge 'A')

11. Discuss and explain (*a*) structural isomerism, (*b*) geometrical isomerism, and (*c*) optical isomerism, using as examples appropriate isomers with the molecular formula C_3H_4ClBr.

(Oxford 'A')

12. Identify the species A to D which have the following compositions:

	A	B	C	D
Protons	1	3	8	11
Neutrons	1	4	8	12
Electrons	0	3	10	10

Draw diagrams to illustrate the shapes of the following covalent molecules: (*a*) methane, (*b*) water, and (*c*) boron trichloride. What explanation can you give for these shapes?

What shapes would you expect the following to adopt: (i) the methyl cation $(CH_3)^+$, (ii) the methyl carbanion $(CH_3)^-$, and (iii) the covalent molecule $BeCl_2$? (Cambridge 'A')

6. The Equilibrium Law

Dynamic equilibrium

WHEN one mole of ethanol and one mole of ethanoic acid are mixed they react together to produce ethyl ethanoate and water according to the equation:

$$C_2H_5OH(l) + CH_3CO_2H(l) \rightarrow CH_3CO_2C_2H_5(l) + H_2O(l)$$

The reaction is a slow one at room temperature, but when there is no further change in the composition of the reaction mixture it is found to contain only two-thirds of a mole of each of the products instead of the mole of each expected from the equation.

When one mole each of ethyl ethanoate and water are mixed they react to form ethanol and ethanoic acid:

$$CH_3CO_2C_2H_5(l) + H_2O(l) \rightarrow C_2H_5OH(l) + CH_3CO_2H(l)$$

Again the reaction does not go to completion. When the composition of the mixture becomes constant it is found to have an identical composition with that described in the opening paragraph, i.e. there are two-thirds of a mole each of ethyl ethanoate and water with one-third of a mole each of ethanol and ethanoic acid.

Many other examples are known of chemical reactions that do not continue to completion but take place to a limited extent. This poses a problem in explanation. If some of the reactant molecules interact to form products, why not all? We can explain this phenomenon in terms of the concept of *dynamic equilibrium*, which postulates that forward and reverse reactions both take place. When, in our example, ethanol and ethanoic acid have reacted to form products these products start to interact to reform the reactants. Because the concentration of products is initially low, the reverse reaction is much slower at first than the forward reaction but after some time, as the concentration of products rises and that of reactants falls, the two rates become equal. The overall composition of the reaction mixture then becomes constant, and is said to have reached a state of dynamic equilibrium.

It is found that a numerical relationship exists between the concentrations of the reactants and products for a mixture in a state of dynamic equilibrium. This relationship is known as the *Equilibrium Law*. Consider the generalized reaction:

$$aA + bB \rightleftharpoons cC + dD$$

In this, A and B are the reactants, C and D are the products, and a, b, c, and d are the stoichiometric numbers of moles of the appropriate substances involved in the reaction. The sign $\rightleftharpoons$ represents a reaction that proceeds to a significant extent when either pure reactants or pure products are mixed resulting in the formation of an equilibrium mixture.

The equilibrium law in this case states:

$$K_c = \frac{[C]^c_{eqm} \times [D]^d_{eqm}}{[A]^a_{eqm} \times [B]^b_{eqm}}$$

In this expression $[A]_{eqm}$ represents the concentration of substance A in the reaction mixture at equilibrium, expressed in mol dm^{-3}. A corresponding notation for the equilibrium concentrations of the other substances involved has been employed. K_c, known as the *equilibrium constant in terms of concentration*, is found to have a constant value for any particular equilibrium system at a specified temperature.

We can calculate K_c at room temperature for the formation of ethyl ethanoate from the information given previously. Let V be the volume, in dm^3, of one mole of ethanol plus one mole of ethanoic acid. We must assume that this volume does not change appreciably during the course of the reaction. Equilibrium is then established, with the concentration changes shown on the next page.

	$C_2H_5OH(l)$ +	$CH_3CO_2H(l) \rightleftharpoons$	$CH_3O_2C_2H_5(l)$ +	$H_2O(l)$
Equilibrium concentration/mol dm^{-3}	$0.33/V$	$0.33/V$	$0.67/V$	$0.67/V$

$$K_c = \frac{[CH_3CO_2C_2H_5]_{eqm} \times [H_2O]_{eqm}}{[C_2H_5OH]_{eqm} \times [CH_3CO_2H]_{eqm}}$$

$$= \frac{0.67/V \times 0.67/V}{0.33/V \times 0.33/V}$$

$$= 4$$

It should be noticed that when the total number of moles of reactants equals the total number of moles of products, the concentration terms in the equilibrium law cancel and K_c is a dimensionless constant.

This equilibrium system has been extensively investigated by Berthelot and St. Gilles, who took varying amounts of alcohol and acid and allowed them to reach equilibrium. The mixtures were then analysed by determining the equilibrium amount of acid by titration with alkali. Within the limits of experimental error, consistent values for K_c were obtained.

Given a value for K_c for such a system, the equilibrium composition that would be obtained from any initial mixture can be calculated, as the example below will make clear.

If 2 moles of ethanol and 1 mole of ethanoic acid are allowed to react to equilibrium, let x be the number of moles of ethyl ethanoate and water formed. Let the volume of the reaction mixture be V dm^3. The calculation then proceeds as shown below.

If this calculation is performed for other initial mixtures of ethanol and ethanoic acid, the result shown in Table 6.1 is obtained.

Table 6.1 Equilibrium compositions of ethanol/ethanoic acid/ethyl ethanoate/water

Initial moles of:		*Equilibrium molar quantities:*	
acid	*alcohol*	*acid*	*ethyl ethanoate*
1	0.5	0.58	0.42
1	1.0	0.33	0.67
1	1.5	0.22	0.78
1	2.0	0.15	0.85
1	2.5	0.12	0.88
1	3.0	0.10	0.90
1	4.0	0.07	0.93
1	5.0	0.06	0.94
1	6.0	0.05	0.95
1	∞	0.00	1.00

Inspection of these figures shows that if a two or three to one molar ratio of alcohol to ethanoic

	$C_2H_5OH(l)$ +	$CH_3CO_2H(l) \rightleftharpoons$	$CH_3CO_2C_2H_5(l)$ +	$H_2O(l)$
Initial molar concentration	$2/V$	$1/V$	$0/V$	$0/V$
Equilibrium molar concentration	$(2-x)/V$	$(1-x)/V$	x/V	x/V

$$K_c = \frac{[CH_3CO_2C_2H_5]_{eqm} \times [H_2O]_{eqm}}{[C_2H_5OH]_{eqm} \times [CH_3CO_2H]_{eqm}} = 4$$

Substituting yields

$$\frac{(x/V)^2}{(2-x)/V \times (1-x)/V} = 4$$

$$\frac{x^2}{(2-x) \times (1-x)} = 4$$

$$x^2 = 4(2-x)(1-x) = 8 - 12x + 4x^2$$

$$3x^2 - 12x + 8 = 0$$

Solving this expression using the general formula for quadratic equations yields: $x = 3.155$ or 0.845. However, as x must lie between 0 and 1, its only acceptable value is 0.845.

Table 6.2 The hydrogen/iodine/hydrogen iodide equilibrium

$[H_2(g)] \times 10^3$	$[I_2(g)] \times 10^3$	$[HI(g)] \times 10^2$	K_c
5.617	0.5936	1.270	48.38
3.841	1.524	1.687	48.61
4.580	0.9733	1.486	49.54
1.696	1.696	1.181	48.48
1.433	1.433	1.000	48.71
4.213	4.213	2.943	48.81

acid is initially taken there is between 85 and 90 per cent conversion. There is little yield advantage in departing further from the stoichiometric proportions. Similar results would be obtained if an excess of ethanoic acid were taken initially. These considerations are of obvious importance in the laboratory and commercial production of chemicals involving equilibrium systems. In such cases it is sensible to take an excess of the cheaper component when this is possible.

When considering equilibria where the total number of moles of reactants differs from the total number of moles of products, the total volume must be taken into account as it does not cancel from the equilibrium expression. An example is shown below.

When mutually dissolved in a suitable solvent, pentene, and ethanoic acid react to form an equilibrium proportion of propyl ethanoate.

$$C_5H_{10} + CH_3CO_2H \rightleftharpoons CH_3CO_2C_5H_{11}$$

It is found that 0.00645 mole of pentene and 0.001 mole of ethanoic acid react in 845 cm^3 of solvent to give 0.000784 mole of pentyl ethanoate. The calculation for K_c is completed as shown below.

Some gaseous equilibria

1. The hydrogen/iodine/hydrogen iodide system

$$H_2(g) + I_2(g) \rightleftharpoons 2HI(g)$$

This system was studied by A. H. Taylor and R. H. Crist in 1941. In some experiments mixtures of hydrogen and iodine in known mole quantities were placed in sealed tubes and held at 457.6 °C to attain equilibrium. The tubes were rapidly cooled to 'freeze' the equilibrium (i.e. to effectively reduce the rates of forward and reverse reactions to zero). The tubes were broken under a solution of potassium iodide and the amount of iodine determined by titration with sodium thiosulphate solution. In another series of experiments hydrogen iodide was placed in the sealed tubes which were again equilibrated at 457.6 °C. Table 6.2 shows the results of three determinations from the first series of experiments, followed by three from the second.

The values of K_c, which show good consistency, have been calculated from the expression:

$$K_c = \frac{[HI(g)]^2_{eqm}}{[H_2(g)]_{eqm} \times [I_2(g)]_{eqm}}$$

It will be observed that as the same number of concentration terms appear in the numerator and denominator, K_c is dimensionless.

$$K_c = \frac{[CH_3CO_2C_5H_{11}]_{eqm}}{[C_5H_{10}]_{eqm}[CH_3CO_2H]_{eqm}}$$

$$= \frac{(7.84 \times 10^{-4})/0.845}{(64.5 \times 10^{-4} - 7.84 \times 10^{-4})/0.845 \times (10 \times 10^{-4} - 7.84 \times 10^{-4})/0.845}$$

$$= 540 \text{ mol}^{-1} \text{ dm}^3$$

The Equilibrium Constant in terms of Partial Pressure, K_p

$$pV = nRT$$

$$p_X = \frac{n}{V} RT = [X]\, RT$$

From the general gas equation and the application of Dalton's Law of Partial Pressures, it can be seen that the partial pressure of a gas is proportional to its concentration in mol dm^{-3}. We can write an equilibrium constant for gaseous equilibria, therefore, using the partial pressures as concentration terms. The equilibrium constant in these cases is given the symbol K_p. Thus for the hydrogen–iodine–hydrogen iodide system:

$$K_p = \frac{p_{HI}^{\,2}}{p_{H_2} \times p_{I_2}}$$

As the same number of partial pressure terms appear in the numerator as in the denominator, *for this system K_p* is dimensionless and has the same numerical value as K_c. In cases where there are different numbers of moles of reactants and products the two equilibrium constants are related by the expression:

$$K_p = K_c(RT)^{\Delta n}$$

where Δn = number of moles of products − numbers of moles of reactants.

N.B.—for this purpose, the partial pressures must be expressed in atmospheres. In this context, therefore, the appropriate value for R is 0.0821 dm^3 atm K^{-1} mol^{-1}.

2. The ammonia equilibrium

Ammonia is made directly from nitrogen and hydrogen in the Haber process by the reaction:

$$N_2(g) + 3H_2(g) \rightleftharpoons 2NH_3(g); \quad \Delta H^{\ominus} = -100 \text{ kJ mol}^{-1}$$

$$K_p = \frac{p_{NH_3}^{\,2}}{p_{N_2} \times p_{H_2}^{\,3}}$$

Consider a mixture of nitrogen and hydrogen consisting initially of 1 mole of nitrogen to 3 moles of hydrogen. Let α be the fraction of the mole of nitrogen converted to ammonia under the particular equilibrium conditions chosen. Let the total pressure be p.

The equilibrium partial pressure of each substance is obtained by allocating to it a fraction of the total pressure proportionate to the fraction of moles of that substance to the total number of moles (see foot of page). These values can be substituted into the equilibrium expression, which on simplification yields:

$$K_p/\text{atm}^{-2} = \frac{4\alpha^2(4 - 2\alpha)^2}{27(1 - \alpha)^4 \times p^2}$$

In this case K_p has dimension as well as magnitude and the fraction converted depends on the total pressure. For this reason the Haber process is carried out under high pressure conditions.

Le Châtelier's Principle

The *Principle of Mobile Equilibrium* was stated by Henri Le Châtelier in 1885. This states that: *if a system is in equilibrium and a change occurs in one of the equilibrium conditions, such as temperature or pressure, the position of equilibrium shifts so as to oppose that change as far as possible.* Although this is a useful and often profitable principle to apply, some caution is necessary in its use. When applied to the ammonia equilibrium considered above, it is seen that the system can best respond to an increase in pressure by producing more ammonia, for by so doing the system occupies a lesser volume. It should be stressed that changing the pressure without alteration to the temperature does not alter the value of the equilibrium constant. Consideration of the equilibrium expression for the ammonia equilibrium given previously shows that, with a constant value for K_p, changes in total pressure can affect the fraction of nitrogen converted.

Le Châtelier's principle can also be used to predict the effect of a temperature change on this system. As the reaction is exothermic, an increase in temperature will oppose the forward reaction and result in a decreased yield of ammonia, although in this situation a change in value of the equilibrium constant is involved (a quantitative relationship between temperature and equilibrium constant is given later in this chapter).

	$N_2(g)$	+	$3H_2(g)$	$\rightleftharpoons$	$2NH_3(g)$	Total
Moles initially	1		3		0	4
Moles at equilibrium	$1 - \alpha$		$3(1 - \alpha)$		2α	$4 - 2\alpha$
Equilibrium pressure/atmosphere	$\frac{1 - \alpha}{4 - 2\alpha} \times p$		$\frac{3(1 - \alpha)}{4 - 2\alpha} \times p$		$\frac{2\alpha}{4 - 2\alpha} \times p$	p

The combined effect of temperature and pressure on the percentage yield of ammonia is shown graphically in Figure 6.1.

The commercial production of such materials as ammonia depends not only on equilibrium considerations of this type but also on the rates of the reactions involved. The need to choose suitable conditions to satisfy both equilibrium and kinetic considerations is discussed in Chapter 10.

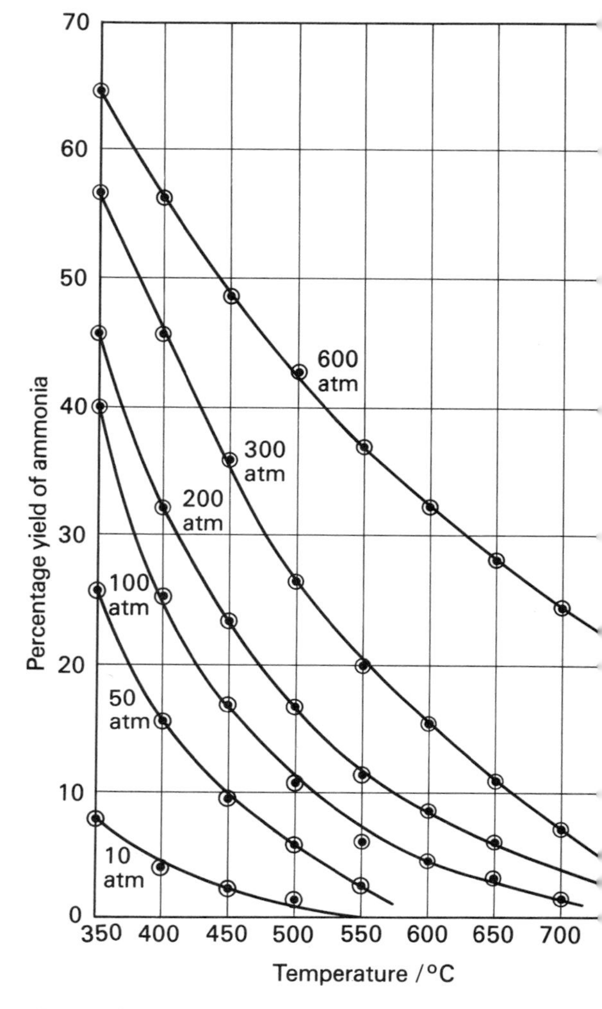

Figure 6.1 Effect of temperature and pressure on the percentage yield of ammonia at equilibrium in the Haber process

3. The dissociation of dinitrogen tetraoxide

Dinitrogen tetraoxide, N_2O_4, exists as a colourless solid at temperatures below −10 °C. Between this temperature and 22 °C, at which it boils, N_2O_4 exists as a liquid consisting of a proportion of nitrogen dioxide molecules, NO_2, formed by dissociation which increases in extent as the temperature rises. By determining the average molar mass of the vapour using a gas syringe heated in a furnace as described in Chapter 1, it can be shown that the vapour consists of a mixture of N_2O_4 and NO_2 molecules. The proportion of the latter steadily increases with temperature until at 140 °C a molar mass of 46 is observed indicating completion of the dissociation process:

$$N_2O_4(g) \rightleftharpoons 2NO_2(g); \quad \Delta H^{\ominus} = 57 \text{ kJ mol}^{-1}$$

Molar mass	92	46

From such measurements, the equilibrium constant at various temperatures can be calculated. Consider an initial one mole of dinitrogen tetraoxide assuming no dissociation and let α (known as the degree of dissociation) be the fraction that dissociates under the conditions described. Let the total external pressure be p. The calculation then proceeds as shown below.

	$N_2O_4(g)$	$\rightleftharpoons$ $2NO_2(g)$	Total
Moles initially	1	0	1
Moles at equilibrium	$1-\alpha$	2α	$1+\alpha$
Pressure at equilibrium/atm	$\frac{1-\alpha}{1+\alpha} \times p$	$\frac{2\alpha}{1+\alpha} \times p$	p

$$K_p = \frac{p_{NO_2}^{\,2}}{p_{N_2O_4}} = \frac{\left(\frac{2\alpha}{1+\alpha} \times p\right)^2}{\frac{1-\alpha}{1+\alpha} \times p} = \frac{4\alpha^2 p}{1-\alpha^2}$$

Thus from experimentally determined values of α, the value of K_p can be calculated. α can be calculated as follows; let the experimental value of average molar mass be M_{av}:

$$M_{av} = \frac{92(1-\alpha) + 46(2\alpha)}{(1+\alpha)}$$

This simplifies and rearranges to give:

$$\alpha = \frac{92 - M_{av}}{M_{av}} = \frac{92}{M_{av}} - 1$$

Table 6.3 Data for the thermal dissociation of dinitrogen tetraoxide at 1 atmosphere pressure

T/K	*Average relative molar mass*	α	K_p/atm
308	72.4	0.271	0.37
318	66.8	0.377	0.60
328	61.2	0.503	0.84
338	56.5	0.628	2.60
348	52.9	0.739	5.72
373	51.4	0.790	6.64
413	46.0	1.000	∞

Homogeneous and heterogeneous equilibria

The equilibrium systems considered so far have all involved substances in one phase only, i.e. either liquid or gas; such equilibria are known as *homogeneous equilibria*. We shall now consider *heterogeneous equilibria*, which are those taking place between substances in different phases.

1. Thermal dissociation of carbonates

Cadmium carbonate is suitable for studies of this type as it shows measurable dissociation at much lower temperatures than calcium carbonate, upon which the classical studies were made. Figure 6.2 shows apparatus suitable for the investigation. After checking that the set-up is leakproof, air is removed from the apparatus via the 3-way stop-cock. Very gentle heat is applied to the can and eventually the manometer records a pressure within the apparatus. This is due to the dissociation reaction:

$$CdCO_3(s) \rightleftharpoons CdO(s) + CO_2(g)$$

Corresponding values of pressure (the partial pressure of carbon dioxide) and temperature are made. We could consider writing the equilibrium expression in the form:

$$K_c = \frac{[CdO(s)]_{eqm} \times [CO_2(g)]_{eqm}}{[CdCO_3(s)]_{eqm}}$$

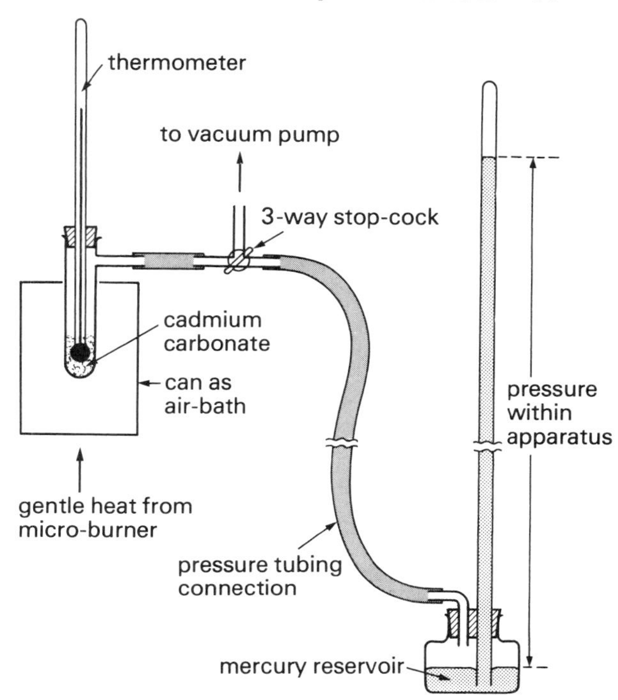

Figure 6.2 Apparatus to study the thermal dissociation of cadmium carbonate

However, the concentrations of cadmium carbonate and cadmium oxide are constant. This is because the quantities present do not affect the concentration (in mol dm^{-3}) for pure solids, this being the molar density of the substances. (For example, if we double the number of moles, we also double the volume.) We therefore write the equilibrium expression with these constants contained within the equilibrium constant:

$$K_c = [CO_2(g)]_{eqm}$$

similarly, $K_p = p_{CO_2}$

Table 6.4 Equilibrium data for the thermal dissociation of cadmium carbonate

Temperature/K	$p_{CO_2}/mmHg$	K_p/atm
535	2.5	0.0033
553	4.0	0.0053
573	6.0	0.0079
581	23	0.030
603	190	0.250

2. The reaction between steam and iron

At temperatures above 100 °C iron and steam react according to the equation:

$$3Fe(s) + 4H_2O(g) \rightleftharpoons Fe_3O_4(s) + 4H_2(g)$$

It must be stressed that equilibrium will only be attained within a closed system, i.e. one in which the substances involved are not allowed to escape. Allowing the hydrogen, for example, to escape in the above reaction will, as predicted by Le Châtelier's principle, cause the forward reaction to continue to produce more.

Using the general rule for heterogenous equilibria that the concentrations of all pure solid and pure liquid phases will be constant, the equilibrium expression is written:

$$K_p = \frac{p_{H_2}{}^4}{p_{H_2O}{}^4}$$

3. Partition of ammonia between water and trichloromethane

Ammonia gas is extremely soluble in water and also dissolves to a lesser extent in trichloromethane. If an aqueous solution of ammonia and an approximately equal volume of trichloromethane are placed in a separating funnel and shaken together, ammonia passes between the two immiscible phases until equilibrium is established:

$$NH_3(CHCl_3) \rightleftharpoons NH_3(aq)$$

If the two layers are run into separate containers, the concentration of ammonia in each can be determined by titration against a standard acid solution. If the amounts removed for the determination are replaced with equal volumes of each solvent and the layers equilibrated again, a further determination of the concentrations can be made as the remaining ammonia partitions itself between the layers. Repeating this procedure allows a series of corresponding pairs of equilibrium concentrations to be determined. The results are shown graphically in Figure 6.3.

The graph takes the form of a straight line connecting the points associated with the higher concentrations. We would theoretically expect this to continue to the origin, for zero concentration in one layer must be matched by zero concentration in the other. The deviation from this in the experimental curve can be explained in terms of the partial ionization of ammonia in the aqueous layer:

$$NH_3(aq) + H_2O(l) \rightleftharpoons NH_4{}^+(aq) + OH^-(aq)$$

Only the unionized ammonia can be partitioned between the solvents. Approximately one ammonia molecule in 10^4 is in the ionic form in a 1 M aqueous solution; however, the proportion increases with dilution and the effect is therefore more prominent at the lower concentrations. The gradient of the straight part of the curve is the equilibrium constant in the expression:

$$K_c = \frac{[NH_3(aq)]_{eqm}}{[NH_3(CHCl_3)]_{eqm}} = 23.3$$

K_c in this context is sometimes called the partition, or distribution, coefficient for ammonia between water and trichloromethane.

4. Partition of ammonia between 0·1 M copper(II) sulphate and trichloromethane

If the procedure described above is repeated but with 0.1 M copper(II) sulphate substituted for water, the results shown graphically in Figure 6.4 are obtained. The intercept on the aqueous concentration axis that would result from extrapolation

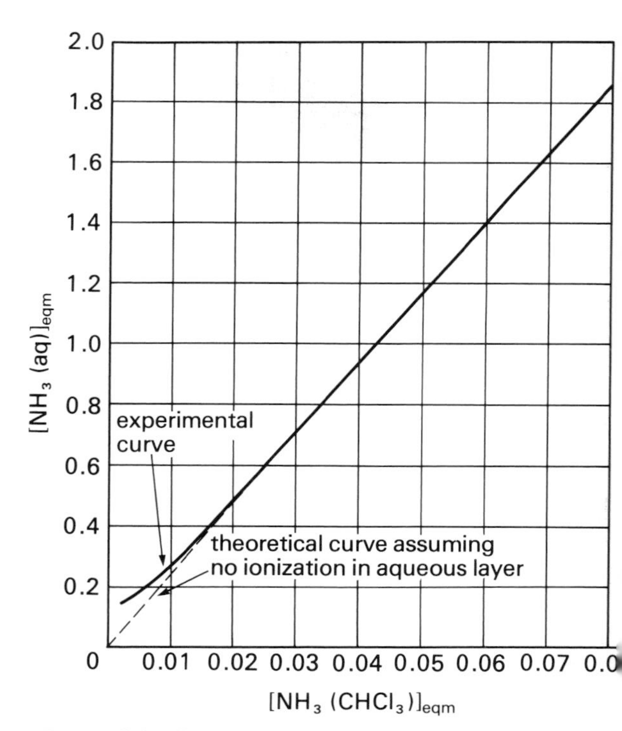

Figure 6.3 The partition of ammonia between water and trichloromethane

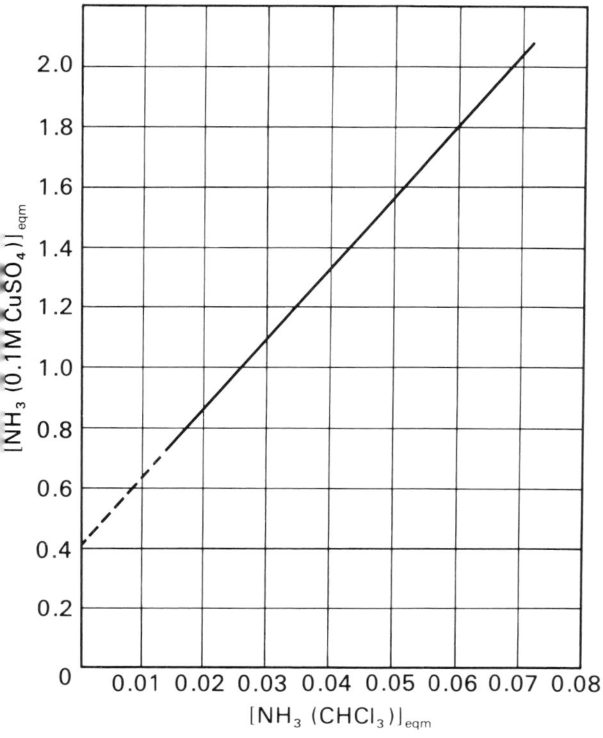

Figure 6.4 The partition of ammonia between 0.1 M copper(II) sulphate and trichloromethane

of the experimental curve is due to combination of part of the aqueous ammonia with the copper(II) ions. Only the uncombined or free ammonia can be partitioned. It will be observed that 0.4 moles of ammonia become associated with 0.1 moles of copper(II) in this way. This observation is consistent with the reaction below occurring:

$$Cu(H_2O)_4{}^{2+}(aq) + 4NH_3(aq) \rightarrow Cu(NH_3)_4{}^{2+}(aq) + 4H_2O(l)$$

5. Partition of ethanoic acid between water and tetrachloromethane

If experiments similar to those described above are carried out to examine the partition of ethanoic acid between water and tetrachloromethane, results as shown in Table 6.5 are obtained.

This unexpected result can be interpreted in terms of the hydrogen bonding between ethanoic acid molecules and solvent molecules which cause them to exist entirely as monomer in solution in water. In tetrachloromethane they exist as an equilibrium mixture of monomer and dimer. Hydrogen bonding is discussed in Chapter 5. In the organic layer the following equilibrium is established:

$$(CH_3CO_2H)_2(CCl_4) \leftrightharpoons 2CH_3CO_2H(CCl_4)$$

Concentration at equilibrium $\quad B(1 - \alpha) \quad 2B \times \alpha$

(where α is the degree of dissociation of the dimer)

$$K_c(\text{dissociation of dimer}) = \frac{(2B \times \alpha)^2}{B(1 - \alpha)}$$

$$2B \times \alpha = \sqrt{K_c(\text{dissociation of dimer}) \times B(1 - \alpha)}$$

Only those molecules in the same state can be partitioned between the solvents. In this case, therefore, only monomer molecules may be so partitioned.

$$K_c(\text{partition coefficient of monomer}) = \frac{\text{concentration of monomer in aqueous layer}}{\text{concentration of monomer in organic layer}}$$

$$= \frac{A}{2B \times \alpha}$$

$$= \frac{A}{\sqrt{K_c(\text{dissociation of dimer}) \times B(1 - \alpha)}}$$

When the degree of dissociation of the dimer is small, as is normally the case, $(1 - \alpha)$ can be approximated to unity. Under these conditions the equation above simplifies to:

$$\frac{A^2}{B} = \text{constant}$$

Table 6.5 Partition of ethanoic acid between water and tetrachloromethane

Concentration of ethanoic acid/mol dm^{-3}			
In aqueous layer(A)	*In tetrachloromethane layer (B)*	$\frac{A}{B}$	$\frac{A^2}{B}$
4.87	0.292	16.7	81
7.98	0.795	11.0	80
10.70	1.410	8.3	81

Some applications of partition of solutes

1. Chromatography

(*a*) *Liquid–liquid chromatography* Since their discovery by Tswett in 1906, chromatographic techniques have become among the most widespread and versatile analytical tools available to the chemist. Liquid–liquid chromatographic techniques include the use of columns packed with a material such as alumina, absorbent paper, and thin layers of materials such as silica gel held on glass plates or celluloid film. These materials hold water, known as the *stationary phase*, and flowing up or down this there is a second liquid known as the *running solvent*. The mixture to be separated is applied to the end of the stationary phase where the running solvent enters. In the case of column chromatography the running solvent is allowed to pass completely down the stationary phase, but with paper and thin-layer it is stopped before reaching the other end. Different components in the mixture have different partition coefficients between the water held on the stationary phase and the running solvent, and are therefore separated. The location of the components can be visualized, if they are not already coloured, by the use of a suitable locating agent. By employing a range of running solvents with varying degrees of molecular polarity, even substances of closely similar chemical structures can be separated and identified.

(*b*) *Gas–liquid chromatography* (*GLC*) In this technique the stationary solvent is a liquid, held on the surface of a finely divided solid such as powdered fire-brick, which is contained in a thin tube some two metres in length placed in a thermostatically controlled oven. The running solvent is a gas which passes continuously at a steady rate through the tube. The liquid sample to be analysed is injected into this stream of gas before the stationary phase is reached. On their passage through the tube, the components in the mixture are separated in terms of their partition coefficients between the running and stationary solvents. The components are detected as they emerge from the far end of the tube—the time between injection and emergence is characteristic for a particular material with a particular column packing and running solvent at a specific temperature.

2. Solvent extraction

When aniline (phenylamine) is prepared by the reduction of nitrobenzene it is separated from the reaction mixture by steam distillation (this technique is discussed in Chapter 7); the distillate so obtained consists of water and aniline. These liquids are largely immiscible and the bulk of the aniline can thus be readily separated. A little aniline, however, remains dissolved in the water and it is sometimes profitable to attempt to extract it. The technique employed is known as *solvent extraction*, in which the aqueous solution is shaken with another immiscible liquid in which the solute is more soluble than it is in water.

Consider such a situation in which the partition coefficient of the solute between water and the other solvent is 4 (the second solvent will be referred to simply as the solvent hereafter). If we have 50 cm^3 of aqueous layer containing 15 g of solute and we shake it with 50 cm^3 of solvent, 12 g will be extracted into the solvent layer. However, if we use the solvent twice in two equal portions of 25 cm^3 the extraction is seen to be more efficient. For the first extraction:

$$\frac{\text{mass of solute in 1 cm}^3\text{ solvent}}{\text{mass of solute in 1 cm}^3\text{ water}} = 4$$

$$\frac{\text{mass of solute in 25 cm}^3\text{ solvent}}{\text{mass of solute in 50 cm}^3\text{ water}} = 2$$

Thus the first extraction yields two-thirds of the solute, i.e. 10 g. This leaves 5 g of solute still in the water. If a further extraction is performed with the remaining 25 cm^3 of solvent two-thirds of this are extracted, i.e. 3.3 g. The total extracted by the second technique is therefore 13.3 g.

Variation of equilibrium constants with temperature

The value of the equilibrium constant for any one reaction varies with temperature according to the equation:

$$\ln K = -\frac{\Delta H^\ominus}{RT} + \text{constant}$$

(The derivation of this equation is given in Chapter 11, page 176.) This equation can be used to calculate $\Delta H^\ominus$ for a reaction when the values of the equilibrium constants are known at at least two different temperatures. If K_1 and K_2 are the equilibrium constants at temperature T_1 and T_2 respectively, we can write the equation as

$$\ln \frac{K_1}{K_2} = -\frac{\Delta H^\ominus}{R}\left(\frac{1}{T_1} - \frac{1}{T_2}\right)$$

Thus if four of the quantities $\Delta H^\ominus$, K_1, K_2, T_1, and T_2 are known, the fifth may be calculated.

Table 6.6 Data relating K_p and temperature for the contact process reaction

Temperature/K	*10^3K/T*	*K_p/atm$^{-\frac{1}{2}}$*	*ln K_p*
800	1.24	31.1	3.43
900	1.11	6.75	1.91
1000	1.00	1.86	0.620
1100	0.90	0.671	−0.400

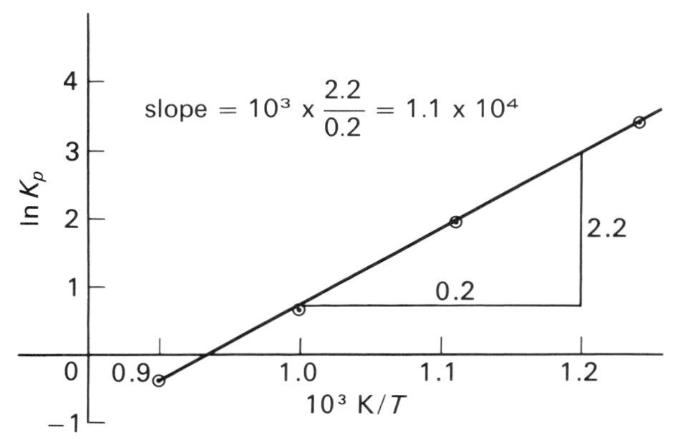

Figure 6.5 Graph of ln K_p against 10^3 K/T for the contact process reaction

For the reaction $SO_2(g) + \frac{1}{2}O_2(g) \rightleftharpoons SO_3(g)$ the data in Table 6.6 have been determined.

These data have been plotted in Figure 6.5. From the equation above it can be seen that the slope of the graph equals $-\Delta H^\ominus/R$:

$$\text{slope} = -\frac{\Delta H^\ominus}{R} = 1.1 \times 10^4 \text{ K}$$

$$\Delta H^\ominus = -(11 \times 10^3 \times 8.3) \text{ J mol}^{-1}$$
$$= -91 \text{ kJ mol}^{-1}$$

Suggestions for further reading

BROWNING, D. R. (ed.) *Chromatography* (London: McGraw-Hill, 1969)

Problems

1. 1 mole of propanoic acid and 0.5 mole of ethanol are mixed and thermostated at 50 °C and the following equilibrium is established:

$$C_2H_5CO_2H(l) + C_2H_5OH(l) \rightleftharpoons C_2H_5CO_2C_2H_5(l) + H_2O(l)$$

At equilibrium, titration with alkali shows the mixture to contain 0.55 mole of unchanged propanoic acid. Calculate K_c.

What will be the composition of the equilibrium mixture formed when 3 moles of propanoic acid and 0.5 mole of ethanol are reacted?

2. 1 mole of ethanol and 0.091 mole of ethanal were mixed at 25 °C in a total volume of 63 cm³. The reaction between them is

$$2C_2H_5OH(l) + CH_3CHO(l) \rightleftharpoons CH_3CH(OC_2H_5)_2(l) + H_2O(l)$$

At equilibrium, 0.0084 mole of ethanal are found unreacted. Calculate K_c for the reaction and state its units.

Concentration/mol dm^{-3}						
Initially			*At equilibrium*			*K_c*
H_2	*I_2*	*HI*	*H_2*	*I_2*	*HI*	
0.0123	0.008 78	0		0.001 06		
0	0	0.0152		0.001 70		

The data in the table refer to the equilibrium:

$$H_2(g) + I_2(g) \rightleftharpoons 2HI(g)$$

at a temperature of 458 °C. Complete the table. Why does an alteration in pressure not affect the equilibrium concentrations in this reaction?

4. At elevated temperatures, phosphorus pentachloride dissociates according to the equilibrium:

$$PCl_5(g) \rightleftharpoons PCl_3(g) + Cl_2(g)$$

At 200 °C and 1 atmosphere pressure, 256 cm^3 of the equilibrium mixture was found to weigh 0.922 g. Calculate (*a*) the average molar mass, (*b*) the degree of dissociation, and (*c*) K_p at this temperature. What are the units of K_p? Calculate K_c.

5.

$[I_2(aq)]_{eqm}$	0.197	0.178	0.146	0.077
$[I_2(CCl_4)]_{eqm}$	17.2	15.6	12.8	6.71

From the results in the table, calculate the partition coefficient of iodine between tetrachloromethane and water.

When iodine dissolves in aqueous potassium iodide solution, the following equilibrium (which lies well to the right) is established:

$$I_2(aq) + I^-(aq) \rightleftharpoons I_3^-(aq)$$

What results would you expect from partition experiments of iodine between tetrachloromethane and aqueous potassium iodide?

6.

[Benzoic acid (aq)]	0.001 00	0.007 30	0.004 60
[Benzoic acid (C_6H_6)]	0.170	0.0090	0.360

From the results in the table concerning the partition of benzoic acid between water and benzene, deduce the molecular state of the acid in these solvents.

7. For the reaction:

$$N_2O_4(g) \rightleftharpoons 2NO_2(g)$$

the data below have been reported:

T/K	300	333	373
K_p/atm	0.168	1.34	6.64

Determine (*a*) $\Delta H^{\ominus}$ for the reaction and (*b*) K_p at 350 K.

8. The oxidation of sulphur dioxide is a reversible reaction:

$$2SO_2 + O_2 \rightleftharpoons 2SO_3; \quad \Delta H^{\ominus} = -197 \text{ kJ}$$

Discuss the effect of (*a*) pressure, (*b*) temperature, and (*c*) a catalyst on the position of equilibrium.

Calculate the value of the equilibrium constant in terms of partial pressures at 800 °C from the following data:

Partial pressures/N m^{-2}

$p(SO_2)$	$p(O_2)$	$p(SO_3)$
10 100	68 800	80 100

If the above equilibrium mixture was obtained by starting with a mixture of sulphur dioxide and oxygen in a sealed vessel at 800 °C, what were the initial partial pressures of these two gases? (Oxford 'A')

9. Write down the equation for the reversible reaction in which carbon monoxide and hydrogen react to form methanol (methyl alcohol). How could it be shown that this system is in *dynamic* equilibrium?

The partial pressures of the above components in an equilibrium mixture at 320 °C are: CO 3.333×10^{-1} atm, H_2 6.666×10^{-1} atm, CH_3OH 9.924×10^{-4} atm. Calculate the value of the equilibrium constant (K_p) at 320 °C.

If one started with a mixture of carbon monoxide and hydrogen in the molar ratio 1:2, what would be the molar percentages of these gases in an equilibrium mixture in which the molar percentage of methanol is 19?

In the industrial synthesis of methanol, carbon monoxide and hydrogen in the molar ratio 1 : 2 are passed over an activated zinc oxide catalyst at 320 °C. Using the value of K_p obtained above, calculate the total pressure (in atmospheres) required to produce a molar yield of 19 per cent methanol.

(Oxford and Cambridge 'A')

10. Define the *standard state* and the *standard enthalpy of formation* of a compound. Explain why the enthalpy of formation is important in chemical equilibrium.

Bodenstein investigated the gaseous equilibrium

$$H_2 + I_2 \rightleftharpoons 2HI$$

at 710 K and found the equilibrium pressures in one experiment to be 1.9, 3.4, and 18 kPa respectively. Calculate the value of the equilibrium constant at this temperature. Starting with 8 and 5.3 kPa pressure of H_2 and I_2 respectively, calculate the partial pressures of H_2, I_2, and HI at equilibrium. What information is necessary to evaluate the equilibrium constant at any other temperature?

(Cambridge Entrance and Awards)

11. Describe how partition between two solvents may be used to find either the formula of a complex ion or whether a carboxylic acid is associated in non-aqueous solution.

500 cm^3 of water containing 4.0 g potassium iodide was shaken with 100 cm^3 of a solution of iodine in benzene. After reaching equilibrium, the layers were separated. On titrating with standard aqueous sodium thiosulphate (0.1 mol dm^{-3}), 10 cm^3 of the benzene layer required 5.1 cm^3 and 50 cm^3 of the aqueous layer required 2.9 cm^3.

If the distribution coefficient of iodine between benzene and water is 130, calculate the equilibrium constant for the reaction:

$$KI + I_2 \rightleftharpoons KI_3$$

(WJEC 'S')

12. (*a*) State the distribution (partition) law.

Solid X is three times as soluble in solvent A as it is in solvent B, and has the same molar mass in both solvents.

What mass of X would be extracted from a solution of 4 grams of X in 12 cm^3 of B by extracting with (i) 12 cm^3 of A and (ii) three successive portions of 4 cm^3 of A?

(*b*) Describe an experiment to illustrate the separation of a mixture by a chromatographic method and explain the principles involved.

(Oxford 'A')

13. For the following reaction:

$$H_2O(g) + CO(g) \rightleftharpoons H_2(g) + CO_2(g)$$

(*a*) Write an expression for the equilibrium constant, K_p, in terms of the partial pressures of the gases involved.

(*b*) Calculate the percentage of hydrogen in the equilibrium mixture when one mole of steam and one mole of carbon monoxide are placed in a vessel and allowed to reach equilibrium at a temperature at which the value of the equilibrium constant is 4.0.

(*c*) Why is the equilibrium constant for this particular reaction not expressed in any specified units?

(*d*) State, with your reasons, whether you would expect an increase of pressure to affect the yield of hydrogen in (*b*). (Oxford 'A')

14. Nitrogen oxide (NO) in the exhaust gases from internal combustion engines causes pollution of the environment. The nitrogen oxide can be removed by reducing it catalytically in the exhaust system with carbon monoxide (also present in exhaust gases).

$$CO(g) + NO(g) \rightleftharpoons CO_2(g) + \tfrac{1}{2}N_2(g); \quad \Delta H^{\ominus} = -374 \text{ kJ mol}^{-1}$$

(*a*) Write down the expression for the equilibrium constant, K_p.

(*b*) What is the effect of temperature and pressure on (i) the position of equilibrium; (ii) the rate of reduction of nitrogen oxide?

(*c*) What is the origin of the nitrogen oxide present in exhaust gas?

(*d*) Explain why a catalyst is necessary in the exhaust system.

(*e*) The exhaust gases from an engine contain 75% nitrogen, 10% carbon dioxide, 2% carbon monoxide by volume. Calculate the equilibrium partial pressure of nitrogen oxide at 770 K and at atmospheric pressure. K_p at 770 K $= 1.0 \times 10^{20}$ atm$^{-\frac{1}{2}}$.

(Oxford and Cambridge 'A')

15. How does the equilibrium constant for a reaction depend upon the temperature?

The equilibrium constant (K_c) at 410 °C for the reaction

$$2NO_2(g) \rightleftharpoons 2NO(g) + O_2(g)$$

is 9.00 mol dm^{-3}. Using thermochemical information from tables, calculate ΔH for the reaction and hence calculate or graphically estimate the equilibrium constant K_c at 800 °C.

Taking the gas constant R as 8.2 J mol^{-1} K^{-1}, calculate K_p for the reaction at 410 °C.

Without using Le Chatelier's principle, explain what happens to the equilibrium concentration of oxygen if (*a*) the partial pressure of O_2 is increased, (*b*) the total pressure of the system is increased. (Nuffield 'S')

7. Vapour Pressure

Vapour pressure and Raoult's Law

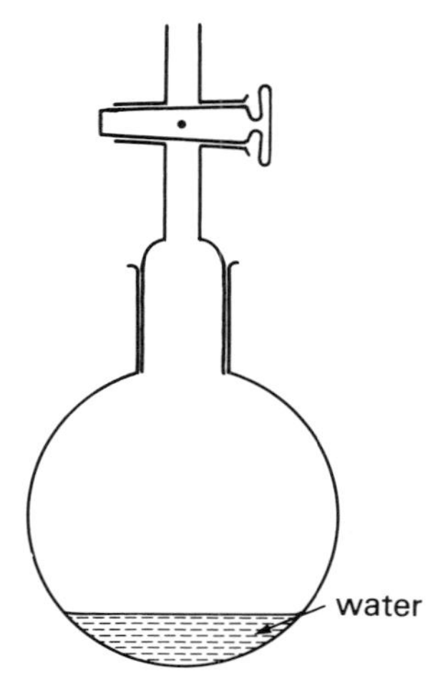

Figure 7.1 Liquid/vapour equilibrium in water

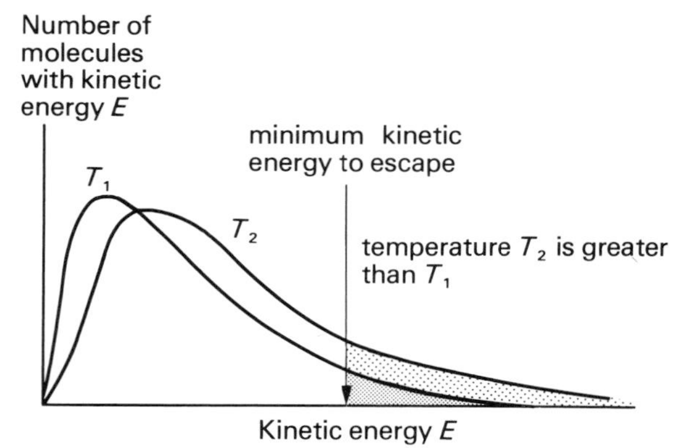

Figure 7.2 Distribution of molecular energies at different temperatures

CONSIDER the situation illustrated in Figure 7.1. The flask containing water is connected to a vacuum pump and evacuated. Molecules of water will evaporate from the surface and form a gaseous phase in the upper part of the flask. After random collisions between some of these molecules among themselves and the side of the flask, some will lose energy and fall back into the liquid phase. In this way an equilibrium is rapidly established:

$$H_2O(l) \rightleftharpoons H_2O(g)$$

The pressure of the gaseous phase so formed is known as the *vapour pressure*. As a certain minimum energy is required before molecules can escape from the surface of the liquid phase, if the temperature of the water is raised the fraction of the molecules with this energy is increased and the vapour pressure rises. This can be more fully understood in terms of the energy distribution of the molecules, shown in Figure 7.2. (An experimental method by which such information is obtained is discussed in Chapter 1, page 12.) The number of molecules with at least the minimum energy required is the shaded portion under the curves. When the mean energy of the molecules equals this minimum escape energy the liquid boils, and the temperature corresponding to this is the normal boiling point of the liquid.

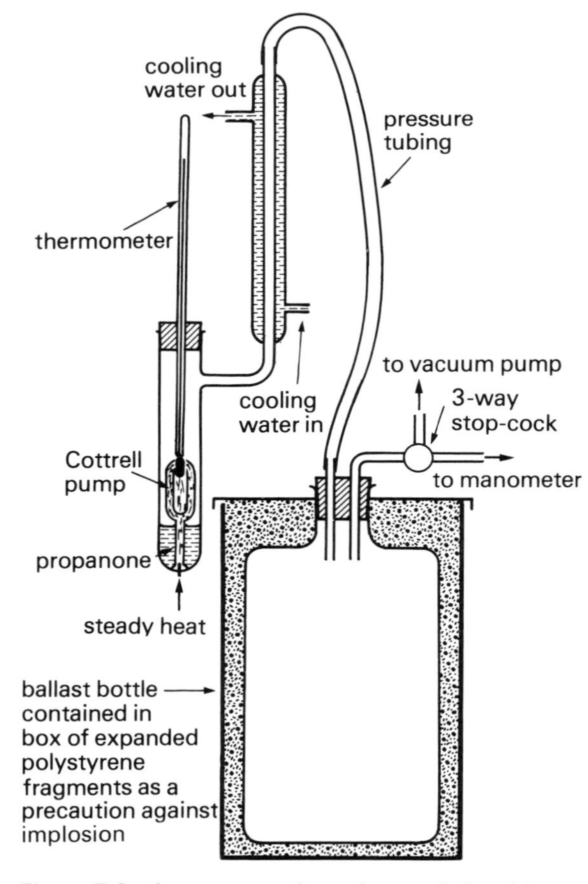

Figure 7.3 Apparatus to investigate relationship between temperature and vapour pressure

The quantitative effect of temperature on vapour pressure can be investigated using the apparatus shown in Figure 7.3. Propanone has been chosen as the liquid because its lower surface tension makes it more suitable than water. The heating causes the liquid to boil and an equilibrium mixture of liquid and vapour passes, by convection, up the Cottrell pump. The thermometer records the temperature of this equilibrium mixture and simultaneously the pressure is read on the manometer. The apparatus is then partially evacuated and the new equilibrium temperature is noted together with the pressure. By repeating this procedure, corresponding values of vapour pressure and temperature can be obtained. The ballast bottle is necessary so that the pressure can be reduced in small steps. The results are shown graphically in Figure 7.4. It will be observed that propanone boils at its normal boiling point of 56 °C under standard atmospheric pressure.

If the experiment is repeated using solutions of a suitable solute, such as naphthalene, dissolved in the propanone, the results shown in Figure 7.5 are obtained. It can be seen that at any particular temperature the naphthalene is lowering the vapour pressure of the propanone. This can be explained in terms of the surface of the solution consisting not of molecules all potentially volatile, as in the case of pure solvents, but having a proportion of completely non-volatile molecules on it. This effect was investigated by the French chemist Raoult towards the end of the last century. His work can be understood by considering the situation shown in Figure 7.6.

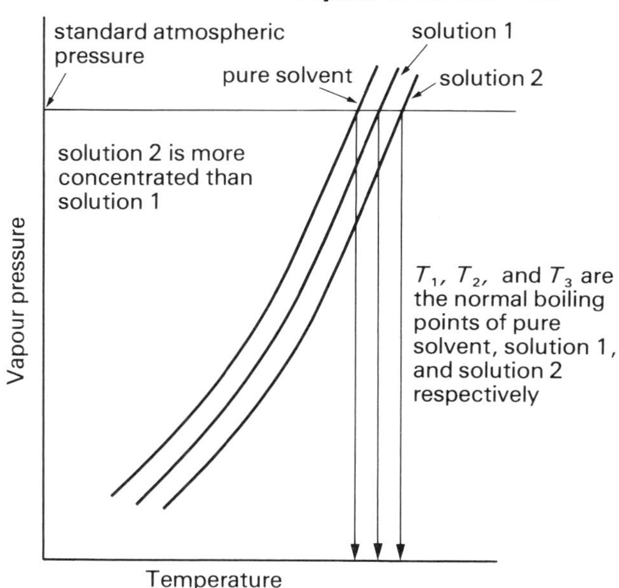

Figure 7.5 The effect of a non-volatile solute such as naphthalene on the vapour pressure of propanone

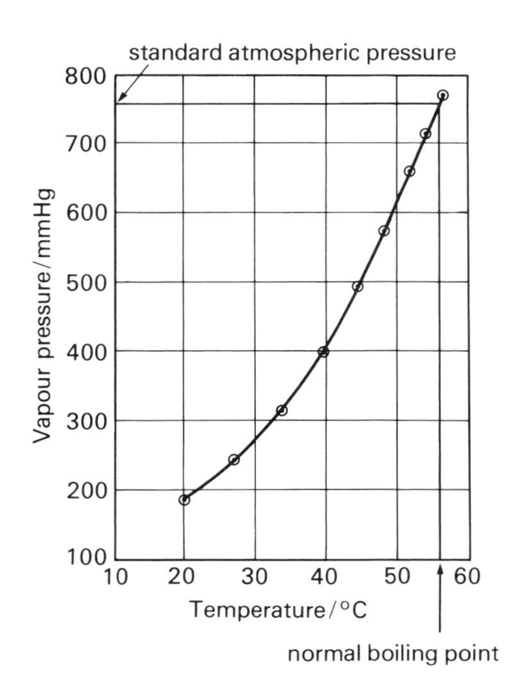

Figure 7.4 Relationship between vapour pressure and temperature for propanone

$(h_1 - h_2)$ represents the vapour pressure of water at room temperature; this is given the symbol p_0. $(h_1 - h_3)$ is the vapour pressure of the sucrose solution at room temperature; this is given the symbol p. $(p_0 - p)$ is known as *the lowering of vapour pressure*. $(p_0 - p)/p_0$ is known as the *relative* lowering of vapour pressure. After quantitative investigation of such situations, Raoult discovered a simple relationship to exist between the relative lowering of vapour pressure and the concentration of the solution expressed as the *mole fraction* of solute. The mole fraction of solute is defined as:

$$\frac{\text{moles of solute}}{\text{moles of solute} + \text{moles of solvent}}.$$

Table 7.1 shows results obtained for aqueous solutions of sucrose at room temperature.

Table 7.1 Effect of dissolved sucrose at room temperature

Mole fraction of sucrose	*Vapour pressure of solution/ mmHg*	*Relative lowering of vapour pressure $(p_o - p)/p_o$*
0.000	17.5	0.000
0.050	16.6	0.051
0.100	15.8	0.097
0.150	14.9	0.150

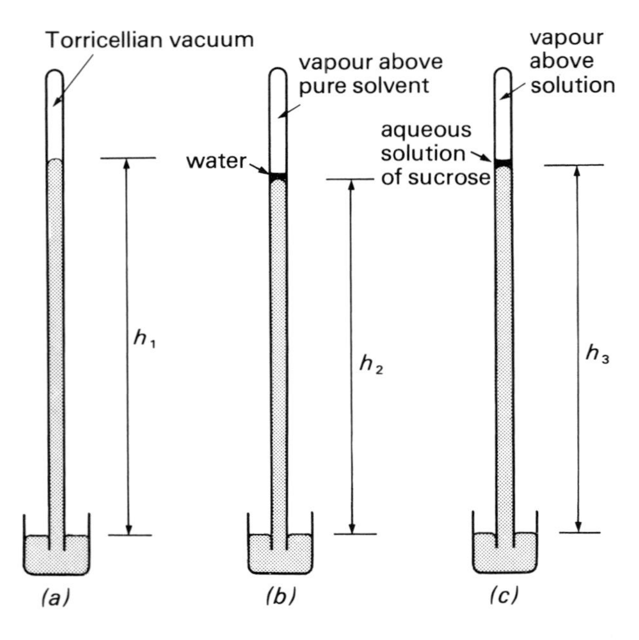

Figure 7.6 The effect of vapour from water and an aqueous solution on the height of a mercury barometer

It can be seen that, within the limits of experimental error, *the relative lowering of vapour pressure is equal to the mole fraction of solute.* This is a statement of Raoult's Law which can also be written mathematically:

$$\frac{p_0 - p}{p_0} = \frac{n_{\text{solute}}}{n_{\text{solute}} + n_{\text{solvent}}} = x_{\text{solute}}$$

where x_{solute} is the mole fraction of solute.

The methods for determining vapour pressures present considerable practical difficulties (the barometric technique described above, for example, involves the introduction of water and solutions into an instrument sensitive enough to read to 0.1 mmHg). It is useful to utilize measurements of the boiling points of liquids instead. Liquids with low vapour pressures at room temperature will have correspondingly high boiling points. We can use the apparatus shown in Figure 7.7 to determine boiling points. In this example we shall be using mixtures of liquids in which both the components are volatile; for different mixtures of hexane and heptane, the results are shown graphically in Figure 7.8.

Knowing Raoult's Law we would expect this result because it predicts that the vapour pressure is affected linearly by mole fraction of solute. However, both components of the mixture are volatile and thus, although each will reduce the vapour pressure of the other, the measured vapour pressure will be the sum of the vapour pressures of

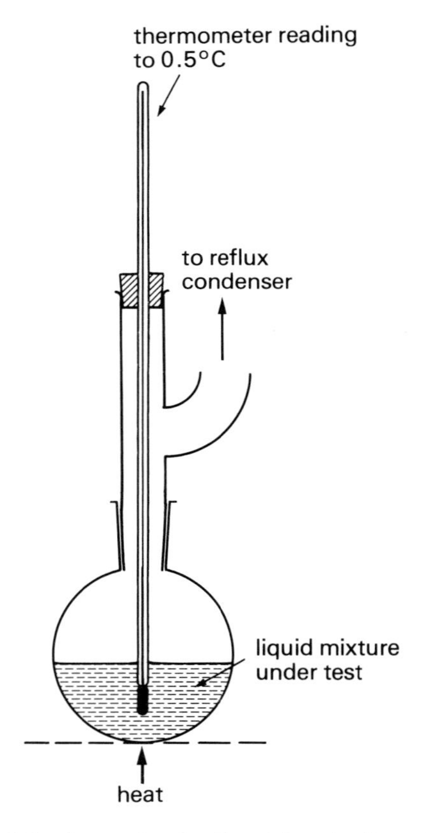

Figure 7.7 Apparatus for the determination of boiling points of liquids

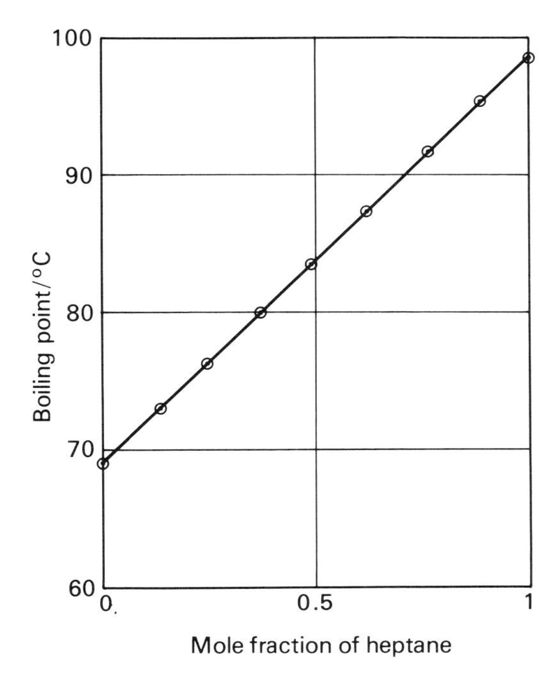

Figure 7.8 Variation of boiling point with molar composition for mixtures of hexane and heptane

Figure 7.9 Vapour pressure relationships for mixtures of hexane and heptane

the two components. This relationship between the individual vapour pressures of the components and that of the mixture is shown in Figure 7.9. The slope is the reverse of that for the boiling point/composition graph, reflecting the inverse relationship involved. Liquid pairs which form mixtures that obey Raoult's Law are said to be '*ideal*'.

Many examples are known of pairs of liquids that form mixtures that deviate from Raoult's Law; such a pair is ethanol and cyclohexane. The result of investigating such mixtures is shown in Figure 7.10, and the mixtures are seen to have higher vapour pressures (i.e. they are more volatile) than is to be expected from Raoult's Law—they are said to exhibit *positive deviation* from the law. Ethanol molecules have forces between them (intermolecular forces) which are reinforced by the permanent dipole each molecule possesses. (See Chapter 5, page 70, for a discussion of dipoles in molecules.) Cyclohexane has no such permanent dipole, and when mixtures of the two liquids are made, dipole interactions between the ethanol molecules tend to be broken up by the interposition of cyclohexane molecules. This explains the increased vapour pressure exhibited by these mixtures, and the reasoning suggests that making mixtures of these liquids should result in a temperature fall. Calorimetric measurements show this to be the case.

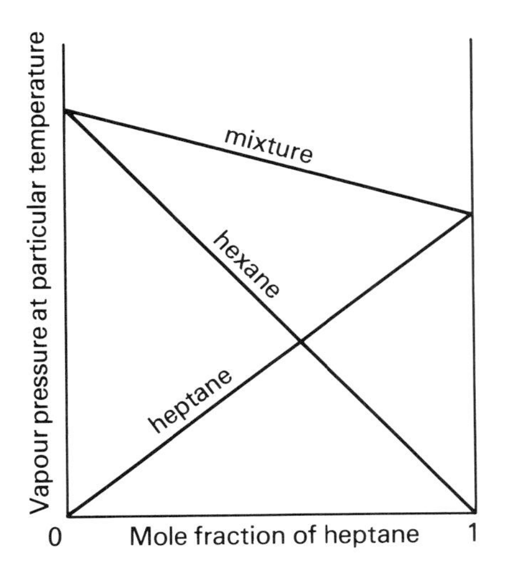

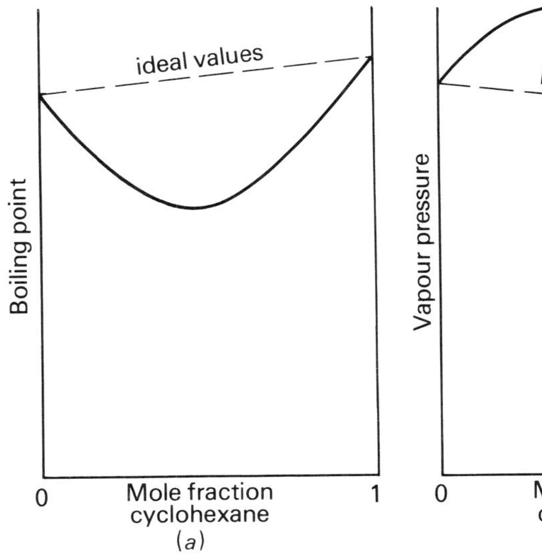

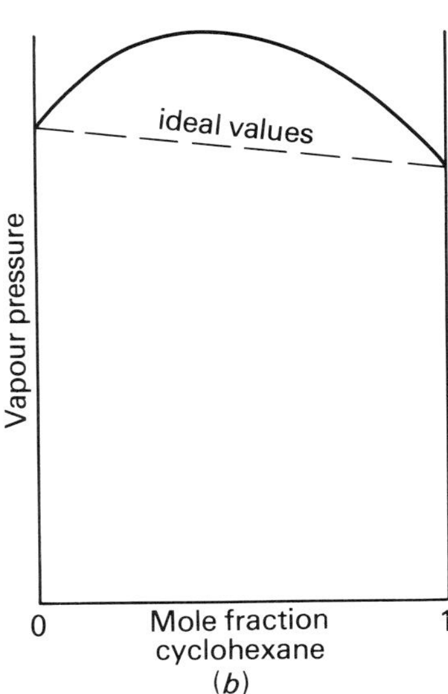

Figure 7.10 (*a*) boiling point/composition and (*b*) vapour pressure/composition curves for mixtures of ethanol and cyclohexane

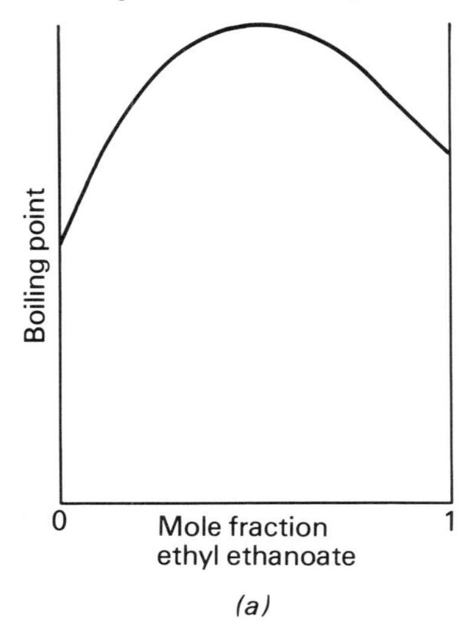

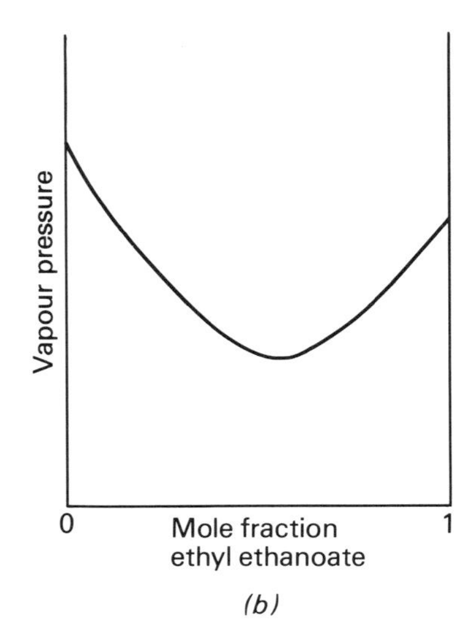

Figure 7.11 (*a*) boiling point/composition and (*b*) vapour pressure/composition curves for mixtures of trichloromethane and ethyl ethanoate

Other pairs of liquids show a *negative deviation* from Raoult's Law. Such a pair is ethyl ethanoate and trichloromethane, for which data are given in Figure 7.11. In this example both the pure liquids have a permanent dipole. Each must therefore break down to some extent the dipole interactions in the other. The energy to do this is more than provided by the new dipole interactions between the molecules in the mixture. In this case we would expect the mixing to be accompanied by a temperature rise, and this is confirmed by calorimetry.

Vapour pressure and molar masses

Consideration of the mathematical statement of Raoult's Law shows that it could be the basis of a method of determining molar masses. If a known mass of solute is dissolved in a known mass of solvent and the vapour pressures of solvent and solution determined, provided the molar mass of one is known, the other can be calculated. Methods formerly employed involving the direct or indirect determination of vapour pressure have now been superseded by the practically easier measurements of elevation of boiling point (*ebullioscopy*) or depression of freezing point (*cryoscopy*) which are related to vapour pressure changes.

Ebullioscopy

The relationship between elevation of boiling point and lowering of vapour pressure when a solute is dissolved in a solvent can be deduced from a consideration of the data in Figure 7.12. If the vapour pressure curves are considered to approximate to straight lines then triangles ADB and AFC are similar. Hence the ratios of the lengths of corresponding sides are equal, i.e.

$$\frac{AB}{AC} = \frac{AD}{AF}$$

In terms of the physical quantities involved, this means that the elevation of the boiling point (ΔT) is proportional to the lowering of vapour pressure. Thus for every solvent there is an ebullioscopic constant, given the symbol K_b. *N.B.*—it is *not* an equilibrium constant. K_b for propanone is 1.67 °C for 1 mole of solute dissolved in 1000 g of solvent. A specimen calculation will make the significance of the constant clear.

2.475 g of a substance is dissolved in 35 g of propanone. The boiling point of the pure solvent was found to be 56.38 °C and that of the solution 56.88 °C. What is the molar mass of the solute?

2.475 g solute in 35 g solvent is equivalent to 2.475 × 100/35 g in 1000 g of solvent. This is 70.71 g of solute.

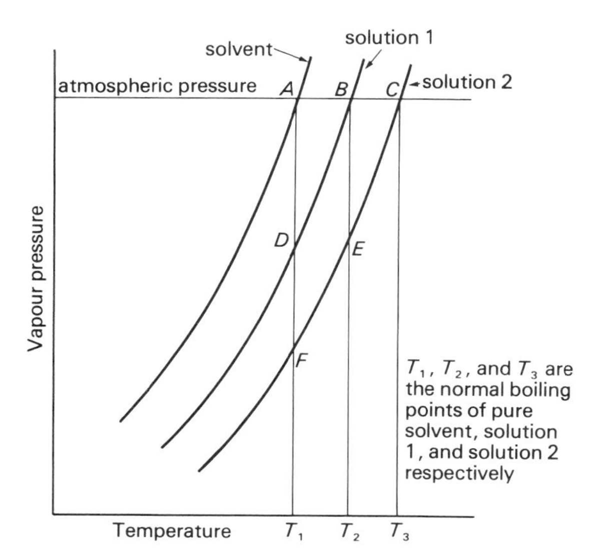

Figure 7.12 Data to deduce the relationship between elevation of boiling point and depression of vapour pressure

A solution of 70.71 g of solute in 1000 g solvent gives $\Delta T = 0.50$ °C.

A solution of $70.71 \times \frac{1.67}{0.50}$ g in 1000 g would give
$\Delta T = 1.67$ °C

Therefore the relative molar mass of solute $= 70.71 \times \frac{1.67}{0.50} = 236$

Cryoscopy

In a similar manner, the depression of freezing point that occurs when a solute dissolves in a solvent can be used to deduce its molar mass. This has the advantages that melting points are easier to determine and that the technique lends itself to use on a micro-scale. A common solvent to use for this purpose is molten camphor. This melts at 177 °C which obviates the need for freezing mixtures. It also has a large value of K_f, the cryoscopic constant, of 40 °C for 1 mole of solute in 1000 g of solvent; this leads to large depressions which can be measured with good precision.

Cryoscopy was pioneered by Rast and the technique is named after him. A specimen calculation will clarify the procedure.

A solution containing 0.150 g of naphthalene in 3.000 g of camphor was found to melt at 161.4 °C. What is the molar mass of naphthalene?

0.150 g of naphthalene in 3 g camphor is equivalent to $0.150 \times \frac{1000}{3}$ g $= 50$ g in 1000 g.

50 g of naphthalene in 1000 g of camphor give $-\Delta T = 15.6$ °C.

$50 \times \frac{40}{15.6}$ g will give $-\Delta T = 40$ °C.

Therefore the relative molar mass of naphthalene

$$50 \times \frac{40}{15.6} = 128.$$

N.B.—samples of camphor from different sources have slightly differing melting points and so a melting point determination should be carried out first. Similarly if it is possible to determine the value of K_f using a substance of known molar mass this should be done.

Table 7.2 Some ebullioscopic and cryoscopic data

Substance	*m.p./°C*	*b.p./°C*	*For 1 mole solute in 1000 g of solvent* K_f	K_b
Water	0	100	1.86	5.2
Benzene	5	80	4.9	2.67
Camphor	177	—	40	—

Ebullioscopic and cryoscopic properties of solutions depend for their magnitude on the molar concentration of solute particles and not on the nature of the solute. They are known as *colligative properties.* A third colligative property, osmosis, will be discussed later in the chapter.

Vapour pressure and distillation

Simple distillation

This is a technique used to separate a liquid which can be vaporized without decomposition from non-volatile substances dissolved in it; suitable apparatus is shown in Figure 7.13.

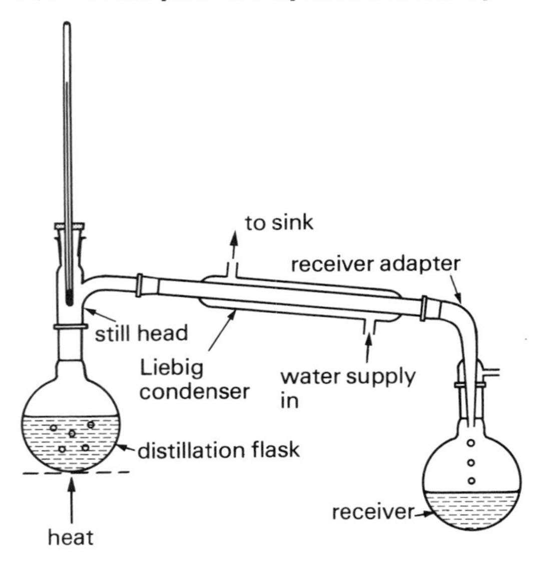

Figure 7.13 Apparatus for simple distillation

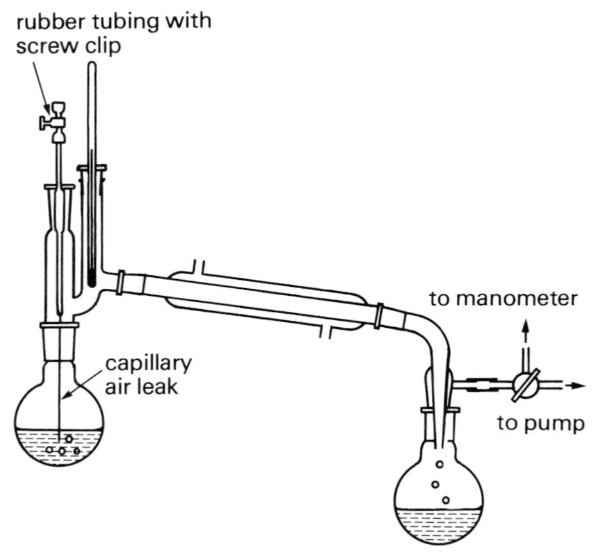

Figure 7.14 Apparatus for distillation under reduced pressure

Distillation under reduced pressure (vacuum distillation)

This method is used to distil liquids that would partly or completely decompose if distilled at atmospheric pressure. Examination of Figure 7.4 shows that, if the pressure is reduced, a liquid will boil below its normal boiling point. Phenylamine (aniline) can be purified in this way. It boils at 184 °C under standard atmospheric pressure, but this can be reduced to 77 °C at 15 mmHg using the apparatus in Figure 7.14. The capillary air leak is to reduce 'bumping', which otherwise accompanies this distillation.

Steam distillation

This technique is used to separate a volatile liquid (which must be immiscible with water) from non-volatile contaminants at a temperature below its normal boiling point. When phenylamine (aniline) is prepared by the reduction of nitrobenzene, the final reaction mixture contains water, phenylamine, and the ionic (non-volatile) side products of the reaction. Phenylamine and water can be separated from this by steam distillation using the apparatus in Figure 7.15. Because phenylamine and water are immiscible and form two separate liquid phases, the vapour pressure of the mixture at any temperature is the sum of the vapour pressures of the individual liquids. This addition of vapour pressures is shown in Figure 7.16. It will be seen that the total vapour pressure equals atmospheric pressure at 97.5 °C, and this is therefore the distillation temperature.

Steam distillation can also serve as a means of determining relative molar masses.

$$p = n\frac{RT}{V} \quad \text{(the general gas equation)}$$

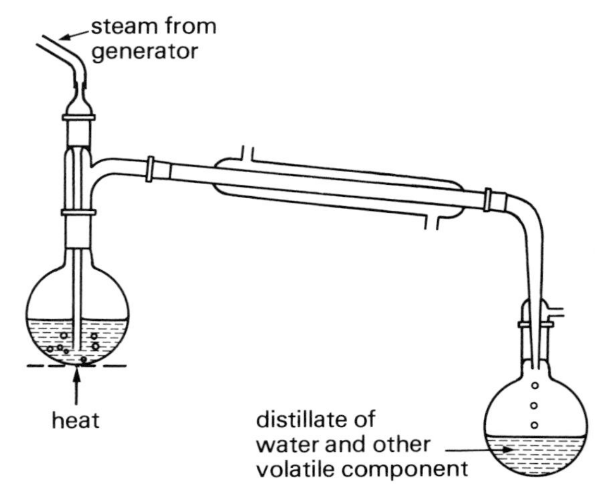

Figure 7.15 Apparatus for steam distillation

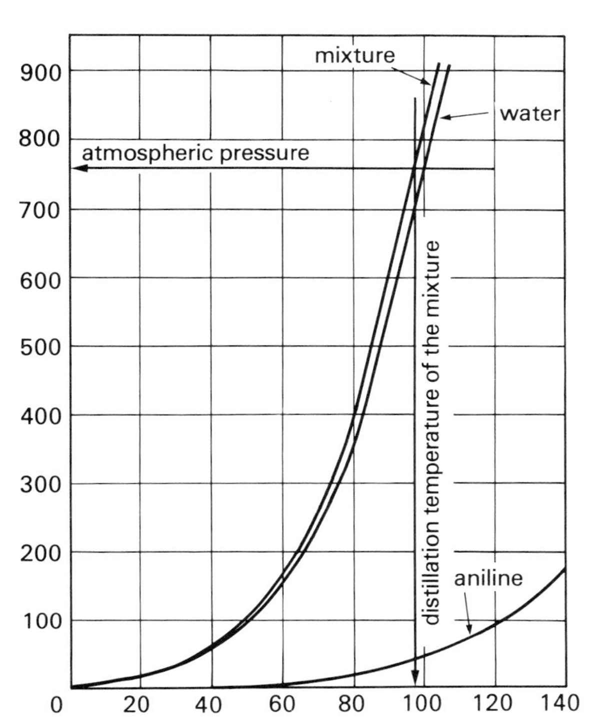

Figure 7.16 Vapour pressure data for aniline, water, and a mixture of the two liquids

Thus water and another liquid undergoing steam distillation each exert a vapour pressure proportional to the mole fraction present in the distillate. This means that

$$\frac{p_{\text{water}}}{p_{\text{liquid } X}} = \frac{\text{moles of water}}{\text{moles of } X}$$

$$= \frac{\text{mass of water}}{18} \times \frac{\text{relative molar mass of } X}{\text{mass of } X}$$

This example should illustrate how the method works in practice: When nitrobenzene is steam distilled under standard atmospheric pressure, the temperature of distillation is 99 °C (at which the vapour pressure of water is 733 mmHg). The distillate contains 80 per cent by mass of water. What value for the molar mass of nitrobenzene do these figures indicate? Substitution yields:

$$\frac{733}{27} = \frac{80}{18} \times \frac{\text{relative molar mass of nitrobenzene}}{20}$$

Relative molar mass of nitrobenzene =

$$\frac{733}{27} \times \frac{18 \times 20}{80} = 122$$

Fractional distillation

This is used to separate mixtures of miscible liquids. For ideal mixtures of liquids such as hexane and heptane (page 97), complete separation can be achieved, but as will be shown later (page 102) this is not always possible with non-ideal mixtures. Data required for an explanation of fractional distillation of an ideal mixture can be obtained using the apparatus shown in Figure 7.13 modified in two important respects. Firstly the thermometer is lowered until it is in a position to read the temperature of the liquid, and secondly heat insulating material is wrapped around the neck of the distillation flask and still head. Mixtures of the two liquids in varying molar proportions are then placed in the flask and a small proportion distilled; in each case the temperature is recorded. The distillates are reserved for future investigation. A graph is then plotted of the temperature at which the liquid mixture boiled against its molar composition. A typical result is shown in Figure 7.17.

The vapour and hence the distillate does not have the same composition as the liquid from which it came, and each distillate must therefore have its composition analysed. This can be done by finding its boiling point and using the data in Figure 7.17 as a calibration curve. After this has been completed the data for the liquid and the data for the vapour can be combined to give the

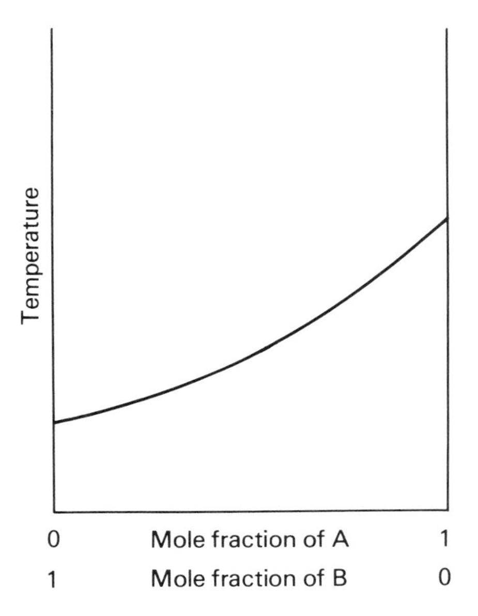

Figure 7.17 Variation in boiling point of mixtures of liquids A and B with molar composition

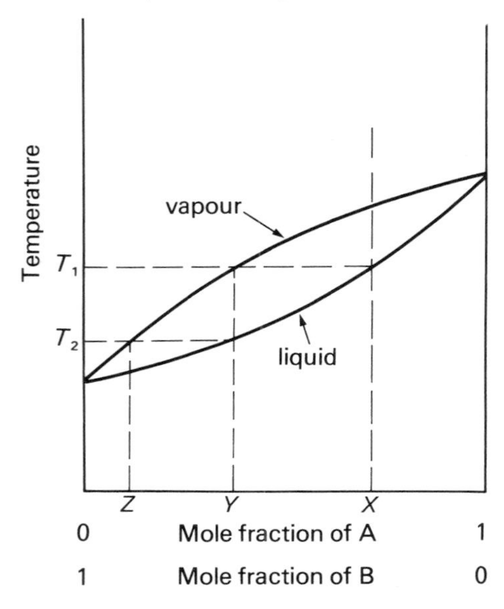

Figure 7.18 Distillation data for ideal liquids A and B

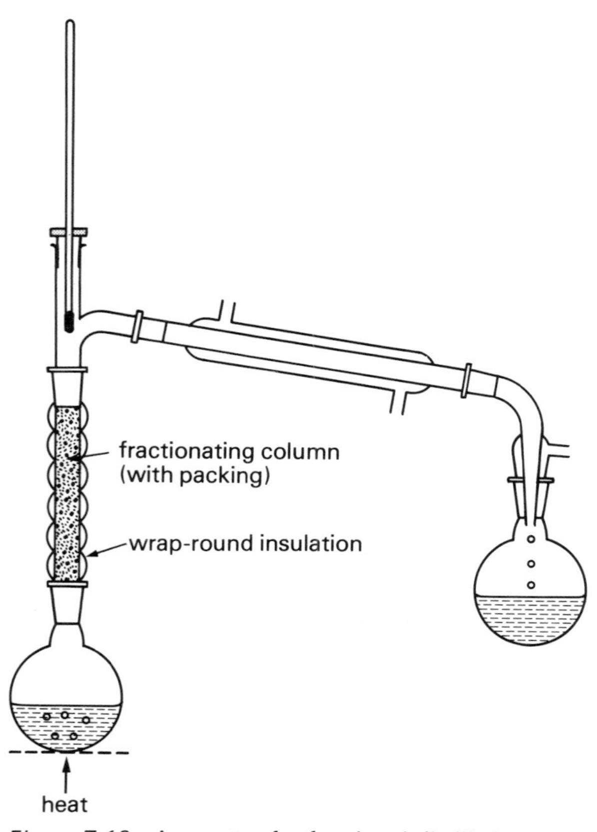

Figure 7.19 Apparatus for fractional distillation

result shown in Figure 7.18. It will be seen that if a mixture of mole composition X is distilled, the liquid boils at temperature T_1 at which vapour forms with composition Y. If the distillate with composition Y is distilled, it boils at temperature T_2 and gives vapour of composition Z. Continual distillations of this type therefore produce distillates successively richer in the more volatile component, B.

To avoid carrying out multiple distillations of this type, a fractionating column is employed; one type is shown in Figure 7.19. As vapour passes up the column it undergoes repeated condensation and re-evaporation, equivalent to a series of separate distillation procedures.

The distillation data for two types of non-ideal liquid mixtures are shown in Figure 7.20. In distilling a mixture with composition X' of a pair of liquids of type (*a*), primary distillation will give a distillate of composition Y'. Successive distillations will result in distillates gradually closer in composition to C_1. Further distillation will not alter the composition of this mixture which is therefore called the constant boiling mixture or *azeotropic mixture*. Distillation of mixtures richer in the more volatile component than C_1 will also eventually yield this constant boiling mixture. A common example of a system of this type is ethanol/water, which forms a constant boiling mixture at 78.2 °C with an ethanol content of 95.6 mole per cent.

Non-ideal liquid mixtures of type (*b*) form azeotropic mixtures having a higher boiling point than either component or any other mixture; examples include trichloromethane/ethyl ethanoate and water/hydrochloric acid. This provides an easy method of preparing standard volumetric solutions

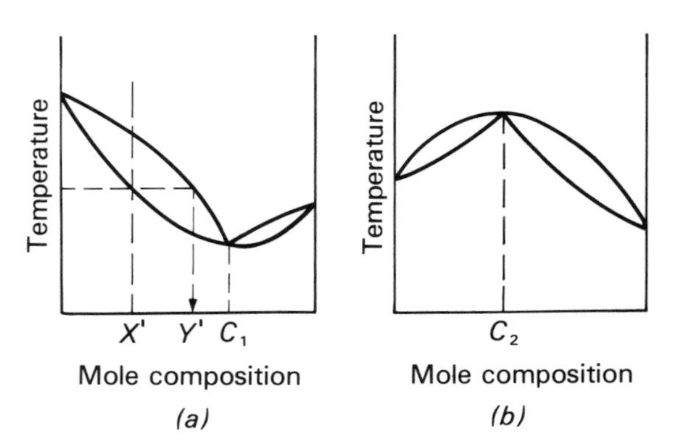

Figure 7.20 Distillation data for two types of non-ideal liquid mixtures

of hydrochloric acid, for to the constant boiling mixture can be added the calculated quantity of water to give a solution of the concentration required.

Fractional distillation is widely used in the chemical industry. A common example is seen in the fractionating towers used for distillation of crude oils at refineries (Figure 7.21); the fractions are run off from the appropriate height of the tower.

Osmosis

The phenomenon of osmosis was first recorded by the Frenchman Abbé Nollet in 1748. He filled a pig's bladder with wine and placed this in water. He observed that the bladder swelled and eventually burst and deduced that water was passing into the bladder. Many animal and plant tissues act in a similar manner to the pig's bladder in the Abbé's experiment and are known as *semi-permeable membranes*. Osmosis may be recognized as the passage of solvent molecules, but not those of solute, through a semi-permeable membrane from solvent to solution or from a solution of low to one of higher concentration.

The concept of a semi-permeable membrane is an ideal one, and real membranes approach this ideal with differing degrees of success. One of the most useful man-made membranes for use in studies on osmosis can be made by placing a solution of copper(II) sulphate inside a porous pot and standing this in a solution of potassium hexacyanoferrate(II), whose ions seep into the pot. When the solutions meet they react to deposit copper(II) hexacyanoferrate(II) within the walls of the pot.

$$2Cu^{2+}(aq) + Fe(CN)_6{}^{4-}(aq) \rightarrow Cu_2Fe(CN)_6(s)$$

An early investigation into the phenomenon is shown in Figure 7.22. An inverted thistle funnel with a semi-permeable membrane blocking the wide end had sucrose solution placed in the bulb. This was then suspended as shown in a vessel of water. The height of the liquid column was seen to rise slowly and then stop at a maximum height. The osmotic activity is supporting a liquid column of height h, and could thus be regarded as exhibiting an *osmotic pressure*.

The Dutch chemist van't Hoff was studying pressure–temperature–volume relationships in gases

Figure 7.21 Fractionating tower for crude oil distillation

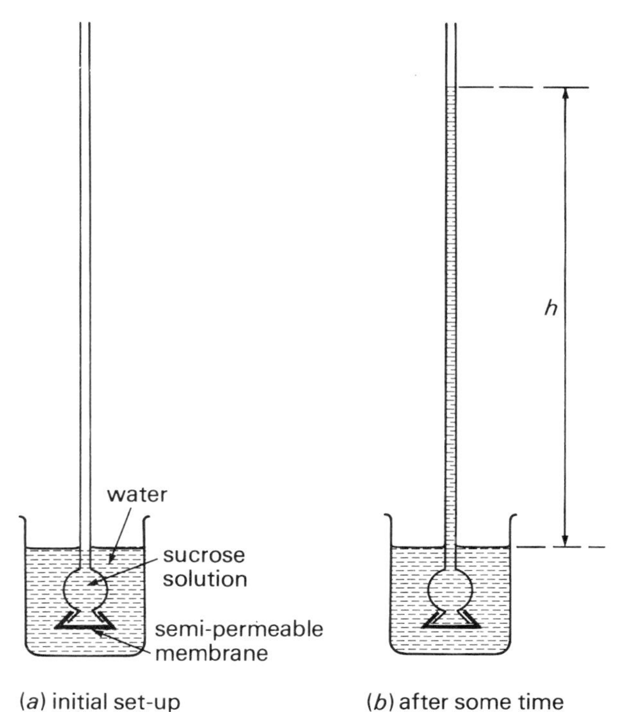

Figure 7.22 Apparatus demonstrating the pressure of osmotic systems

(as discussed in Chapter 1). He suggested that the behaviour of osmotic systems and gases were analogous. He was able to show by calculation that osmotic systems obeyed the general gas equation with the same value for R within the limits of experimental error.

Table 7.3 Osmotic pressure data for aqueous solutions of sucrose at 20 °C

[sucrose]/ mol dm^{-3}	p/atm.	$\frac{p}{T\,[\text{sucrose}]} = R$/dm^3 atm. K^{-1} mol^{-1}
0.1	2.59	0.0884
0.2	5.06	0.0863
0.3	7.61	0.0866
0.4	10.14	0.0865
0.5	12.75	0.0870
0.6	15.39	0.0875

Consideration of the magnitude of the pressures quoted in Table 7.3 shows that these results could not have been obtained using apparatus of the type shown in Figure 7.22 without involving columns of liquid over 150 metres in height. In addition, analysis reveals that the osmotic pressure of the solution balanced by the hydrostatic pressure of the liquid in the tube is not that of the original solution but that of the solution after dilution by the necessary inflow of water. Van't Hoff's calculations were based on measurements obtained a decade previously with an improved osmometer devised in 1877 by the German botanist Pfeffer. A still further refinement in osmometer design due to the Earl of Berkeley and E. G. Hartley and used by them between 1906 and 1909 allowed osmotic pressures of up to 150 atmospheres to be measured under conditions that obviated the disadvantages outlined above. Their apparatus is shown schematically in Figure 7.23.

The pressure, p, applied to the piston on the right-hand side of the apparatus is adjusted until it exactly opposes the osmotic flow. The osmotic pressure is therefore equal and opposite to this externally applied pressure.

In recent times measurement of osmotic pressure has proved a useful way of determining the molar masses of polymers; because of their high molar masses, cryoscopic and ebullioscopic effects are very small and cannot be accurately measured. Mass spectrometry measurements are also difficult as it is not possible to get a high concentration of ions representing the unbroken molecule. A calculation illustrates the principle.

A 1.5 per cent solution of polyvinyl chloride in solution in dioxan was found to have an osmotic pressure of 0.0026 atmospheres at 20 °C. What is the molar mass of polyvinyl chloride.

$$pV = \frac{m}{M_r}RT$$

$$\text{Relative molar mass } (M_r) = \frac{mRT}{pV}$$

$$= \frac{1.5 \times 0.0821 \times 293}{0.0026 \times 0.1}$$

$$= 1.39 \times 10^5$$

Reverse osmosis

Consideration of the principle of Berkeley and Hartley's osmometer shows that it would be possible to reverse the osmotic flow by opposing the osmotic pressure with a greater applied pressure on the solution side of the membrane. This is known as *reverse osmosis* and its use has been suggested as a means of desalination of water.

Biological significance of osmosis

The single-celled protozoan, Amoeba, is among the simplest known of animals. Several species are known, some inhabiting fresh and some salt water environments. Their outer membrane acts as a

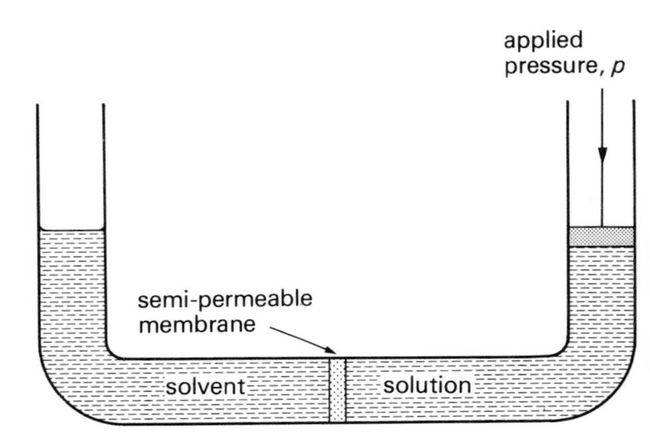

Figure 7.23 Schematic diagram of Berkeley and Hartley's osmometer

semi-permeable membrane and those species with fresh water habitats have to contend with continual inflow of water through this membrane (endosmosis). They have a special organ, the contractile vacuole, to pump out the water entering this way. This organ is missing in sea water varieties of amoeba as the osmotic potentials of the cell contents and the environment are equal (the two solutions are said to be *isotonic*).

The fact that the body fluids of land animals, including man, are approximately isotonic with sea water is powerful evidence that land creatures evolved from sea-living ancestors.

Osmosis is of great importance in the life of plants, it being the chief means of water uptake via the root system. The osmotic pressure within plant cells can be determined by the method of plasmolysis described by the botanist H. de Vries in 1884. Thin sections of plant tissue (the epidermis of leaves is suitable) are placed in a series of sucrose solutions ranging in concentration from 0.1 to 1 M. Each is examined under the microscope for plasmolysis, the phenomenon in which the normal pressure of the cell contents (known as the 'turgor') is reduced by osmotic outflow (exosmosis) to such an extent that the cytoplasm shrinks and detaches itself from the cellulose cell wall, as shown in Figure 7.24. As there are slight variations in osmotic potential between the cells, the mean osmotic pressure is taken to be that when half the cells show plasmolysis. In an experiment of this type, half the cells plasmolysed at room temperature (20 °C) in a 0.6 M solution of sucrose.

$$pV = nRT$$

$$p = \frac{n}{V} RT$$

$$= [\text{sucrose}]\, RT$$

$$= 0.6 \times 0.0821 \times 293$$

$$= 14.4 \text{ atmospheres}$$

Anomalous results for measurements of colligative properties

(*a*) *Association in solution* When association of solute molecules in the solvent employed takes place, the value for the molar mass of the solute determined by ebullioscopy, cryoscopy, or osmometry is that for the associated molecule. When the molar mass of ethanoic acid, for example, is determined by cryoscopy using benzene as a solvent, a value of about 120 is obtained as compared to the 60 expected from the formula CH_3CO_2H. This is due to dimerization by hydrogen bonding (page 71).

(*b*) *Electrolytes* When the molar masses of electrolytes are determined using the colligative properties of their solutions, the results are very different from those expected from their formulae. Van't Hoff introducted the quantity, given the symbol *i*:

$$i = \frac{\text{observed value of molar mass}}{\text{molar mass calculated from formula}}$$

This *i* factor was found to vary from electrolyte to electrolyte and also to depend on the concentration of the solution studied.

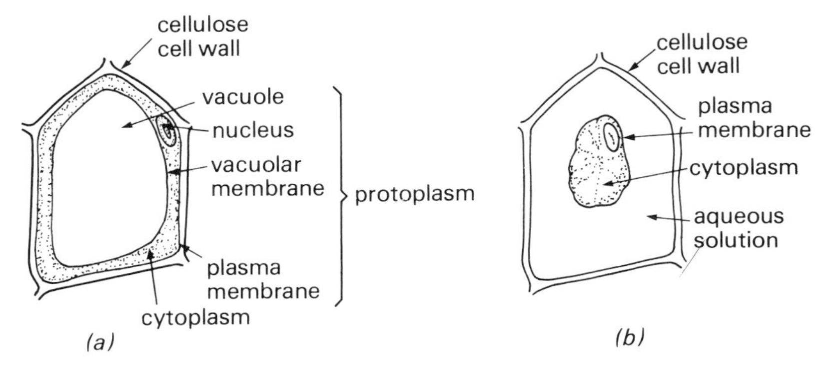

Figure 7.24 Plant cells in (*a*) turgid and (*b*) plasmolysed form

Table 7.4 Van't Hoff *i* factors for some electrolyte solutions

Concentration/mol dm^{-3}	NaCl	$CuSO_4$	K_2SO_4	$K_3Fe(CN)_6$
0.005	1.95	1.54	2.77	3.51
0.01	1.92	1.45	2.70	3.31
0.05	1.88	1.22	2.45	3.01
0.10	1.85	1.12	2.32	2.85
0.20	1.82	1.03	2.17	2.69
0.40	1.80	0.97	2.04	—
1.00	1.79	0.93	—	—

Van't Hoff observed that these values for the *i* factor were consistent whichever of the techniques, ebullioscopy, cryoscopy, or osmometry, was used to measure the molar mass. The Swedish chemist, Svente Arrhenius, in 1887, interpreted these results in terms of the dissociation of electrolytes into ions. When these results were first considered the main interest in the problem seemed to be why the *i* factor for sodium chloride, for example, was more than 1. In terms of modern theory the main interest is in why the factor is less than two. Although we now believe that such strong electrolytes are completely ionized in solution, the results suggest that there is interionic attraction within the solution which prevents the ions exerting their full independence. As would be expected, this effect increases with concentration as the ions are then closer together. The corresponding factors for the other binary electrolyte (i.e. one in which one mole of substance gives a total of two moles of ions) in the table, copper sulphate, are lower than those for sodium chloride because each of the ions carries a double charge thereby exhibiting increased interionic attraction. The ternary and quaternary electrolytes have *i* factors approaching 3 and 4 as expected.

Problems

1. When 2.9 g of urea is dissolved in 150 cm^3 of water, the freezing point is depressed by 0.60 °C. K_f for water is 1.86 °C for 1 mole in 1000 g. Calculate the molar mass of urea.
2. 1.455 g of benzoic acid was dissolved in 80 g of ethanol. The boiling point of the solution was 78.97 °C while that of pure ethanol is 78.80 °C. K_b for ethanol is 1.15 °C for 1 mole in 1000 g. Calculate the molar mass of benzoic acid in this solvent.
3. A 0.0416 M aqueous solution of potassium chloride boils at 100.041 °C. K_b for water is 0.52 °C for 1 mole in 1000 g. Calculate the *i* factor for potassium chloride at this concentration.

		Temperature/°C					
		0	20	40	60	80	100
Vapour pressure/ mmHg	Water	4.6	17.5	55.3	149	355	760
	Octane	2.9	10.4	30.8	77.5	175	354

4. Water and octane are immiscible. Use the information in the table to plot (on the same axes) their individual vapour pressure/temperature curves, together with the total vapour pressure/temperature curve for the two. Deduce the temperature at which octane would steam distil under standard pressure and calculate the percentage by mass of octane in the distillate (the molar mass of octane is 106).
5. A solution of 1.271 g dm^{-3} of phenol is found to have an osmotic pressure of 0.324 atmosphere at 15 °C. What is the molar mass of phenol?
6. Pfeffer found that a solution of gum arabic having a concentration of 164 g dm^{-3} exerted an osmotic pressure of 1193 mmHg at 15.6 °C. What value for the molar mass of gum arabic do these figures suggest?
7. 9 g of non-volatile methyl stearate was dissolved in 100 g of a solvent *X*. The vapour pressure of the solvent at 20 °C is 440 mmHg and that of the solution is 431 mmHg at the same temperature. If the relative molar mass of methyl stearate is 298, calculate the relative molar mass of *X*.
8. Ethanol (b.p. 78.5 °C) and water form a constant boiling mixture having a boiling point of

78.2 °C and a composition of 95.6 per cent ethanol.

(*a*) Define the term *constant boiling mixture.*

(*b*) Sketch, and fully label, the boiling point/composition diagram for ethanol and water.

(*c*) An ethanol/water mixture shows positive deviation from Raoult's Law. Explain and account for this and state the law.

(*d*) What intermolecular changes take place when ethanol is added to water?

(*e*) State qualitatively the result of distilling (i) a mixture containing 75 per cent ethanol, (ii) a mixture containing 97.5 per cent ethanol.

(*f*) State, with reasons, which *one* of the following pairs of substances most closely obeys Raoult's Law: (i) $C_2H_5NH_2$ and $C_6H_5NH_2$, (ii) CH_3COCH_3 and $CH_3COC_2H_5$, (iii) C_6H_6 and $C_6H_5CH_3$. (Associated Examining Board 'A')

9. Describe the use of freezing point measurements in the determination of relative molecular masses.

0.289 g of a carboxylic acid depresses the freezing point of 100 g of benzene by 0.147 K; 0.313 g of the same acid depresses the freezing point of 100 g of water by 0.097 K. Evaluate the relative molecular mass of the acid in each solvent and comment on your results. How would you expect the results to change as each solution is made progressively more dilute?

[1 mol of added solute depresses the freezing point of 1 kg of benzene by 5.10 K and that of 1 kg of water by 1.86 K.]

(Oxford Entrance and Awards)

10. Among the techniques used by chemists for the isolation and separation of substances from complex systems or reaction mixtures are (*a*) simple distillation, (*b*) fractional distillation, (*c*) solvent extraction, (*d*) chromatography, and (*e*) crystallization.

Outline the physico-chemical principles underlying **four** of the above techniques. Illustrate the usefulness of each technique you have chosen by means of one example of an actual application. The application may relate either to an industrial or to a laboratory situation. (London 'A')

11. What do you understand by osmosis and osmotic pressure? Describe briefly (i) the measurement of osmotic pressure and (ii) any one actual or possible application of osmosis.

Calculate the osmotic pressure of a 0.0500 M solution of sodium chloride at 25 °C, assuming that inter-ionic effects are negligible at this dilution. (Assume that 1 atmosphere = 1.01 $\times 10^5$ N m^{-2} and that 1 mole of gas occupies 22.4 dm^3 at s.t.p.) (Oxford 'A')

12. Draw a boiling point/composition diagram for a mixture of two volatile miscible liquids which do not give rise to a constant boiling mixture (ideal mixture). State the law obeyed by such mixtures. Discuss, with the aid of a diagram, what happens when a typical mixture of such liquids is fractionally distilled and sketch the apparatus you would use for the distillation.

What is the advantage of distilling liquids under reduced pressure? Steam distillation is essentially distillation of one component of a non-miscible pair of liquids under reduced pressure. Explain this statement. (WJEC 'A')

8. The Ionic Theory

An electric current consists of a movement of electrically charged particles through the conductor. In the case of metals and graphite, the charged particles concerned are the delocalized electrons associated with these structures, as discussed in Chapter 5. The movement of these electrons does not cause changes in the chemical nature of the conductor. Certain chemical compounds, either when molten or when dissolved in water, are good electrical conductors. In these cases, however, the passage of the electric current results in chemical reactions occurring where the electricity enters and leaves the melt or solution. Michael Faraday introduced the terms *electrolyte* for compounds that behaved in this way and *non-electrolyte* for those that did not. To explain the chemical changes associated with electrolytic conductance, it is necessary to postulate the existence of charge carriers other than electrons in them. Faraday proposed the name *ion* for these charge carriers. He also named the metallic or graphite conductors connecting the electrolyte with the applied voltage the *electrodes*; that connected to the positive was called the *anode* and that to the negative the *cathode*. Ions moving towards the anode (and therefore carrying a negative charge) were called *anions* and those travelling towards the cathode (bearing positive charge) *cations*.

In 1805, Baron von Grotthaus suggested that the molecules of electrolytes are very polar and that the applied electrical potential aligns them in a chain. The electrical field was then supposed to cause the molecules at the end of the chain to dissociate into ions which were then discharged at the electrodes. It was soon observed, however, that even the smallest applied voltage could result in an electrical current flowing and that thus no energy was required to dissociate the molecules as would be required by the Grotthaus theory.

In 1857, Rudolph Clausius proposed that in electrolytes there is a small equilibrium proportion of the particles in the ionized form even in the absence of an applied voltage. Between 1882 and 1886 Julius Thomsen published data on the enthalpies of neutralization of acids and bases, involving techniques discussed in Chapter 4. He showed that the neutralization of one mole of strong acid by one mole of strong alkali gave constant enthalpy values of -57.7 kJ, no matter which acid or alkali pair was involved. This led Svante Arrhenius in 1887 to suggest that this was due to the almost complete dissociation into ions of these electrolytes and that their interaction was always due to the fundamental reaction:

$$H^+(aq) + OH^-(aq) \rightarrow H_2O(l); \quad \Delta H^{\ominus} = -57.7 \text{ kJ mol}^{-1}$$

As he was developing this theory, Arrhenius became aware of the interpretation van't Hoff was placing on the abnormal values for molar masses of electrolytes determined from methods depending on their colligative properties, as discussed in Chapter 7. He noted that the van't Hoff i factor approached closely, as more and more dilute solutions of each strong electrolyte were considered, the number of ions that would be formed if complete dissociation took place. He wrote to van't Hoff:

> 'It is true that Clausius assumed only a minute quantity of dissolved electrolyte is dissociated, and that all other physicists and chemists had followed him; but the only reason for this assumption, as far as I can understand, is a strong feeling of aversion to a dissociation at so low a temperature, without any facts against it being brought forward. At extreme dilution all salt molecules are completely dissociated.'

Measurement of the electron densities from X-ray diffraction studies on solid electrolytes such as sodium chloride have shown that even in the solid state the substance exists entirely of ions —'molecules of sodium chloride' do not exist. The aversion Arrhenius ascribed to scientists to accept the completely ionic nature of strong electrolytes is confirmed in the words of Sir Lawrence Bragg writing about his experiments on the X-ray diffraction investigation of sodium chloride:

> '. . . chemists had talked of common salt, sodium chloride, as being composed of "molecules" of NaCl. My very first crystal determination showed that there are no molecules of NaCl. The atoms are arranged like the black and white squares of a chessboard, though in three dimensions. Each atom of sodium has six

atoms of chlorine around it at the same distance, and each atom of chlorine has correspondingly six atoms of sodium around it. Some chemists at that time were very upset indeed about this discovery and, as I well remember, begged me to find that there was just a slight approaching of one atom of sodium to one of chlorine so that they could be regarded as a properly married pair.'

From this discussion it should be seen that strong electrolytes can be regarded as being completely in the ionic form in solutions of about 1 M concentration or less. Some, like sodium chloride, are completely ionic at all concentrations and even in the solid state. Others, such as hydrogen chloride, are non-ionic when anhydrous, but in the aqueous solutions quoted are completely ionic.

Further development of these concepts requires the quantitative investigation of electrolysis, which began with the two laws enunciated by Faraday in 1833.

Faraday's laws of electrolysis

If a suitable electrolyte solution containing silver ions is electrolysed between silver electrodes, it is found that after some time the mass of the anode has decreased and that of the cathode has increased by the same amount. By investigating the effect of changing each of the possible variables (temperature, current, concentration, surface area of electrodes, etc.) one at a time, Faraday discovered that the quantity of material transferred from anode to cathode depended solely on the charge in coulombs that had been passed. This quantity of charge is calculated from the relationship:

$$\text{charge/coulomb} = \text{current/amp} \times \text{time/sec}$$

The results of experiments of this type are shown in Figure 8.1. He summarized his findings in these investigations in his first law of electrolysis. This states (in modern wording): *the mass of substance liberated at an electrode during electrolysis is proportional to the quantity of electrical charge that has passed.*

It will be seen from Figure 8.1 that 1 g of silver is deposited on the cathode by the passage of 895 coulomb. Therefore,

charge required to deposit 1 mole (108 g) of

$$\text{silver} = 895 \times \frac{108}{1} = 96\,700 \text{ C}$$

The modern accepted value for this quantity is 96 487 C mol^{-1}, and it is known as a *faraday*.

Apparatus for investigating electrolysis, consisting of a container for the electrolyte fitted with electrodes and possibly with facilities for collecting any gaseous products, are called *voltameters*. Voltameters which produce quantitative results for the charge passed and the mass of product liberated are known as *coulometers*.

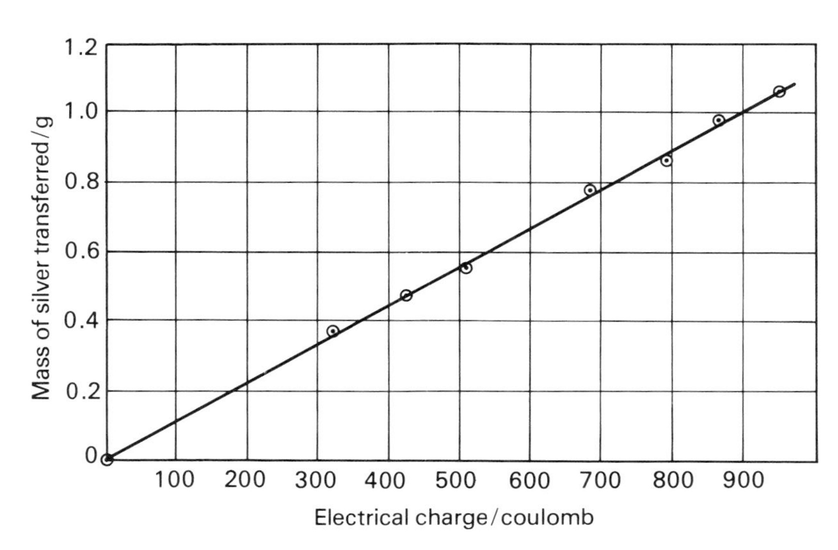

Figure 8.1 Data to confirm Faraday's first law of electrolysis

Using suitable coulometers, information similar to that for silver may be obtained for other electrolytes and the ions of the elements of which they are composed.

The results in Table 8.1 exemplify Faraday's second law of electrolysis which states (in modern wording): *the quantity of electrical charge required to liberate* 1 *mole of atoms of any element by electrolysis is always a whole number of faradays*. The relationship between the value of this whole number of faradays and the charge on the ion is illustrated by the examples below (the sign of the charge is known from the electrode of liberation).

Ag^+	+	e^-	→	Ag
(1 mole ions)		(1 mole electrons)		(1 mole atoms)
Al^{3+}	+	$3e^-$	→	Al
(1 mole ions)		(3 moles electrons)		(1 mole atoms)
I^-	−	e^-	→	$\frac{1}{2}I_2$
(1 mole ions)		(1 mole electrons)		(1 mole atoms)

The faraday can thus be identified with the charge associated with Avogadro's number of electrons.

Sometimes useful information of this type can be deduced from loss of mass of the anode in a suitable coulometer. If, for example, a mixed electrolyte solution of sodium chloride and sodium hydroxide is electrolysed between copper electrodes at 80 °C, 63.5 g of copper (1 mole) is stripped from the anode by the passage of 1 faraday. This implies that the anode reaction is:

$$Cu(s) - e^- \rightarrow Cu^+(aq)$$

From experiments such as these the existence of different ions of the same metal has been demonstrated. This is discussed further in Chapter 12.

Theories of acidity

By 1663, Robert Boyle had introduced the term *acid* to describe substances which shared the following properties:

(*a*) sourness to the taste,
(*b*) having an effect on the colour of certain vegetable dyes,
(*c*) ability to react with some metals to produce hydrogen.

The term *alkali* was given those substances that had the opposite effect on the dyes, felt 'soapy' in water, and which would neutralize the acids forming salts and water.

In 1777, Antoine Lavoisier defined an acid as the oxide of a non-metallic element dissolved in water and an alkali as an aqueous solution of a metallic oxide. Lavoisier's interpretation is not completely satisfactory because not all non-metallic oxides give acidic solutions in water and not all substances that dissolve to give acidic solutions (known as acids anhydride) are non-metallic oxides. The Venn diagram in Figure 8.2 clarifies the relationship between the sets of non-metallic oxides and acids anhydride.

In 1754 the term *base* was introduced by Rouelle to describe substances that reacted with an acid to give a salt and water. The term alkali is reserved for soluble bases.

Arrhenius in 1884 showed how the ionic theory could be applied to explain the nature of acids and alkalis. According to his theory, an acidic solution was one that contained more hydrogen ions than hydroxide ions and an alkaline solution one containing more hydroxide ions than hydrogen ions. In neutral substances, like water, there were equal numbers of hydrogen and hydroxide ions. He

Table 8.1 Coulometric data for selected elements

Element	Electrode at which liberated	Coulombs to liberate 1 mole of atoms	Faradays to liberate 1 mole of atoms	Charge on the ion
Aluminium	cathode	289 500	3	+3
Copper		193 000	2	+2
Hydrogen		96 500	1	+1
Lead		193 500	2	+2
Silver		96 500	1	+1
Zinc		193 000	2	+2
Chlorine	anode	96 500	1	−1
Iodine		96 500	1	−1
Oxygen		193 000	2	−2

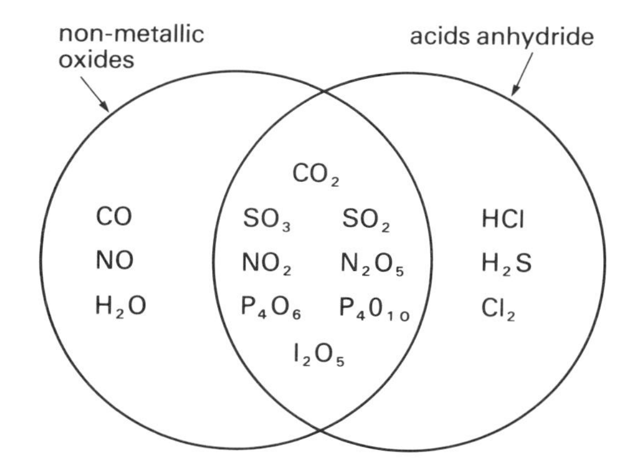

Figure 8.2 Venn diagram relating the sets of non-metallic oxides and acids anhydride

explained the formation of hydrochloric acid from hydrogen chloride by the equation:

$$HCl(g) + aq \rightarrow H^+(aq) + Cl^-(aq)$$

If this reaction is carried out by wetting the bulb of a thermometer with water and dipping this into a gas jar of hydrogen chloride, it is apparent that the reaction is exothermic. The equation above, however, implies a bond-breaking process which, as discussed in Chapter 4, would be an endothermic reaction. In addition, if the hydrogen ion really existed as a single proton, as is implied above, it would have almost infinite polarizing power by virtue of its high charge to surface area ratio, and thus could not exist in solution. Solutions of hydrogen chloride in some other solvents, such as toluene, do not conduct an electric current and do not show the chemical properties expected of acids such as action on indicator paper. This suggests that water plays a much more active role in the formation of hydrochloric acid than the Arrhenius theory implies. J. N. Brønsted, of Denmark, and T. M. Lowry, of England, independently proposed an alternative theory of acidity in 1923 to overcome the objections to the Arrhenius theory and incorporate the new knowledge then available. The Lowry–Brønsted theory defines an acid as a proton (or hydrogen ion) donor and a base as a proton (or hydrogen ion) acceptor. In terms of this theory the reaction occurring when hydrogen chloride dissolves in water is

$$H{:}\underset{H}{\ddot{O}}{:} + H{:}\ddot{Cl}{:} \rightarrow \left[H{:}\underset{H}{\ddot{O}}{:}H\right]^+ + \left[{:}\ddot{Cl}{:}\right]^-$$

water + hydrogen chloride → hydronium ion + chloride ion

Hydrogen chloride is seen to fulfil the definition of an acid required by the Lowry–Brønsted theory in that it donates a proton to the water which, acting as a base, accepts it. The exothermic nature of the reaction is due to the energy required to break the hydrogen/chloride bond being less than that evolved in the formation of the new hydrogen/oxygen bond. The objection to the existence of isolated protons—that they would have unacceptable charge densities—is overcome, as the charge will now be spread over the much greater area of the hydronium ion. The failure of solutions of hydrogen chloride, in solvents such as toluene, to exhibit the reactions regarded as characteristic of acids is due to the lack of electron 'lone pairs' on the solvent molecules, preventing them from acting as bases. Further evidence for the Lowry–Brønsted theory comes from the observation that chloric(VII) acid–1–water ($HClO_4.H_2O$) and ammonium chlorate(VII) are isomorphous. Diffraction studies show these crystals to contain the H_3O^+ and NH_4^+ ions in corresponding positions in the crystal lattice.

Conjugate acids and bases

Ethanoic acid is regarded as a weak acid because in solutions of around 1 M concentration it is dissociated only to a limited extent.

$$\underset{\text{(acid 1)}}{CH_3CO_2H(aq)} + \underset{\text{(base 2)}}{H_2O(l)} \rightleftharpoons \underset{\text{(base 1)}}{CH_3CO_2^-(aq)} + \underset{\text{(acid 2)}}{H_3O^+(aq)}$$

In terms of the Lowry–Brønsted theory, the reverse reaction is also an acid-base reaction. In such a system, ethanoic acid molecules and ethanoate ions are said to be *conjugate acids and bases.* Water molecules and hydronium ions are related in a similar manner. Because the equilibrium is balanced well to the left, we can deduce that the hydronium ion is a stronger acid than ethanoic acid and that the ethanoate ion is a stronger base than water. Hydrochloric acid is regarded as a strong acid because in similar solutions it is virtually completely dissociated into ions.

$$\underset{\text{(acid 1)}}{HCl(aq)} + \underset{\text{(base 2)}}{H_2O(l)} \rightleftharpoons \underset{\text{(base 1)}}{Cl^-(aq)} + \underset{\text{(acid 2)}}{H_3O^+(aq)}$$

From the observed balance of this system, it can be seen that water is a stronger base than the chloride ion.

Of the three bases involved in the systems above, therefore, ethanoate ion is the strongest and chloride ion the weakest. An aqueous solution of sodium ethanoate is found to have an alkaline pH whereas one of sodium chloride is neutral. This is because the ethanoate ion is a sufficiently strong base to remove protons from some of the water molecules, causing the latter to act as an acid:

$$CH_3CO_2^-(aq) + H_2O(l) \rightleftharpoons CH_3CO_2H(aq) + OH^-(aq)$$

The increase in concentration of hydroxide ions that results from this accounts for the observed alkaline pH. This is an example of a phenomenon known as *salt hydrolysis*.

Ammonia is a weak base, and when the equilibrium below is established in aqueous solution it lies well to the left:

$$\underset{\text{(base 1)}}{NH_3(aq)} + \underset{\text{(acid 2)}}{H_2O(l)} \rightleftharpoons \underset{\text{(acid 1)}}{NH_4^+(aq)} + \underset{\text{(base 2)}}{OH^-(aq)}$$

The conjugate acid of ammonia, the ammonium ion, is thus a stronger acid than water. This accounts for the observation that aqueous solutions of ammonium salts, such as ammonium chloride, have acidic pH values:

$$NH_4^+(aq) + H_2O(l) \rightarrow NH_3(aq) + H_3O^+(aq)$$

There is further discussion of salt hydrolysis in Chapters 9 and 12.

Measuring the conductance of electrolyte solutions

G. S. Ohm in 1827 enunciated his well-known Law of Electrical Conductance: *the electrical current flowing through a conductor is directly proportional to the potential difference applied and is inversely proportional to its resistance.* The unit of electrical resistance is the ohm, given the symbol Ω. Ohm's law may be stated in terms of an equation:

$$\text{current/amp} = \frac{\text{potential difference/volt}}{\text{resistance/ohm}}$$

The resistances of metallic conductors can be measured using a Wheatstone bridge, the principle of which is shown in Figure 8.3. The galvanometer reading is zero, no current flows through it, and the potentials at A and B must be equal. For this to be the case the potential drop down the 5-ohm resistance must be the same as that down the 25-ohm resistance, and that down the 30-ohm resistance the same as that down the unknown resistance, x. From this it follows:

$$\frac{5}{30} = \frac{25}{x}$$

or

$$x = 150 \text{ ohm}$$

The resistance, R, of a uniform resistor is proportional to its length, l, and inversely proportional to its area of cross-section, A.

$$R = \frac{\rho l}{A}$$

The proportionality constant, ρ, is known as the *resistivity* of the material. In electrolyte studies we are more usually concerned with the reciprocal of the resistivity, known as the *conductivity* and given the symbol γ.

A problem arises if the measurement of the electrical conductance of an electrolyte solution is attempted using a Wheatstone bridge. Gases are often evolved at one or both electrodes, and so an alternating current with a frequency of about 1000 Hz must be used. The simple galvanometer must then be replaced with a detector that will respond to a.c. Modern conductivity meters of the type shown in Figure 8.4 often contain a solid-state diode rectification circuit around the meter to accomplish this.

Consider a uniform cube of material of 1 metre sidelength. The length and area of cross-section of this cube are both unity and hence the electrical resistance measured across opposite faces will be the resistivity of the material in Ω m. The electrical conductivity, being the reciprocal of resistivity, has units $\Omega^{-1}\ m^{-1}$ ($S\ m^{-1}$). It would obviously be impractical to use such an enormous quantity of material. For experimental purposes cells of the type shown in Figure 8.5 are employed. The conducting

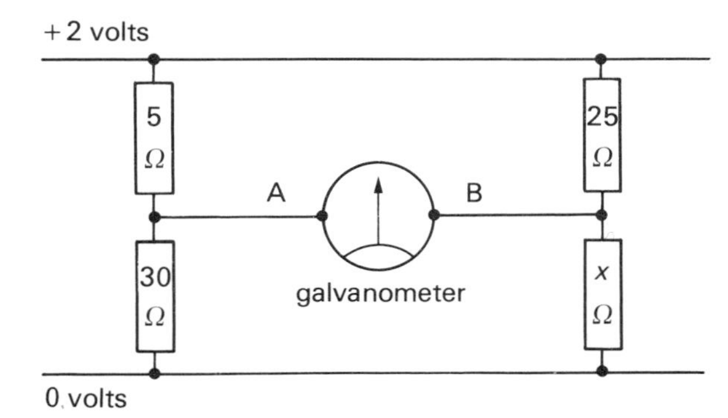

Figure 8.3 The Wheatstone bridge principle (values for voltage and resistance are hypothetical)

pathway through the solution is defined by the dimensions of the electrodes and their distance apart. For each cell it is necessary to determine a cell constant which, when multiplied by the measured conductance, gives the conductivity of the solution:

electrical conductivity $(\gamma)/\Omega^{-1}\ m^{-1} =$ conductance$/\Omega^{-1} \times$ cell constant$/m^{-1}$

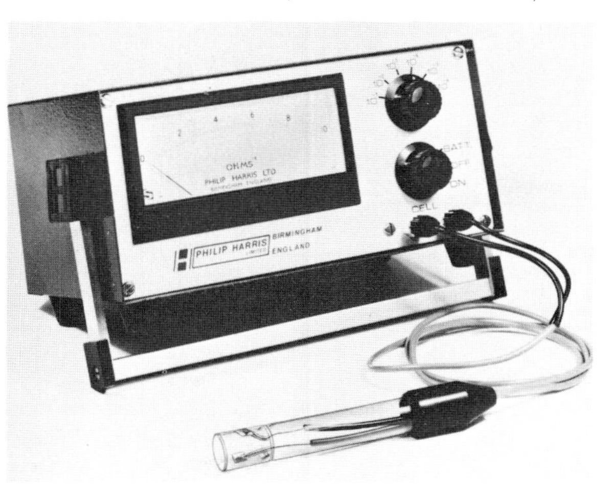

Figure 8.4 A commercial conductivity meter

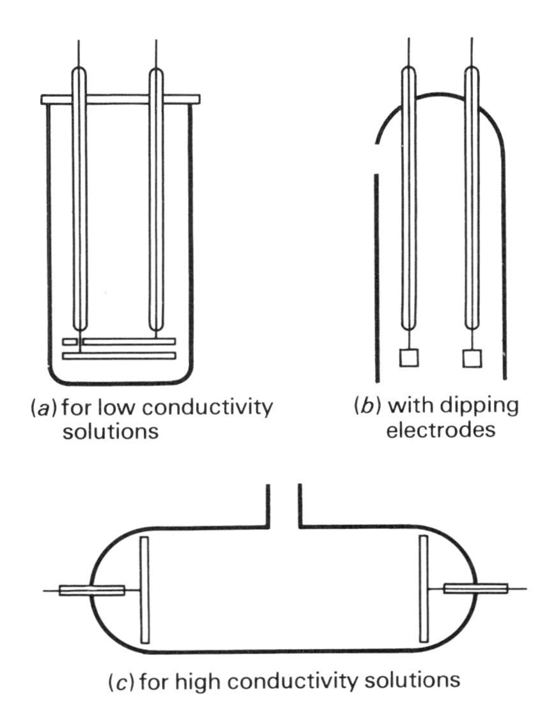

(*a*) for low conductivity solutions

(*b*) with dipping electrodes

(*c*) for high conductivity solutions

Figure 8.5 Types of conductivity cell

The value of the cell constant is most conveniently found by using the cell with a 1.00 M solution of potassium chloride, which is known to have a conductivity of 12.9 $\Omega^{-1}\ m^{-1}$ at 25 °C

The surfaces of the electrodes are coated with platinum black, finely divided platinum, which helps to reduce polarization by catalysing the reverse of the reactions in which gases are formed. The a.c. conductance bridge described above is named after its originator, Friedrich Kohlrausch, who used such an instrument to make many conductivity measurements between 1869 and 1880.

To understand the significance of such conductance measurements two new quantities must be defined:

Dilution, V, is the volume, in dm^3, which contains one mole of solute. It is therefore the reciprocal of the concentration of a solution in mol dm^{-3}, and has the units $dm^3\ mol^{-1}$.

Molar conductivity, Λ is defined by the equation:

$$\Lambda = \gamma/\Omega^{-1}\ m^{-1} \times V/dm^3\ mol^{-1} \times 10^{-3}\ dm^{-3}\ m^3$$

The 10^{-3} term is necessary to convert the units of dilution into basic SI units. The molar conductivity thus has units $\Omega^{-1}\ m^2\ mol^{-1}$.

The results of such measurements on solutions of the strong electrolytes hydrochloric acid, sodium chloride, and sodium ethanoate are shown in Figure 8.6. One would expect the conductivity of such solutions to be proportional to the number of ions in a given volume. Because the number of ions halves when the dilution is doubled, the molar conductivity should be independent of dilution in these cases. Inspection of Figure 8.6 shows this to be largely true at dilutions greater than 200 $dm^3\ mol^{-1}$. In more concentrated solutions, the ions are sufficiently close to exert a dragging effect on those of opposite charge travelling in the other direction. The value of molar conductivity for each of these electrolytes is tending towards a maximum, which represents the condition where the ions are sufficiently separated to travel independently and without exerting a drag on each other. This maximum value, known as the *molar conductivity at infinite dilution*, Λ_∞, is shown on the extreme right of Figure 8.6.

Corresponding data for the weak electrolyte, ethanoic acid, are shown in Figure 8.7. Although the dilution plotted as abscissa has the same scale as Figure 8.6, the values of molar conductivity plotted as ordinate have much smaller values. It is clear that the variation of molar conductivity with

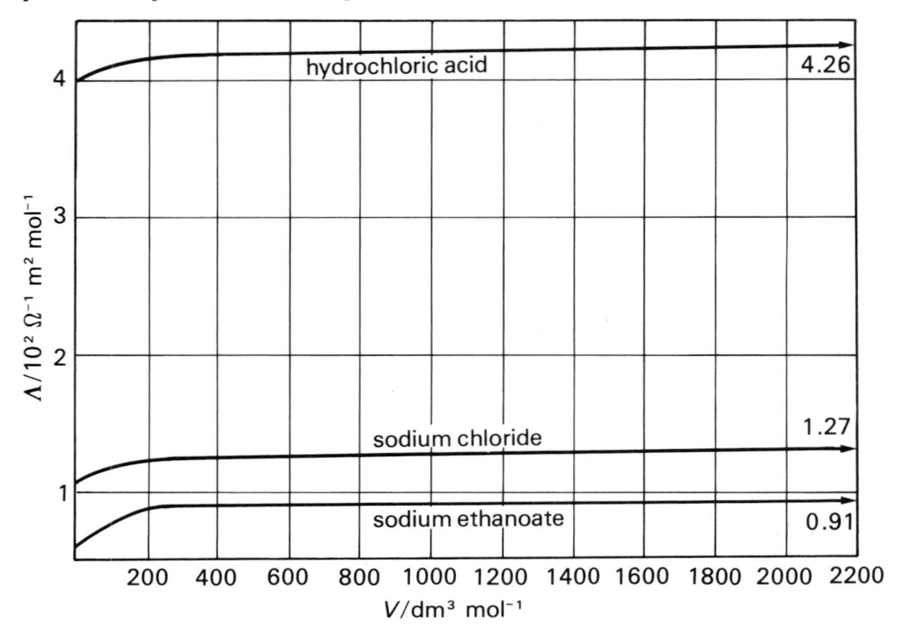

Figure 8.6 Variation of molar conductivity with dilution for some strong electrolytes at 298 K

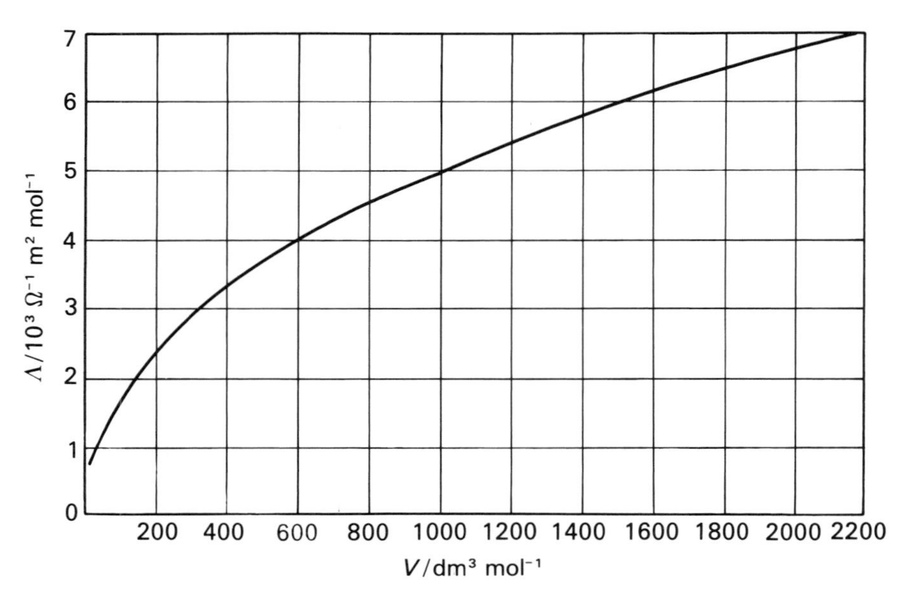

Figure 8.7 Variation of molar conductivity with dilution for ethanoic acid at 298 K

dilution for weak electrolytes is due to factors other than interionic drag. The effect is best understood in terms of the dilution law derived by Wilhelm Ostwald. For a weak binary electrolyte such as ethanoic acid, there is an equilibrium between the molecules of unionized acid and its ions:

	$CH_3CO_2H(aq)$	$\rightleftharpoons CH_3CO_2^-(aq)$	$+ H^+(aq)$
Concentration assuming no dissociation	$\frac{1}{V}$	0	0
Concentration assuming degree of dissociation α/mol dm^{-3}	$\frac{(1-\alpha)}{V}$	$\frac{\alpha}{V}$	$\frac{\alpha}{V}$

$$K_c = \frac{[H^+(aq)]_{eqm}\,[CH_3CO_2^-(aq)]_{eqm}}{[CH_3CO_2H(aq)]_{eqm}} = \frac{(\alpha/V)^2}{(1-\alpha)/V} = \frac{\alpha^2}{(1-\alpha)\,V}$$

For electrolytes as weak as ethanoic acid, α is small compared with 1 and $(1 - \alpha)$ approximates to unity. Hence, to a good approximation:

$$K_c = \frac{\alpha^2}{V}$$

When small, the degree of dissociation is thus proportional to $\sqrt{V}$.

Ostwald's Dilution Law can be used to calculate the equilibrium constant for the dissociation of weak binary electrolytes, such as ethanoic acid, provided the degree of dissociation, α, is known at particular dilutions. Arrhenius proposed the relationship:

$$\alpha = \frac{\Lambda_V}{\Lambda_\infty}$$ (where Λ_V is the molar conductivity at dilution V and Λ_∞ is that at infinite dilution)

(The validity of this relationship is best confirmed by considering the values of Λ_V which would correspond to values of α such as 0.5 and 1.)

Unfortunately it is not possible to extrapolate a value of Λ_∞ for weak electrolytes with any accuracy, as inspection of Figure 8.7 shows. It is, however, possible to obtain a value indirectly by the application of Kohlrausch's Law of Independent Migration of Ions. By 1875, Kohlrausch had examined the molar conductivity values of many strong electrolytes and concluded that *at infinite dilutions* the ions of opposite charge are so far apart that they exert negligible 'drag' on each other. Thus the molar conductance at infinite dilution is the sum of the molar conductances at infinite dilution of the cations and anions:

$$\Lambda_\infty = \lambda_\infty^+ + \lambda_\infty^-$$

Kohlrausch's law is supported by evidence such as the following:

The constant difference between the potassium and sodium salts of the same acid is readily explained in terms of the conductivity of the potassium ion being $0.23 \times 10^2\ \Omega^{-1}\ m^2\ mol^{-1}$ greater than that of the sodium ion when the same anion is involved. The differences between the conductivities of nitrate and chloride ions is seen to be $0.05 \times 10^2\ \Omega^{-1}\ m^2\ mol^{-1}$.

Application of the Kohlrausch law to hydrochloric acid, sodium chloride, and sodium ethanoate gives:

$$\Lambda_\infty(HCl) = \lambda_\infty(H^+) + \lambda_\infty(Cl^-)$$
$$\Lambda_\infty(NaCl) = \lambda_\infty(Na^+) + \lambda_\infty(Cl^-)$$
$$\Lambda_\infty(CH_3CO_2Na) = \lambda_\infty(Na^+) + \lambda_\infty(CH_3CO_2^-)$$

These can be combined to give:

$$\Lambda_\infty(HCl) + \Lambda_\infty(CH_3CO_2Na) - \Lambda_\infty(NaCl)$$
$$= \lambda_\infty(H^+) + \lambda_\infty(Cl^-) + \lambda_\infty(Na+) + \lambda_\infty(CH_3CO_2^-) - \lambda_\infty(Na^+) - \lambda_\infty(Cl^-)$$
$$= \lambda_\infty(H^+) + \lambda_\infty(CH_3CO_2^-)$$
$$= \Lambda_\infty(CH_3CO_2H)$$

Substituting values from Figure 8.6:

$$\Lambda_\infty(CH_3CO_2H)/10^2\ \Omega^{-1}\ m^2\ mol^{-1} = 4.26 + 0.91 - 1.27 = 3.90$$

This value can be used together with the molar conductivity of ethanoic acid at a series of dilution to give the degree of dissociation and thence the dissociation constant (see Table 8.2).

The quantitative aspects of acid–base behaviour are discussed in more detail in Chapter 9.

$\Lambda_\infty/10^2\ \Omega^{-1}\ m^2\ mol^{-1}$

KCl	1.50	KNO_3	1.45	NaCl	1.27	KCl	1.50
NaCl	1.27	$NaNO_3$	1.22	$NaNO_3$	1.22	KNO_3	1.45
	$\Delta = 0.23$		$\Delta = 0.23$		$\Delta = 0.05$		$\Delta = 0.05$

Table 8.2 Conductance and equilibrium data for ethanoic acid at 298 K

Dilution, V/dm³ mol⁻¹	*Molar conductivity/ Ω^{-1} m² mol⁻¹*	*Degree of dissociation* (α)	$K_c/10^{-5}$ *mol dm⁻³*
35 700	2.10	0.539	1.76
6 530	1.12	0.287	1.77
973	0.482	0.123	1.78
414	0.322	0.0825	1.79
169	0.210	0.0537	1.80
77.9	0.144	0.0368	1.80
20.0	0.0736	0.0188	1.81

Conductance changes during acid–alkali titrations

These can be investigated using the apparatus shown in Figure 8.8. Measurements of conductance are made when 1 cm³ portions at a time of 1 M alkali is added to the 0.1 M acid. (The concentration difference is to minimize the diluting effect that would obscure the significance of the conductivity changes if solutions of equal concentration were used.) Various combinations of strong and weak acids and alkalis can be investigated in this manner. Some typical results are shown in Figures 8.9 and 8.10.

1 *Strong acid/strong base titration* (e.g. hydrochloric acid and sodium hydroxide). The conductivity falls between A and B as the rapidly moving, and therefore highly conductive, hydrogen ions are replaced by the slower sodium ions:

$$H^+(aq) + OH^-(aq) \rightleftharpoons H_2O(l)$$

When virtually all the hydrogen ions have been used up at B, the conductivity rises to C. Addition of further sodium hydroxide increases the ionic concentration.

2 *Weak acid/strong base titration* (e.g. ethanoic acid and sodium hydroxide). The conductivity rises gradually between D and E as the ion population is increased by reaction between ethanoic acid molecules and hydroxide ions:

$$CH_3CO_2H(aq) + OH^-(aq) \rightarrow CH_3CO_2^-(aq) + H_2O(l)$$

It rises more steeply between E and F as hydroxide ion is a better conductor than ethanoate ion.

3 *Strong acid/weak base titration* (e.g. hydrochloric acid and ammonia). The conductivity falls between G and H as the hydrogen ions are removed by reaction with ammonia molecules:

$$H^+(aq) + NH_3(aq) \rightarrow NH_4^+(aq)$$

The conductivity falls slightly between H and J as the ionization of the excess ammonia solution added according to the equation:

$$NH_3(aq) + H_2O(l) \rightleftharpoons NH_4^+(aq) + OH^-(aq)$$

is suppressed by the ammonium ions already present. The addition of excess solution, therefore, decreases the ion concentration by dilution.

4 *Weak acid/weak base titration* (e.g. ethanoic acid and ammonia). The conductivity first falls between K and L as the small concentration of hydrogen ions from the dissociation of ethanoic acid is reduced still further by the addition of ammonia. Between L and M the conductivity rises as this effect is offset by the increase in other ions formed. There is little increase in conductivity between M

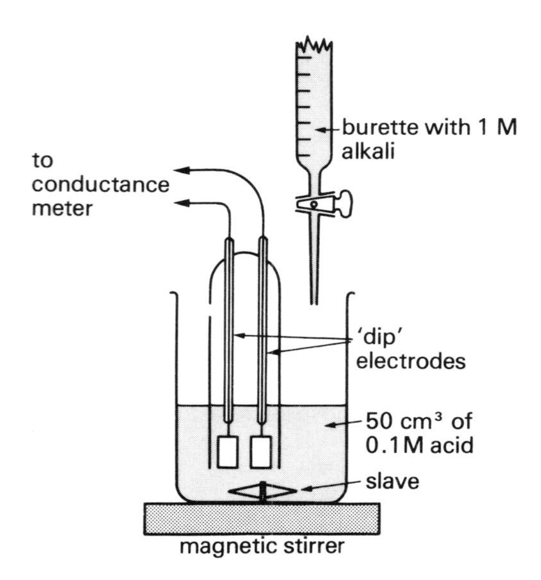

Figure 8.8 Apparatus to investigate conductivity changes during acid/alkali titrations

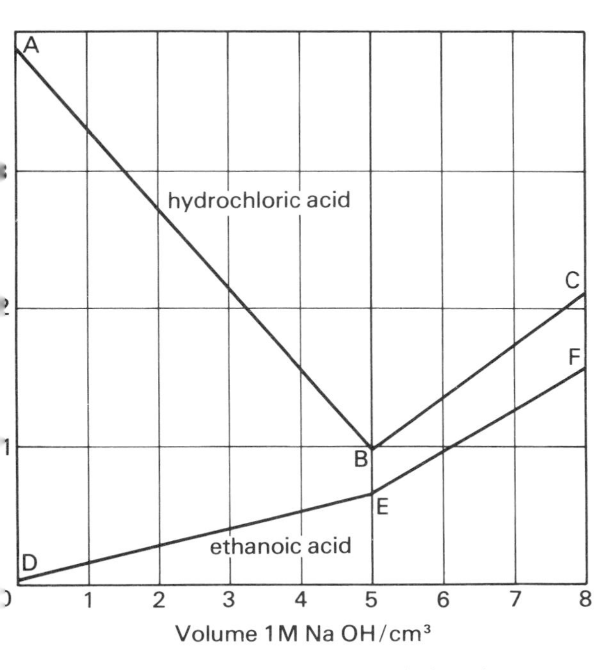

Figure 8.9 Conductivity changes during the titration of 0.1 M hydrochloric and ethanoic acid with 1 M sodium hydroxide

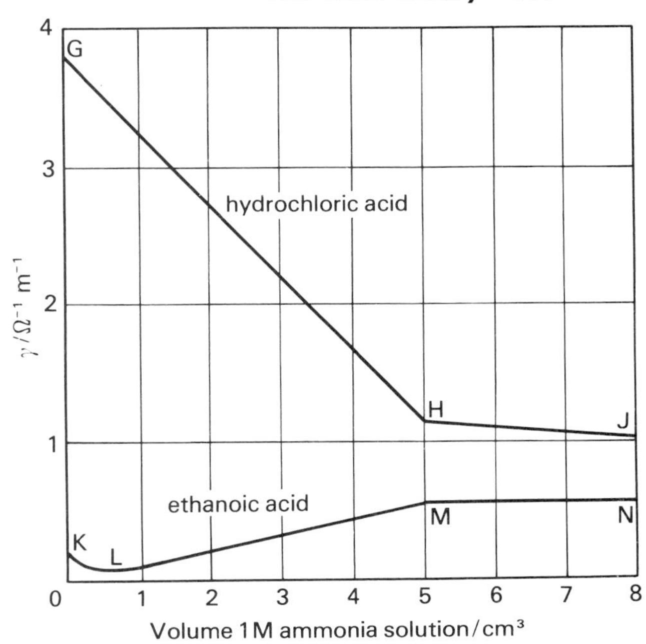

Figure 8.10 Conductivity changes during the titration of 0.1 M hydrochloric and ethanoic acid with 1 M ammonia solution

and N as the ionization of ammonia solution is suppressed by the ammonium ions already present, as discussed previously.

The quantitative aspects of acid–base chemistry are considered in more detail in Chapter 9.

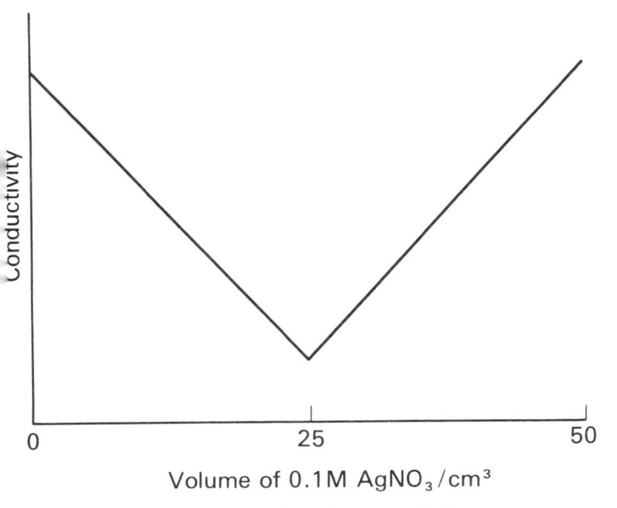

Figure 8.11 Conductivity changes during a titration involving precipitation

Conductimetric titrations

Following conductivity changes in the course of a titration provides a method of monitoring the end-point. This method is not restricted to acid–alkali titrations; Figure 8.11 shows the change of conductivity when 0.1 M silver nitrate solution is added to 25 cm^3 of 0.1 M $Pt(NH_3)_3Cl_2$. As the silver ions react with the chloride ions according to the equation:

$$Ag^+(aq) + Cl^-(aq) \rightarrow AgCl(s)$$

it can be inferred that only one of the chlorine atoms is in the form of free chloride ions and that the structure of the compound is $[Pt(NH_3)_3Cl]^+.Cl^-$.

The migration of ions

The high molar conductivities at infinite dilution of acids is due to the extremely rapid velocity of the hydrogen ion. This results from the breaking and reformation of hydrogen bonds by means of which charge may be passed rapidly through the solution, rather like a relay race, without actual movement of the atoms; this is illustrated in Figure 8.12. The velocities of ions vary considerably be-

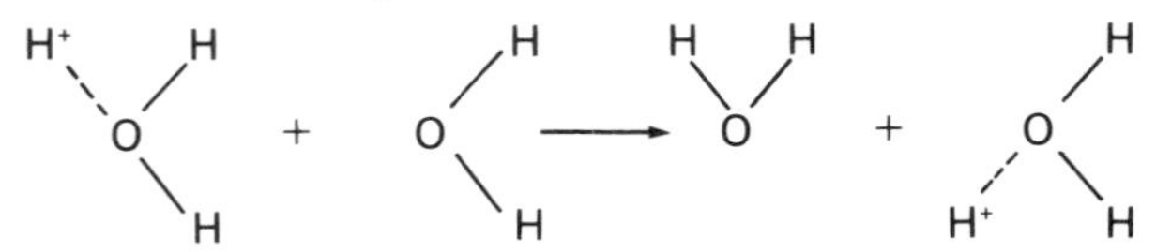

Figure 8.12 Electrical conductance by hydrogen ions

cause of differences in size, charge, and extent of hydration.

The velocity of migration of ions can be studied using the moving-boundary apparatus devised by Sir Oliver Lodge and illustrated in Figure 8.13. The anode is of cadmium and the cathode of silver coated with silver chloride. If the tube is filled with hydrochloric acid and an acid–alkali indicator; during operation cadmium chloride solution is formed by electrolysis at the anode. This neutral solution follows the hydrochloric acid solution up the tube and the boundary between the two solutions is marked by the indicator. When it is not possible to use an indicator in this way to mark the boundary, it can be detected by some physical difference between the solutions such as refractive index.

As different ions have different mobilities, the anion and cation in an electrolyte solution undergoing electrolysis will carry unequal shares of the total current. The fraction of the current carried by the individual ions is known as the *transport* (or transference) *number* for that ion. Transport numbers can be determined using the apparatus shown in Figure 8.14, first employed by Hittorf in 1853, which involves electrolysing a suitable solution and then analysing the contents of the three separate compartments. A sample calculation will make the method clear.

A silver nitrate solution containing 0.010 g of silver ion per cm^3 was electrolysed for 90 minutes at 0.08 A. The anode compartment was then emptied and the 30 cm^3 of solution it contained was found to contain 0.550 g of silver ion. Silver electrodes were employed.

Solution in anode compartment contains initially:

$$0.010 \times 30 = 0.300 \text{ g of silver;}$$

therefore gain in anode compartment

$$= 0.550 - 0.300$$
$$= 0.250 \text{ g;}$$

$$\text{charge passed} = 90 \times 60 \times 0.08 = 432 \text{ C}$$

Because 96 500 coulombs are required to liberate 1 mole of silver atoms (108 g), 432 coulombs liberate

$$108 \times \frac{432}{96\,500} = 0.480 \text{ g of silver from the cathode.}$$

If this had not occurred, the anode compartment would not have gained 0.250 g of silver but would have lost (0.480 − 0.250) g = 0.230 g of silver.

Transport number of Ag^+

$$= \frac{\text{loss of silver ion from anode compartment}}{\text{mass of silver transferred between electrodes}}$$

$$= \frac{0.230}{0.480} = 0.480$$

Transport number of NO_3^-

$$= 1 - 0.480 = 0.520$$

It is possible to check these results by an independent analysis of the contents of the cathode compartment. The concentrations in the central compartment must remain unaltered by the electrolysis for the determination to be valid.

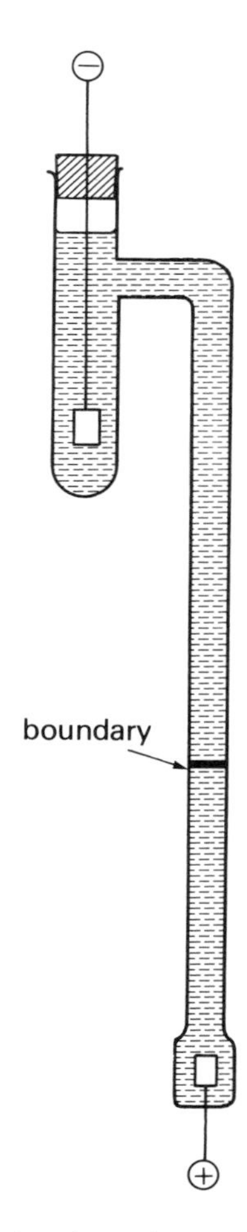

Figure 8.13 Moving-boundary apparatus

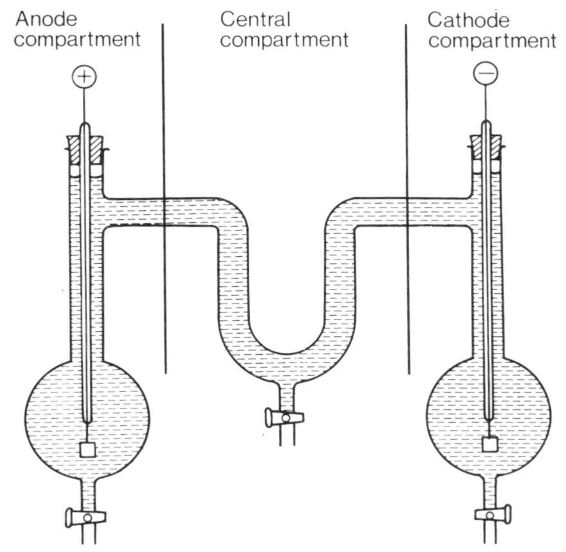

Figure 8.14 Hittorf's apparatus for the determination of transport numbers of ions

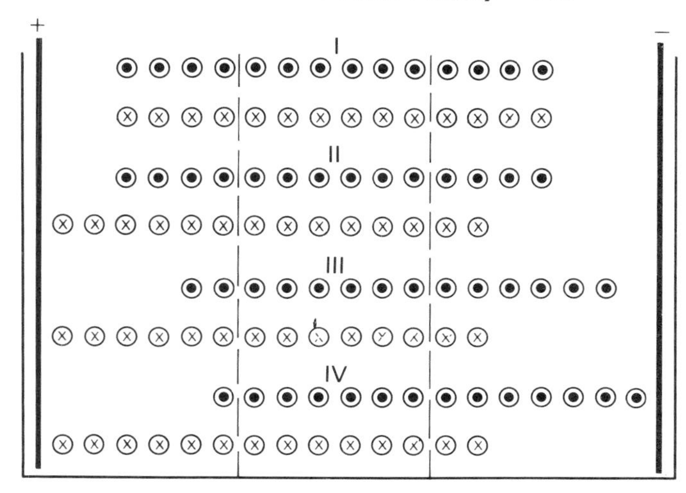

Figure 8.15 Ostwald's hypothetical experiment to explain the effect of differential ionic velocities on the concentrations around the electrodes

The effect of unequal velocities of ions on the concentrations around the electrodes can be understood in terms of the hypothetical experiment due to Ostwald shown in Figure 8.15. The cations are represented by circles with dots and the anions by circles with crosses. The cell is divided into three compartments by porous diaphragms. Before current passes, the concentrations of the ions are uniform throughout the solution as in I. If we imagine that only the anions move when the circuit is completed, the situation when the anion chain has moved two steps towards the anode is as in II. Each unpaired ion in both the anode and cathode compartments is now supposed to be discharged. It will be seen that although no cations have migrated, an equal number of anions and cations have been discharged. In addition, the concentration in the anode compartment remains unchanged but that in the cathode compartment has halved. Supposing that the anions and cations travel with the same velocity, the situation when each chain of ions moves two steps towards the appropriate electrode is shown in III. In this case four of each type of ion will have been discharged leaving the concentrations in the anode and cathode compartments diminished by equal amounts. Finally, if the ratio of velocities of cations to anions is 3:2, when the former move three steps towards the cathode, the latter will have moved two steps towards the anode as shown in IV. Thus when five ions of each type have been discharged, the concentration in the cathode compartment has diminished by two ions while that in the anode compartment by three. The change in concentration in each compartment is therefore proportional to the velocity of the ion leaving it.

Transport numbers of the ions constituting a salt can be used in conjunction with the molar conductivities of salts to assign molar conductivities at infinite dilution for individual ions. The molar conductivity at infinite dilution for silver nitrate at 25 °C is $1.33 \times 10^{-2}\ \Omega^{-1}\ m^2\ mol^{-1}$. Using the previously calculated value of 0.480 for the transport number of the silver ion in silver nitrate, the molar conductivity of silver ion at infinite dilution, $\lambda(Ag^+)_\infty$, can be determined:

$$\begin{aligned}\lambda(Ag^+)_\infty &= 0.480 \times 1.33 \times 10^{-2} \\ &= 6.38 \times 10^{-3}\ \Omega^{-1}\ m^2\ mol^{-1}\end{aligned}$$

Table 8.3 Molar conductivities of selected ions/ $10^3\ \Omega^{-1}\ m^2\ mol^{-1}$ at 298 K

Cation		Anion	
H^+	35.0	OH^-	19.8
K^+	7.36	Br^-	7.84
NH_4^+	7.34	I^-	7.68
Ag^+	6.38	Cl^-	7.63
Na^+	5.01	NO_3^-	6.92
Li^+	3.87	HCO_3^-	4.45

Tabulation of the data in terms of the individual ions is much more economical than tabulating the corresponding data for salts.

Decomposition potentials

If an electrolyte solution is electrolysed between platinum electrodes, when a low voltage is applied little current flows in the circuit. As the voltage is increased there is often a temporary increase in current flow; however, as the products accumulate, particularly gases, an opposing voltage (known as a *back e.m.f.*) is established and causes the current to diminish. For each electrolyte there is a critical voltage above which a relatively large current flows, known as the *decomposition potential.* This phenomenon can be investigated using the circuit shown in Figure 8.16. Using this apparatus, corresponding values of current and voltage may be obtained; when these are plotted, as in Figure 8.17, the decomposition potential of the electrolyte can be determined.

It will be observed that those electrolytes in which the electrode products are oxygen and hydrogen have decomposition potentials close to −1.7 volts. This is because identical electrode reactions are involved, namely:

Anode: $4OH^-(aq) - 4e^- \rightarrow 2H_2O(l) + O_2(g)$
Cathode: $4H^+(aq) + 4e^- \rightarrow 2H_2(g)$

When more dilute solutions of hydrochloric acid are electrolysed, chlorine is replaced by oxygen as the anode product and the decomposition potential falls into the pattern above.

Decomposition potentials control the deposition of products during electrolysis and if mixtures of electrolytes are electrolysed, the one with lowest decomposition potential can be plated out first. If a solution which is 1 M with respect to both silver nitrate and lead nitrate, for example, is electrolysed at a little less than 1 volt, silver is formed on the cathode. If the voltage is then raised to about 2 volts, lead is then deposited. It must be stressed that the decomposition potentials vary with concentration and, particularly as these change during the course of deposition, this consideration must always be borne in mind.

Table 8.4 Decomposition potential/volt for some electrolytes in 1 M solutions

Electrolyte	*Decomposition potential*	*Anode product*	*Cathode product*
Nitric acid	−1.69	oxygen	hydrogen
Sulphuric acid	−1.67	oxygen	hydrogen
Hydrochloric acid	−1.31	chlorine	hydrogen
Sodium hydroxide	−1.69	oxygen	hydrogen
Potassium hydroxide	−1.67	oxygen	hydrogen
Cadmium nitrate	−1.98	oxygen	cadmium
Copper(II) sulphate	−1.49	oxygen	copper
Lead(II) nitrate	−1.52	oxygen	lead
Nickel(II) sulphate	−2.09	oxygen	nickel
Silver nitrate	−0.70	oxygen	silver

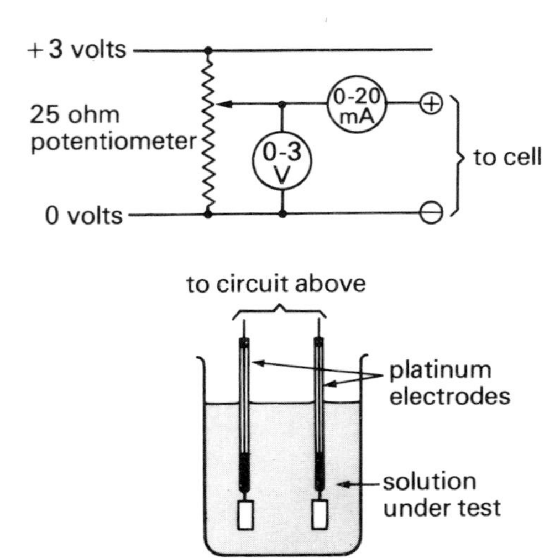

Figure 8.16 Apparatus for determination of decomposition potentials

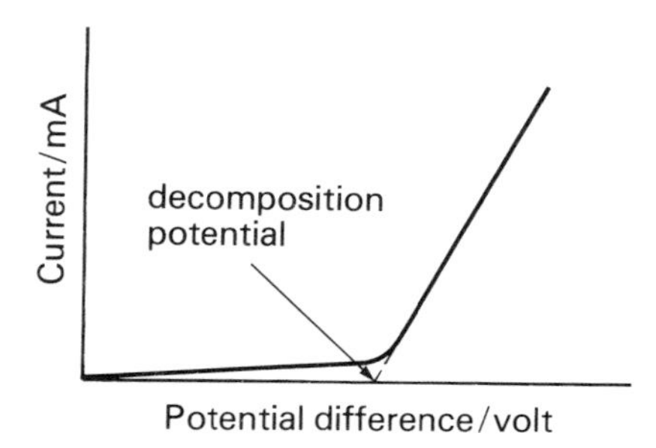

Figure 8.17 Determination of decomposition potential from current/voltage curves

Polarography

In this analytical technique, current/voltage curves are obtained between mercury electrodes for the electrolyte solution under study. Apparatus of the type shown in Figure 8.18 can be employed. The cathode is seen to be a constantly replaced mercury drop. This continuously presents a clean metal surface to the solution and, unlike a solid metal electrode, has no previous history of electrode reactions. In addition, hydrogen is liberated at a much greater voltage at a mercury cathode than at one made of platinum. This effect, known as the *hydrogen overpotential*, may be due to the slow attainment of equilibrium at the electrodes (i.e. slow transfer of electrons between the ions in the solution and the electrode surface) but no completely satisfactory explanation has yet been advanced. The hydrogen overpotential makes it possible to study the cathodic reduction of a wide range of metal ions and organic molecules by polarography. Solutions in the order 10^{-4} M are normally employed in a background electrolyte solution such as 0.1 M potassium chloride. A small amount of a gelatine or similar solution is added to reduce the mobility of the ions.

A typical result of polarography (known as a polarogram) is shown in Figure 8.19. A quantity known as the *half-wave potential* can be deduced from such curves as shown. The half-wave potential, determined under the specified conditions, is characteristic of the ion or molecule undergoing reduction. The amplitude of the wave is proportional to the concentration of the species concerned and thus the technique can be used to yield quantitative results.

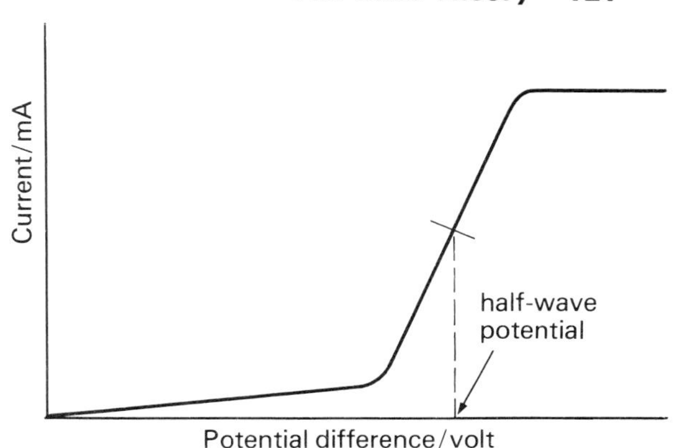

Figure 8.19 A polarograph result (polarogram)

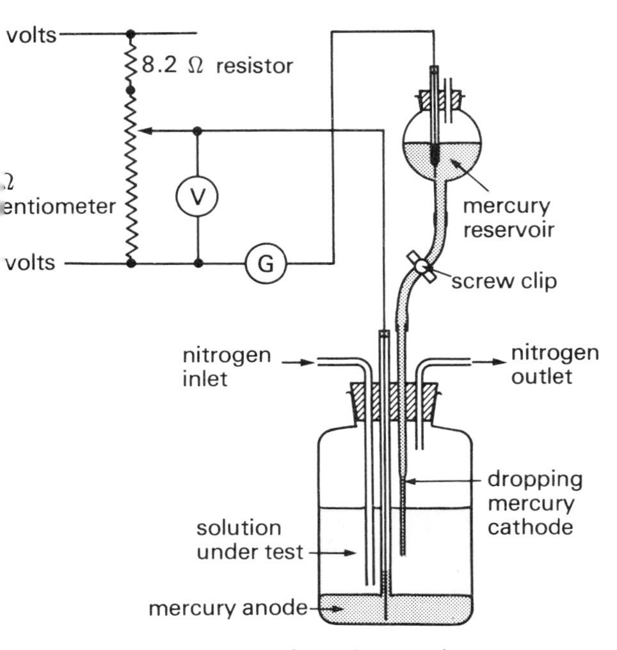

Figure 8.18 Apparatus for polarography (polarograph)

Table 8.5 Some polarographic half-wave potentials for a background medium of 0.1 M potassium chloride at 298 K

Ion	*Half-wave potential/volt*
Cd^{2+}	−0.60
Ce^{2+}	−1.40
Cr^{3+}	−0.90
Cu^{2+}	+0.04
Ni^{2+}	−1.10
Pb^{2+}	−0.40
Zn^{2+}	−1.00

Oxidation and reduction

The term oxidation was originally introduced into chemistry to describe reactions in which an element combined with oxygen to form its oxide, for example:

$$Pb(s) + \tfrac{1}{2}O_2(g) \rightarrow PbO(s)$$

Reduction was regarded as the opposite of this, as when hydrogen is passed over heated lead oxide:

$$PbO(s) + H_2(g) \rightarrow Pb(s) + H_2O(l)$$

Lead oxide can be melted and when a pair of electrodes is inserted into the melt and connected to a d.c. supply, electrolysis occurs with the formation of lead metal at the cathode and oxygen gas at the anode. From this it can be seen that the fundamental changes taking place during the oxidation and reduction reactions are, as far as the lead is concerned:

$$Pb \underset{\text{reduction}}{\overset{\text{oxidation}}{\rightleftharpoons}} Pb^{2+} + 2e^-$$

Oxidation/reduction reactions, or *redox reactions*, can thus be interpreted in terms of the ionic theory as electron transfer reactions. Oxidation is equivalent to loss of electrons and reduction to gain of electrons.

Some metals form more than one type of ion. Iron, for example, forms two series of compounds, one containing Fe^{2+} and the other Fe^{3+} ions. In the terms outlined above, iron is in a more oxidized state in the latter ion than in the former. It is said to exhibit oxidation number (or *oxidation state*) of +2 in Fe^{2+} compounds and +3 in Fe^{3+} compounds. The simple ions of the non-metals carry a negative charge and thus chlorine in chloride ion (Cl^-) can be assigned oxidation number −1 and sulphur in sulphide ion (S^{2-}) oxidation number −2. In more complicated ions and molecules, the following rules should be applied to assign oxidation numbers to the elements they contain:

1. Elements in the free uncombined state have oxidation number zero.
2. Oxygen generally has oxidation number −2. (An important exception to this is in the case of true peroxides. In these compounds there are oxygen-to-oxygen bonds and each oxygen atom has oxidation number −1.)
3. Hydrogen generally has oxidation number +1. (An important exception to this is in the case of hydrides of the more electropositive metals such as sodium. In these hydrogen exists as H^- ion and has oxidation number −1.)
4. The sum of the oxidation numbers of all atoms constituting an ion is equal to the charge on the ion, or in the case of molecules, equal to zero.

If we wish to assign an oxidation number to sulphur in the SO_4^{2-} ion, for example, rules 2 and 4 may be applied. SO_4^{2-} contains four oxygen atoms each with oxidation number −2. As the charge on the ion is 2−, the oxidation number of the sulphur atom must be +6.

The oxidation number concept can be used to give systematic names to inorganic compounds. $FeSO_4$, for example, contains Fe^{2+} and SO_4^{2-} ions and its full systematic name is iron(II) sulphate.

Ion exchange

Several natural and synthetic resins are known which contain ions so lightly bonded that if an electrolyte solution is passed through the resin, an equivalent number of the ions of the same charge in the solution are exchanged for ions in the resin. A well-known example of this is the 'Permutit' process for water softening. Water is 'hard' when it contains Ca^{2+} ions (and to a lesser extent Mg^{2+} ions) which react with anions in soap, forming an insoluble scum. In the Permutit process the hard water is passed through a column on which sodium ions are loosely held and these are exchanged for the offending calcium and magnesium ions:

$$Ca^{2+}(aq) + 2Na^+(resin) \rightleftharpoons Ca^{2+}(resin) + 2Na^+(aq)$$

As the reaction is an equilibrium one, the column can be regenerated by passing a solution of sodium chloride down it.

Water of considerable purity can be obtained by passing tap-water through a cation exchange column which replaces its cation content by an equivalent number of hydrogen ions. The water is then passed through an anion exchange column which replaces its anions with hydroxide ions. The hydrogen and hydroxide ions then react to give water.

$$H^+(aq) + OH^-(aq) \rightarrow H_2O(l)$$

Water produced by this process is known as *deionized or demineralized water*. Non-ionic impurities, if present, are not removed. The method is particularly useful for the preparation of conductivity water for making up the solutions used in measurements on electrolyte conductivity.

Avogadro's number, *L*

A value for Avogadro's number can be deduced from the results of quantitative measurements of the masses of substances liberated during electrolysis in combination with the result of Millikan's experiment to determine the charge on an electron.

$$L = \frac{\text{charge associated with 1 mole of electrons}}{\text{charge carried by 1 electron}}$$

$$= \frac{96\,487 \text{ C mol}^{-1}}{1.60 \times 10^{-19} \text{ C}} = 6.03 \times 10^{23} \text{ mol}^{-1}$$

Suggestions for further reading

MACDONALD, D. K. C., *Faraday, Maxwell, and Kelvin*, Heinemann Science Study Series No. 28. (London: Heinemann Educational Books, 1965)

Problems

1. In a series of experiments with an iodine coulometer, known currents were passed for measured times through solutions of potassium iodide, the anode reaction being:

$$2I^-(aq) - 2e^- \rightarrow I_2(aq)$$

The iodine liberated was estimated by titration with 0.400 M sodium thiosulphate solution with which it reacts according to the equation:

$$I_2(aq) + 2S_2O_3^{2-}(aq) \rightarrow 2I^-(aq) + S_4O_6^{2-}(aq)$$

Complete the table of results:

Experiment	Current/mA	Time/min	Charge/C	Volume of thiosulphate solution/cm³	Mass of iodine/g
A	50	30		23.3	
B	75	25		29.2	
C	100	20		31.1	
D	125	12		23.3	
E	150	12		28.0	

Interpret these results in terms of Faraday's laws of electrolysis.

2. A copper(II) sulphate solution was electrolysed in the apparatus shown in Figure 8.14 (page 119). 0.229 g of copper was deposited at the cathode. Before electrolysis a sample of the solution from the anode compartment was found to contain 1.195 g of copper(II) ion and after electrolysis a similar volume contained 1.360 g. The concentration in the central compartment remained unaltered. Use these results to calculate the transport numbers of $Cu^{2+}(aq)$ and $SO_4^{2-}(aq)$.

3. Describe briefly, with explanations, *what you would observe* when the following systems are electrolysed:

(*a*) fused sodium hydroxide, using platinum electrodes,
(*b*) concentrated sodium chloride solution, using carbon electrodes,
(*c*) a mixed aqueous solution of copper(II) chloride and copper(II) bromide, using platinum electrodes.

When a current of 3 amps flow for 1 hour, 3.55 g of copper are deposited on the cathode from an aqueous solution of copper(II) sulphate in an electrolytic cell. If one mole of copper contains 6×10^{23} atoms, calculate the charge on an electron. State any laws of electrolysis which your calculation assumes. (JMB 'A')

4. What do you understand by (*a*) the electrolytic conductivity and (*b*) the molar conductivity of an electrolyte? Draw rough curves to show how the latter varies with dilution for (i) a strong electrolyte and (ii) a weak electrolyte, and briefly explain the causes of these differences.

How may the limiting molar conductivity of a weak electrolyte such as ethanoic (acetic) acid be deduced from conductivity measurements? (Oxford 'A')

9. Ionic Equilibria

Electrode potentials

If a small square of silver foil is counted for background radioactivity and then placed in a silver nitrate solution which contains some silver ions of the radio-active isotope silver-110 and its radioactivity is remeasured after about one hour, the count rate is found to have substantially increased. This is direct evidence for the establishment of the equilibrium:

$$Ag(s) + aq \rightleftharpoons Ag^{+}(aq) + e^{-}$$

It is reasonable to assume that similar equilibria are established in the case of other metals and solutions of their ion; however, the extent to which this occurs will differ with each metal and this will affect the position of the equilibrium:

$$M(s) + aq \rightleftharpoons M^{n+}(aq) + ne^{-}$$

The electrical potential developed on a strip of metal partially dipping into a solution of its ions is thus a quantitative measure of the extent of this equilibrium. Consider the situation illustrated in Figure 9.1. As the tendency for copper to lose electrons in this manner is greater than that of silver, electrons will pass from the copper to the silver electrode. The voltmeter will move anticlockwise and register a voltage corresponding to the *difference* between their electrode potentials. In this example a voltage of 0.46 volts is registered.

To avoid having to make diagrams of such cells or use wordy descriptions, the following cell notation has been conventionally accepted:

$$Cu(s) \mid Cu^{2+}(aq) \,\vdots\, Ag^{+}(aq) \mid Ag(s); \quad E = +0.46\ V$$

The solid lines represent boundaries between phases and the dashed line represents the ionic bridge. The sign of E is conventionally taken as that of the right-hand-side electrode. From the direction of electron movement it can be seen that the cell reaction taking place in this example is:

$$Cu(s) + 2Ag^{+}(aq) \rightarrow Cu^{2+}(aq) + 2Ag(s)$$

(with $-2e^{-}$ from Cu(s) to Cu^{2+}(aq), and $+2e^{-}$ from $2Ag^{+}$(aq) to 2Ag(s))

A useful generalization from this is that for a cell:

$$A \mid B \,\vdots\, C \mid D$$

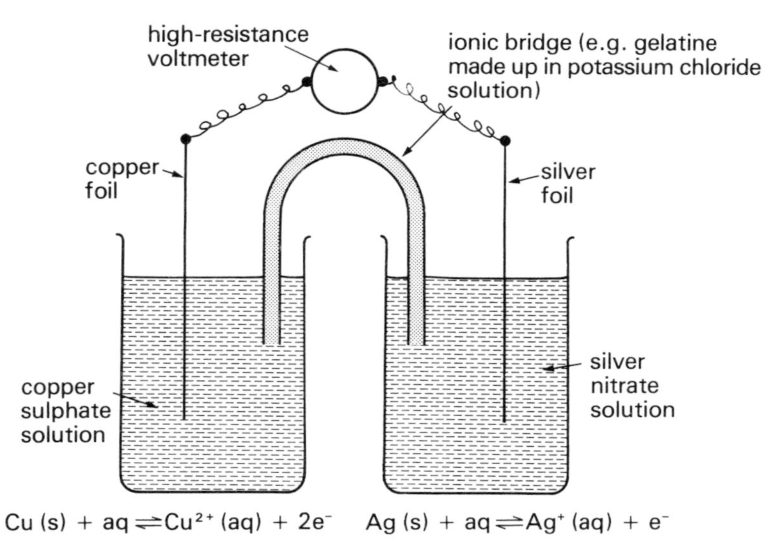

Figure 9.1 Cell consisting of copper and silver electrodes

If E is positive, the cell reaction is

$$A + C \rightarrow B + D;$$

If E is negative, the cell reaction is

$$B + D \rightarrow A + C.$$

This will be referred to later as the *ABCD rule.*

The effect of concentration on cell potentials

The silver/copper cell discussed above is readily constructed, as shown in Figure 9.2. If a series of such cells is assembled, each containing 1 M copper(II) sulphate but with differing concentrations of silver nitrate, a series of corresponding values of cell potential and silver ion concentration can be obtained. The results can then be plotted, as in Figure 9.3. The intercept on the y axis corresponds to the cell in which the concentrations of both copper and silver ions are unity. As explained in Chapter 4, when such measurements are made on systems involving unit concentrations of reactants this can be denoted using the standard state superscript. Thus $E^{\ominus}$ for the silver/copper cell is 0.46 V. The intercept on the x axis corresponds to the concentration of silver ions that are in equilibrium with 1 M copper ions because under these concentration conditions both sides of the cell have the same potential.

$$[Ag^+(aq)]_{eqm} = \lg^{-1}\ (-7.8)$$
$$= 1.6 \times 10^{-8}$$

Together with the unit concentration of copper ions, this can be inserted into the equilibrium expression to determine K_c for the reaction:

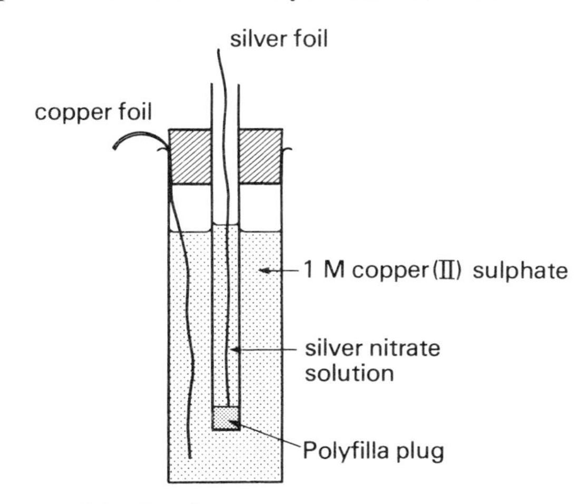

Figure 9.2 Cell for investigating the effect of concentration on potential

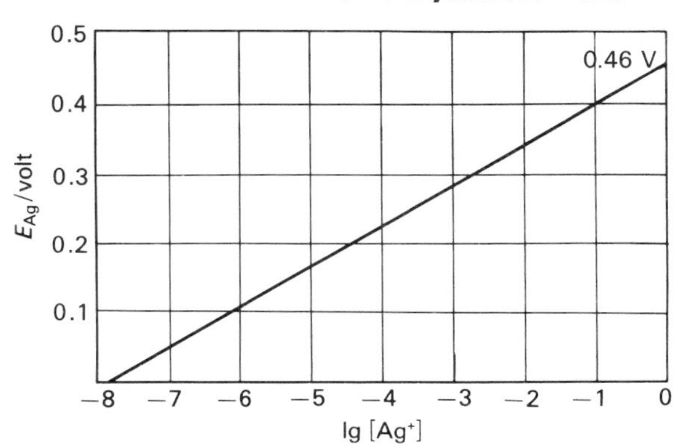

Figure 9.3 Effect of concentration on a cell potential

$$K_c = \frac{[Cu^{2+}(aq)]_{eqm}}{[Ag^+(aq)]^2_{eqm}}$$
$$= \frac{1}{(1.6 \times 10^{-8})^2} = 3.9 \times 10^{15}\ dm^3\ mol^{-1}$$

This large value for the equilibrium constant means that, for all practical purposes, the reaction proceeds to completion. [In a more accurate treatment of such effects, the quantity known as *activity* should be used instead of concentration. This is explained at the end of this chapter.]

Standard electrode potentials

The experiment explained above shows that the values of electrode potentials depend on concentration. For such quantities to be meaningfully tabulated, therefore, it is necessary to define a standard concentration and 1 M has been agreed upon for this purpose. The intercept on the y axis in Figure 9.3 thus gives the Standard Electrode Potential of silver *relative to* copper. The copper electrode, however, is not the internationally agreed standard for this purpose. The International Union of Pure and Applied Chemistry (IUPAC) specify that Standard Electrode Potentials shall be quoted as measured against the standard hydrogen electrode. An experimental arrangement that can be used for this purpose is shown in Figure 9.4. The electrode reaction at the hydrogen electrode is:

$$H_2(g) + aq \rightleftharpoons 2H^+(aq) + 2e^-$$

and the notation for such cells is

$$Pt[H_2(g)]\ |\ 2H^+(aq)\ \vdots\ M^{n+}(aq)\ |\ M(s)$$

in which M^{n+} are the ions of the metal M. If the hydrogen is at 1 atmosphere pressure and the hydrogen ion and metal ion concentrations both unity, the measured potential is the *Standard Electrode Potential* of the metal M.

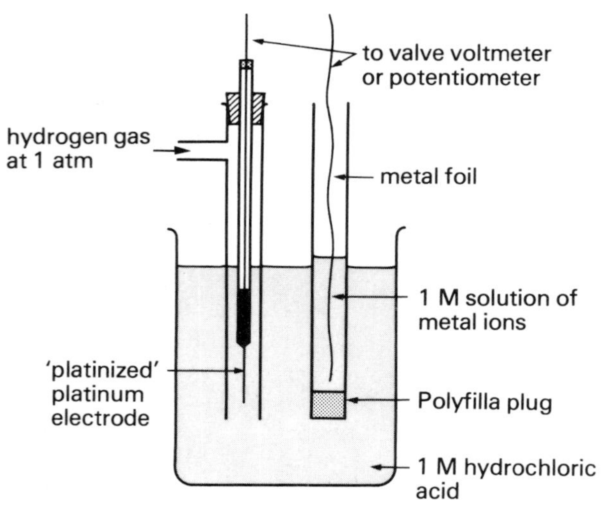

Figure 9.4 Apparatus to measure Standard Electrode Potentials

Electrode potentials of some other redox systems

So far we have only considered electrode processes involving a metal (reduced form) in contact with a solution of its ions (oxidized form). We can extend this to cover other redox systems such as:

$$\underset{\text{(reduced form)}}{2I^-(aq)} \rightleftharpoons \underset{\text{(oxidized form)}}{I_2(aq) + 2e^-}$$

$$\underset{\text{(reduced form)}}{Fe^{2+}(aq)} \rightleftharpoons \underset{\text{(oxidized form)}}{Fe^{3+}(aq) + e^-}$$

The cell notation introduced previously can be stated in the general form:

reduced form | oxidized form ⋮ oxidized form | reduced form

This can be used to cover the more complex examples we are now considering. Such measurements can be made in the apparatus shown in Figure 9.5. If the right-hand-side electrode compartment, for example, contains a solution which is 1 M with respect to both Fe^{2+} and Fe^{3+}, the cell notation is:

$$Pt[H_2(g)] \mid 2H^+(aq) \vdots Fe^{3+}(aq), Fe^{2+}(aq) \mid Pt; \quad E^{\ominus} = +0.75\ V$$

Factors affecting the magnitude of electrode potentials

The reaction between a metal and a solution of its ions that gives rise to the establishment of an electrode potential can be regarded as made up of the following processes:

Table 9.1 Some standard electrode potentials at 29

	$E^{\ominus}$/
$Li^+ + e^- \rightarrow Li$	−3
$K^+ + e^- \rightarrow K$	−2
$Ca^{2} + 2e^- \rightarrow Ca$	−2
$Na^+ + e^- \rightarrow Na$	−2
$Mg^{2} + 2e^- \rightarrow Mg$	−2
$Al(OH)_4^- + 3e^- \rightarrow Al + 4OH^-$	−2
$Al^{3+} + 3e^- \rightarrow Al$	−1
$Zn(OH)_4^{2-} + 2e^- \rightarrow Zn + 4OH^-$	−1
$V^{2+} + 2e^- \rightarrow V$	−1
$Mn^{2+} + 2e^- \rightarrow Mn$	−1
$Te + 2e^- \rightarrow Te^{2-}$	−0
$SO_4^{2-} + H_2O + 2e^- \rightarrow SO_3^{2-} + 2OH^-$	−0
$Se + 2e^- \rightarrow Se^{2-}$	−0
$Zn^{2+} + 2e^- \rightarrow Zn$	−0
$Cr^{3+} + 3e^- \rightarrow Cr$	−0
$S + 2e^- \rightarrow S^{2-}$	−0
$Fe^{2+} + 2e^- \rightarrow Fe$	−0
$Cr^{3+} + e^- \rightarrow Cr^{2+}$	−0
$Cd^{2+} + 2e^- \rightarrow Cd$	−0
$Ni^{2+} + 2e^- \rightarrow Ni$	−0
$V^{3+} + e^- \rightarrow V^{2+}$	−0
$Sn^{2+} + 2e^- \rightarrow Sn$	−0
$Pb^{2+} + 2e^- \rightarrow Pb$	−0
$2H_3O^+ + 2e^- \rightarrow H_2 + 2H_2O$	0
$Cu^{2+} + e^- \rightarrow Cu^+$	+0
$SO_4^{2-} + 2H^+ + 2e^- \rightarrow SO_3^{2-} + H_2O$	+0
$Sn^{4+} + 2e^- \rightarrow Sn^{2+}$	+0
$Cu^{2+} + 2e^- \rightarrow Cu$	+0
$VO^{2+} + 2H^+ + e^- \rightarrow V^{3+}$	+0
$O_2 + 2H_2O + 4e^- \rightarrow 4OH^-$	+0
$I_2 + 2e^- \rightarrow 2I^-$	+0
$Fe^{3+} + e^- \rightarrow Fe^{2+}$	+0
$Ag^+ + e^- \rightarrow Ag$	+0
$Hg^{2+} + 2e^- \rightarrow Hg$	+0
$2Hg^{2+} + 2e^- \rightarrow Hg_2^{2+}$	+0
$NO_3^- + 4H_3O^+ + 3e^- \rightarrow NO + 6H_2O$	+0
$Br_2 + 2e^- \rightarrow 2Br^-$	+1
$4H_3O^+ + O_2 + 4e^- \rightarrow 6H_2O$	+1
$Cr_2O_7^{2-} + 8H_3O^+ + 3e^- \rightarrow Cr^{3+} + 12H_2O$	+1
$MnO_2 + 4H_3O^+ + 2e^- \rightarrow Mn^{2+} + 6H_2O$	+1
$Cl_2 + 2e^- \rightarrow 2Cl^-$	+1
$Au^{3+} + 3e^- \rightarrow Au$	+1
$PbO_2 + 4H_3O^+ + 2e^- \rightarrow Pb^{2+} + 6H_2O$	+1
$ClO_3^- + 6H_3O^+ + 6e^- \rightarrow Cl^- + 9H_2O$	+1
$MnO_4^- + 8H_3O^+ + 5e^- \rightarrow Mn^{2+} + 12H_2O$	+1
$Pt^{2+} + 2e^- \rightarrow Pt$	+1
$MnO_4^- + 4H_3O^+ + 3e^- \rightarrow MnO_2 + 6H_2O$	+1
$Co^{3+} + e^- \rightarrow Co^{2+}$	+1
$Pb^{4+} + 2e^- \rightarrow Pb^{2+}$	+1
$O_3 + 2H_3O^+ + 2e^- \rightarrow O_2 + 3H_2O$	+1
$F_2 + 2e^- \rightarrow 2F^-$	+2

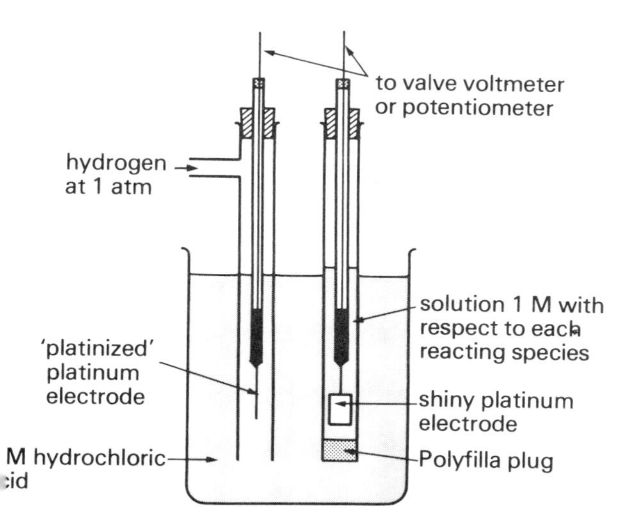

Figure 9.5 Measuring the Standard Electrode Potentials of other redox systems

1. The sublimation enthalpy of the metal: $M(s) \rightarrow M(g)$
2. The 1st and 2nd ionization energies: $M(g) - 2e^- \rightarrow M^{2+}(g)$
3. The hydration enthalpy of the gaseous ions: $M^{2+}(g) + aq. \rightarrow M^{2+}(aq)$

It is instructive to compare the magnitudes of these quantities for two metals of very different reactivities such as copper and zinc.

	Copper	*Zinc*
Sublimation enthalpy/kJ mol^{-1}	339	131
1st + 2nd ionization energies /kJ mol^{-1}	2750	2610
Hydration enthalpy of M^{2+}(aq)/kJ mol^{-1}	−2069	−2013
$\Delta H^\ominus$/kJ mol^{-1} for $M(s) + aq \rightarrow M^{2+}(aq) + 2e^-$	1020	728

It will be seen that the major difference between the two metals is in the values of their sublimation enthalpies. The overall difference in the standard enthalpy of formation of hydrated ions results in values of $E^\ominus$ for copper of +0.35 V and $E^\ominus$ for zinc of −0.76 V.

Reversibility of electrode reactions

It is important to stress at this juncture the reversibility of electrode processes. If an electrode consisting of copper metal dipping in a solution of Cu^{2+}(aq), for example, is opposed in a circuit by an electrode pushing out electrons at a higher negative potential, it operates by using these electrons to convert copper ions to the metal. If, however, it is connected to an electrode of lower negative potential, it operates by forming copper ions in solution from atoms of the metal. As the cell is operating, therefore, concentration changes occur around the electrodes. This results in changes in the cell potential which can be accounted for in terms of the Nernst equation discussed on page 128.

Calculating cell potentials from $E^\ominus$ values

The Daniell cell has the notation:

$$Cu(s) \mid Cu^{2+}(aq) \,\vdots\, Zn^{2+}(aq) \mid Zn(s)$$

$E^\ominus$ values for copper and zinc are +0.35 and −0.76 V respectively; both are quoted for when acting as a right-hand-side electrode against a standard hydrogen electrode. The standard cell potential of the Daniell cell is found by reversing the sign of the copper electrode potential (which is here the left-hand-side electrode) and adding it to the zinc electrode potential, giving a value of −1.11 volts.

Using $E^\ominus$ values to predict the feasibility of redox reactions

Consider the electrode systems:

$$\vdots\, Mg^{2+}(aq) \mid Mg(s); \quad E^\ominus = -2.40 \text{ V}$$
$$\vdots\, Pb^{2+}(aq) \mid Pb(s); \quad E^\ominus = -0.13 \text{ V}$$

These can be combined to make the cell

$$Pb(s) \mid Pb^{2+}(aq) \,\vdots\, Mg^{2+}(aq) \mid Mg(s); \quad E^\ominus = -2.27 \text{ V}$$

As the magnesium electrode is the more negative, electrons will be lost from it (oxidation occurs) and transferred to the lead electrode (reduction occurs).

Reaction at magnesium electrode:

$$Mg(s) + aq \rightarrow Mg^{2+}(aq) + 2e^-$$

Reaction at lead electrode:

$$Pb(s) + aq \rightarrow Pb^{2+}(aq) + 2e^-$$

Therefore the overall reaction is:

$$Mg(s) + Pb^{2+}(aq) \rightarrow Mg^{2+}(aq) + Pb(s)$$

To avoid the routine of this cumbersome reasoning, the '*anticlockwise*' *rule* can be applied. Electrode potentials of the appropriate systems are written with the more negative above the other, and anticlockwise arrows are then drawn as illustrated below; these will be found to indicate the direction of change:

$$Mg^{2+}(aq) \underset{-2e^-}{\overset{+2e^-}{\rightleftharpoons}} Mg(s) \quad E^\ominus = -2.40\ V$$

$$Pb^{2+}(aq) \underset{-2e^-}{\overset{+2e^-}{\rightleftharpoons}} Pb(s) \quad E^\ominus = -0.13\ V$$

This rule is the more convenient to apply when values of electrode potentials are given. When a specific cell notation is given, however, the *ABCD* rule discussed previously is more useful.

The Nernst equation

We have previously considered the effect of changing the silver ion concentration on the potential of the silver/copper cell. If this experiment is repeated using a standard hydrogen electrode and solutions of varying silver ion concentrations, the result shown in Figure 9.6 would be obtained. Both graphs have the same slope, but the intercept on the y axis differs by the Standard Electrode Potential of copper. The linear law can be applied to the graph involving the standard hydrogen electrode:

$$\begin{array}{ccccc} y & = & c & + & mx \\ E_{electrode} & & E^\ominus_{electrode} & + & 0.06 \lg [Ag^+(aq)] \end{array}$$

This is one form of the Nernst equation. It can be stated in the more general form:

$$E_{electrode} = E^\ominus_{electrode} + \frac{RT}{zF} \ln \frac{[\text{oxidized form}]}{[\text{reduced form}]}$$

[where R is the gas constant (8.3 J K^{-1} mol^{-1}), T is the temperature in kelvin, z is the oxidation number change between oxidized and reduced forms and F is the faraday (96 500 coulomb).] In terms of $\log_{10}$, the equation can be written:

$$E_{electrode} = E^\ominus_{electrode} + \frac{2.30RT}{zF} \log_{10} \frac{[\text{oxidized form}]}{[\text{reduced form}]}$$

In the case of metal/metal ion systems, the reduced form is the metal and it is found that this term can be omitted from the Nernst equation. The Nernst equation in the form deduced from the graph cannot be used to derive the more general form; a

Figure 9.6 Variation of cell potential with concentration of silver ion measured against (*a*) standard hydrogen electrode and (*b*) standard copper electrode

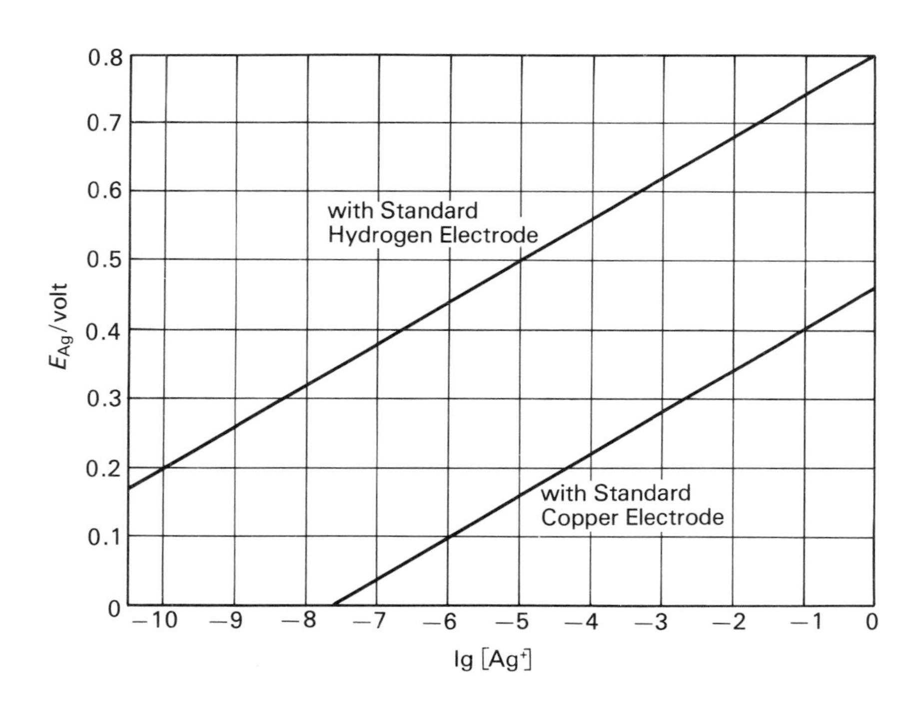

more rigorous derivation will be found at the end of the chapter. We can, however, show that the factor 2.30 RT/zF is consistent in magnitude and dimension with the slope of the experimentally obtained graph:

$$\frac{2.30 \times (8.3\ \text{J K}^{-1}\ \text{mol}^{-1}) \times (298\ \text{K})}{1 \times 96\,500\ \text{C mol}^{-1}} = 0.06\ \text{J C}^{-1} = 0.06\ \text{V}$$

N.B.—The presence of a temperature term in the Nernst equation shows that for accurate work a standard temperature must be specified when $E^{\ominus}$ values are quoted. 298 K has been accepted for this purpose.

Solubility products

If excess of a sparingly soluble salt, such as silver chloride, is shaken with water a saturated solution is soon formed, in which dissolved silver and chloride ions are in equilibrium with the remaining solid salt:

$$AgCl(s) + aq \rightleftharpoons Ag^{+}(aq) + Cl^{-}(aq)$$

The equilibrium constant, K_c, for this process can be written as previously:

$$K_c = \frac{[Ag^{+}(aq)]_{eqm} \times [Cl^{-}(aq)]_{eqm}}{[AgCl(s)]_{eqm}}$$

However, as increasing the mass of solid silver chloride will be accompanied by a corresponding increase in volume of silver chloride, it will be observed that alterations in the quantity of solid silver chloride will not affect the *concentration* of this component. The concentration of silver chloride as the solid is thus always constant; it is in fact the density of this material in mol dm^{-3}. For such systems it is more convenient to use the solubility product than the equilibrium constant. In our example this is:

$$K_s = [Ag^{+}(aq)]_{eqm} \times [Cl^{-}(aq)]_{eqm}$$

The *solubility product*, K_s, is thus the product of the equilibrium constant and the constant concentration of solid silver chloride.

$$C_mA_n(s) + aq \rightleftharpoons mC^{n+}(aq) + nA^{m-}(aq)$$

For the generalized sparingly soluble salt containing m moles of cation to n moles of anion, when it dissolves in water according to the equation above, the solubility product expression can be written:

$$K_s = [C^{n+}(aq)]^m_{eqm} \times [A^{m-}(aq)]^n_{eqm}$$

Determination of solubility product

Solubility products can be determined in a variety of ways.

(*a*) *By direct titration* This method can be used when a sufficiently sensitive titrimetric technique is available to determine the low concentrations involved. When a precipitation reaction is the basis of the titration, the solubility product of the precipitated material must be several orders of magnitude smaller than that of the test material. For example, 25 cm^3 of saturated lithium carbonate solution are found to react with 16.2 cm^3 of 1 M hydrochloric acid:

$$CO_3^{2-}(aq) + 2H^{+}(aq) \rightarrow CO_2(g) + H_2O$$

1 mole of carbonate ion reacts with 2000 cm^3 of 1 M HCl,

0.0081 mole of carbonate ion reacts with 16.2 cm^3 of 1 M HCl.

25 cm^3 of saturated lithium carbonate contains 0.0081 mole CO_3^{2-}

1 dm^3 of saturated lithium carbonate contains 0.324 mole CO_3^{2-}

$$[CO_3^{2-}(aq)]_{eqm} = \frac{[Li^{+}(aq)]_{eqm}}{2}$$

1 dm^3 of saturated lithium carbonate contains 0.648 mole Li^{+}

$$K_s = [Li^{+}(aq)]^2_{eqm} \times [CO_3^{2-}(aq)]_{eqm} = 0.136\ \text{mol}^3\ \text{dm}^{-9}$$

(*b*) *From ion exchange* This method is suitable only when the salt is sufficiently soluble so that when it is passed through a cation exchange column, and the cation it contains is exchanged for hydrogen ions, the concentration of hydrogen ions in the emergent liquid is large enough to be determined by titration. For example, 25 cm^3 of a saturated solution of calcium sulphate were passed through a cationic exchange column:

$$Ca^{2+}(aq) + 2H^{+}(resin) \rightarrow Ca^{2+}(resin) + 2H^{+}(aq)$$

The emergent hydrogen ions, after washing through with additional water, required 22.4 cm^3 of 0.01 M sodium hydroxide solution.

22.4 cm^3 of 0.01 M sodium hydroxide react with 2.24×10^{-4} mole of H^{+}.

25 cm^3 of saturated calcium sulphate contain 1.12×10^{-4} mole of Ca^{2+}

$$[Ca^{2+}(aq)]_{eqm} = [SO_4{}^{2-}(aq)]_{eqm}$$
$$[Ca^{2+}(aq)]_{eqm} = 1.12 \times 10^{-4} \times 40 \text{ mol dm}^{-3}$$
$$= 4.48 \times 10^{-3} \text{ mol dm}^{-3}$$
$$K_s = [Ca^{2+}(aq)]_{eqm} \times [SO_4{}^{2-}(aq)]_{eqm}$$
$$= 2.01 \times 10^{-5} \text{ mol}^2 \text{ dm}^{-6}$$

(*c*) *From measurement of cell potentials* We have seen that the concentrations of metal ions affect the potentials of cells; this effect can be used to measure extremely low concentrations of silver ions, as are encountered in solutions of salts such as silver iodide. When a silver electrode is immersed in saturated silver chloride solution, the electrode potential at 298 K is found to be 0.51 V. The Standard Electrode Potential of silver is 0.80 V. The Nernst equation applied to the silver electrode is:

$$E = E^{\ominus} + 0.06 \log_{10} [Ag^+(aq)]$$
$$0.51 = 0.80 + 0.06 \log_{10} [Ag^+(aq)]$$
$$0.51 = 0.80 + 0.06 \log_{10} [Ag^-(aq)]$$
$$\log_{10} [Aq^+(aq)] = \frac{0.51 - 0.80}{0.06}$$
$$= -4.8$$

$$[Cl^-(aq)]_{eqm} = [Ag^+(aq)] = 1.6 \times 10^{-5}$$
$$K_s = [Ag^+(aq)]_{eqm} \times [Cl^-(aq)]_{eqm}$$
$$= 2.6 \times 10^{-10} \text{ mol}^2 \text{ dm}^{-6}$$

(*d*) *From measurement of electrolytic conductance* The conductance values required for this method are obtained as described in Chapter 8 where the necessary theory is also discussed. A specimen calculation will demonstrate the method.

At 298 K the conductivity of a saturated aqueous solution of silver chloride is $2.68 \times 10^{-4}\ \Omega^{-1}\ m^{-1}$, while that of the conductivity water in which it was made up is $0.86 \times 10^{-4}\ \Omega^{-1}\ m^{-1}$. The molar conductances at infinite dilution of aqueous silver nitrate, hydrochloric acid, and nitric acid are 1.33×10^{-2}, 4.26×10^{-2}, and $4.21 \times 10^{-2}\ \Omega^{-1}\ m^2\ mol^{-1}$ respectively.

The conductivity of the saturated silver chloride solution is the sum of that of the dissolved silver chloride and of the water:

$$\gamma_{(AgCl)} = \gamma_{(solution)} - \gamma_{(water)}$$
$$= (2.68 - 0.86) \times 10^{-4}\ \Omega^{-1}\ m^{-1}$$
$$= 1.82 \times 10^{-4}\ \Omega^{-1}\ m^{-1}.$$

From Kohlrausch's law:

$$\Lambda_{\infty(AgCl)}$$
$$= \Lambda_{\infty(AgNO_3)} + \Lambda_{\infty(HCl)} - \Lambda_{\infty(HNO_3)}$$
$$= (1.33 + 4.26 - 4.21) \times 10^{-2}\ \Omega^{-1}\ m^2\ mol^{-1}$$
$$= 1.38 \times 10^{-2}\ \Omega^{-1}\ m^2\ mol^{-1}$$

These are related to the dilution, $V/\text{dm}^3\ \text{mol}^{-1}$, by the equation:

$$\Lambda_\infty/\Omega^{-1}\ m^2\ mol^{-1}$$
$$= \gamma/\Omega^{-1}\ m^{-1} \times V/\text{dm}^3\ \text{mol}^{-1} \times 10^{-3}\ \text{dm}^{-3}\ \text{m}^3$$

$$V/\text{dm}^3\ \text{mol}^{-1} = \frac{\Lambda_\infty/\Omega^{-1}\ m^2\ mol^{-1}}{\gamma/\Omega^{-1}\ m^{-1} \times 10^{-3}\ \text{dm}^{-3}\ \text{m}^3}$$
$$= \frac{1.38 \times 10^{-2}}{1.82 \times 10^{-4} \times 10^{-3}}$$
$$= 7.58 \times 10^4$$

The concentration of silver chloride in the saturated solution, in mol dm^{-3}, is the reciprocal of the dilution, in $\text{dm}^3\ \text{mol}^{-1}$:

$$[Ag^+(aq)]_{eqm} = [Cl^-(aq)]_{eqm}$$
$$= \frac{1}{7.58 \times 10^4} \text{ mol dm}^{-3}$$
$$= 1.32 \times 10^{-5} \text{ mol dm}^{-3}$$
$$K_s = [Ag^+(aq)]_{eqm} \times [Cl^-(aq)]_{eqm}$$
$$= (1.32 \times 10^{-5})^2 = 1.74 \times 10^{-10} \text{ mol}^2 \text{ dm}^{-6}$$

A considerable divergence will be observed between this result and that obtained using measurements of cell potentials in (*c*). Similar discrepancies will be found in comparing the values of solubility products for the same salt using tables from different sources. The extremely small values quoted for some solubility products should be treated with some caution. The solubility product of mercury(II) selenide, for example, is quoted in one data book as 1.0×10^{-59} mol^2 dm^{-6}.

$$K_s(HgSe) = [Hg^{2+}(aq)]_{eqm} \times [Se^{2-}(aq)]_{eqm}$$

In a saturated solution of the salt the concentrations of mercury(II) and selenide ions are equal, so

$$[Hg^{2+}(aq)]^2_{eqm} = [S^{2-}(aq)]^2_{eqm} = 1.0 \times 10^{-59}$$
$$[Hg^{2+}(aq)]_{eqm} = 3.16 \times 10^{-30} \text{ mol dm}^{-3}$$

Taking Avogadro's number as 6.02×10^{23} mol^{-1}, we can calculate the number of mercury(II) ions per dm^3:

Table 9.2 Some values for solubility product at 298 K

Compound	$K_s/(mol\ dm^{-3})^{m+n}$	*Compound*	$K_s/(mol\ dm^{-3})^{m+n}$
AgBr	5.0×10^{-13}	CuS	6.0×10^{-36}
AgCl	1.8×10^{-10}	$Fe(OH)_2$	2.2×10^{-15}
AgI	8.3×10^{-17}	FeS	6.0×10^{-18}
Ag_2CrO_4	1.6×10^{-12}	$Fe(OH)_3$	2.5×10^{-39}
$Al(OH)_3$	1.0×10^{-32}	HgI_2	2.5×10^{-26}
$AlPO_4$	6.3×10^{-19}	$MgCO_3$	1.0×10^{-5}
$BaCO_3$	5.0×10^{-9}	MgF_2	6.4×10^{-9}
$BaCrO_4$	1.2×10^{-12}	$Mg(OH)_2$	1.0×10^{-11}
$BaSO_4$	1.0×10^{-10}	$MnCO_3$	8.8×10^{-11}
$Bi(OH)_3$	4.0×10^{-31}	$Mn(OH)_2$	1.7×10^{-13}
$CaCO_3$	4.8×10^{-9}	MnS	7.1×10^{-16}
CaC_2O_4	1.3×10^{-9}	$Ni(OH)_2$	1.0×10^{-15}
$Ca(OH)_2$	5.5×10^{-6}	NiS	1.0×10^{-24}
$Ca_3(PO_4)_2$	1.0×10^{-29}	$PbCO_3$	1.0×10^{-13}
$CaSO_4$	1.0×10^{-5}	$PbCl_2$	1.6×10^{-5}
$CdCO_3$	5.2×10^{-12}	$PbCrO_4$	1.6×10^{-14}
$Cd(OH)_2$	1.2×10^{-14}	$Pb(OH)_2$	1.2×10^{-15}
CdS	1.0×10^{-28}	PbS	7.1×10^{-29}
$CoCO_3$	8.0×10^{-13}	$PbSO_4$	1.6×10^{-8}
$Co(OH)_2$	1.3×10^{-15}	$SrCO_3$	1.1×10^{-10}
CoS	2.0×10^{-25}	$Sr(OH)_2$	3.2×10^{-4}
$Cr(OH)_3$	6.0×10^{-31}	$SrSO_4$	3.2×10^{-7}
$CrPO_4$	2.4×10^{-23}	$ZnCO_3$	6.0×10^{-11}
CuCl	1.9×10^{-7}	$Zn(OH)_2$	2.0×10^{-17}
$Cu(IO_3)_2$	1.3×10^{-7}	ZnS	1.6×10^{-23}
$Cu(OH)_2$	6.0×10^{-20}		

Number of mercury(II) ions

$$= 3.16 \times 10^{-30}\ \text{mol dm}^{-3} \times 6.02 \times 10^{23}\ \text{mol}^{-1}$$
$$= 1.90 \times 10^{-6}\ \text{dm}^{-3}$$

It is difficult to conceive of the physical signigicance of this value!

Solubility product calculations

Calculate the solubility, in g dm^{-3}, at 298 K of silver chloride (*a*) in water and (*b*) in 0.1 M hydrochloric acid.

(*a*) From Table 9.2,

$$K_s\ (\text{AgCl})/\text{mol}^2\ \text{dm}^{-6} = 1.8 \times 10$$

In a saturated aqueous solution of silver chloride,

$$[\text{Ag}^+(\text{aq})]_{\text{eqm}} = [\text{Cl}^-(\text{aq})]_{\text{eqm}}$$
$$K_s = [\text{Ag}^+(\text{aq})]_{\text{eqm}} \times [\text{Cl}^-(\text{aq})]_{\text{eqm}}$$
$$= [\text{Ag}^+(\text{aq})]^2_{\text{eqm}}$$
$$= 1.8 \times 10^{-10}$$
$$[\text{Ag}^+(\text{aq})]_{\text{eqm}} = \sqrt{(1.8 \times 10^{-10})}$$
$$= 1.34 \times 10^{-5}$$

This is also the concentration, in mol dm^{-3}, of chloride ion and of silver chloride overall. The molar mass of silver chloride is 143.5 g mol^{-1}:

Solubility of silver chloride

$$= 143.5 \times 1.34 \times 10^{-5}\ \text{g dm}^{-3}$$
$$= 1.9 \times 10^{-3}\ \text{g dm}^{-3}$$

(*b*) The concentration of chloride ions in the final solution comes so overwhelmingly from the hydrochloric acid that the contribution from the silver chloride can be justifiably ignored.

Thus $[\text{Cl}^-(\text{aq})]_{\text{eqm}} = 0.1$,

$$[\text{Ag}^+(\text{aq})]_{\text{eqm}} = \frac{1.8 \times 10^{-10}}{0.1} = 1.8 \times 10^{-9}$$

Solubility of silver chloride

$$= 143.5 \times 1.8 \times 10^{-9}\ \text{g dm}^{-3}$$
$$= 2.6 \times 10^{-7}\ \text{g dm}^{-3}$$

The much-reduced solubility of salts in a solution containing one of its ions compared to its solubility in water is the reason why such solutions are used for the initial washing of precipitates in gravimetric analysis.

The student is warned to look out for situations in which the formation of a complex ion opposes the effect of such a 'common ion'. Figure 9.7 shows the solubility of lead chloride in solutions of varying chloride ion concentration. As the chloride ion concentration increases, the solubility at first decreases because of the 'common-ion' effect but this becomes increasingly opposed by the formation of a complex ion:

$$PbCl_2(s) + 2Cl^-(aq) \rightleftharpoons PbCl_4{}^{2-}(aq)$$

This results in the turnover point on the graph and the eventual increase in solubility.

Solubility products in volumetric analysis

When titrating solutions of halide ions with standard silver nitrate, potassium chromate solution is sometimes used as an indicator. Red silver chromate(VI) is precipitated to indicate the end-point of the titration only when virtually all the halide has been precipitated as silver halide. In a typical titration of this type, 25 cm^3 of approximately 0.1 M chloride solution is placed in a conical flask and a few drops of potassium chromate(VI) solution added. 0.1 M silver nitrate solution is then added until a red precipitate first forms. From Table 9.2, K_s $(Ag_2CrO_4) = 1.6 \times 10^{-12}$. Assuming the concentration of chromate(VI) ions in the titration flask to remain constant at 0.01 M, this value can be substituted in the solubility product expression to determine the concentration the silver ions must exceed if silver chromate is to be precipitated:

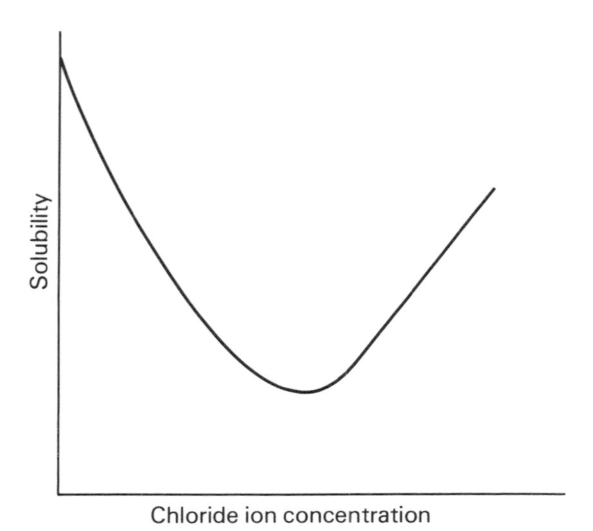

Figure 9.7 Variation in solubility of lead chloride in solutions of varied chloride ion concentration

$$[Ag^+(aq)]^2 = \frac{K_s}{0.01}$$

$$= \frac{1.6 \times 10^{-12}}{0.01} = 1.6 \times 10^{-10}$$

$$[Ag^+(aq)] = 1.3 \times 10^{-5}$$

At the exact end-point of the titration (i.e. when silver and chloride ions are present in the stoichiometric proportions) as we have calculated previously, the silver ion concentration is 1.34×10^{-5}. This may just precipitate silver chromate; however, one drop *after* the equivalence point the silver ion concentration rises dramatically. Assuming a drop to have a volume of 0.1 cm^3 and the total volume in the flask to be 50 cm^3, the silver ion concentration is $0.1 \times \frac{0.1}{50} = 2 \times 10^{-4}$ which will certainly precipitate silver chromate(VI). One drop *before* the equivalance point, the chloride ion concentration is 2×10^{-4}; thus

$$[Ag^+(aq)] = \frac{1.8 \times 10^{-10}}{2 \times 10^{-4}} = 9 \times 10^{-7}$$

This concentration is far too small to precipitate silver chromate.

Cells

Reference electrodes

We have seen how the standard hydrogen electrode is used as a reference electrode in the measurement of Standard Electrode Potentials. This electrode is cumbersome to set up and is often replaced by other electrodes as a reference source. The calomel electrode is commonly used for this purpose and its construction is shown in Figure 9.8. Its Standard Electrode Potential is +0.24 at 298 K. This refers, of course, to the cell:

$$Pt[H_2(g)] \mid 2H^+(aq) \; \vdots \; Hg_2{}^{2+}(aq) \mid 2\,Hg(l); \quad E^\ominus = +0.24\text{ V}$$

As the voltage depends on the concentration of $Hg_2{}^{2+}$ (aq), this is kept constant by the use of a saturated solution of the very sparingly soluble salt, Hg_2Cl_2.

Another reference electrode that is used particularly in the potentiometric measurement of pH (discussed in detail later in this chapter) consists of silver wire with a silver chloride coating dipping into a saturated solution of potassium chloride.

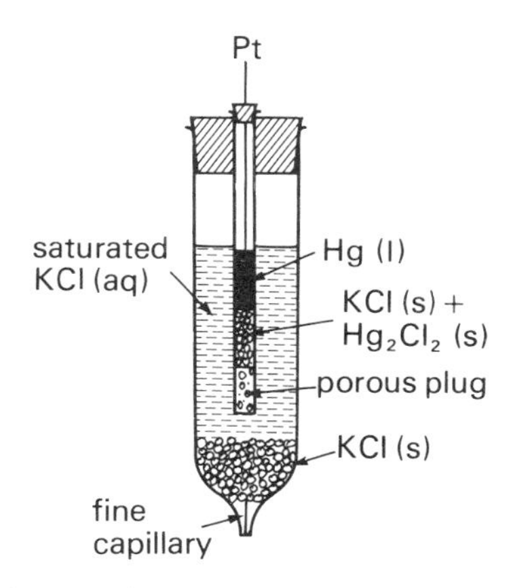

Figure 9.8 The Calomel electrode

Voltage standards

The internationally accepted voltage standard is that of the Weston cadmium cell (Figure 9.9). This has a potential of 1.0188 V at 298 K.

The Mallory mercury cell provides a cheaper but less accurate voltage standard which is adequate for many electrochemical purposes. This has a potential of 1.34 V at 298 K.

Effect of current demand on a cell potential

The construction of a working Daniell cell is shown in Figure 9.10. This can be connected into the circuit illustrated in Figure 9.11 and the effect of drawing different quantities of current on its potential can be investigated by varying the resistance across its terminals via the rheostat. A high resistance voltmeter must be employed in this circuit or considerable current will flow from the cell through the voltmeter and not be registered on the ammeter. The results of such an investigation are shown in Figure 9.12. A maximum potential of 1.1 V is given by the cell when zero current is withdrawn. For this reason cell potentials must be measured as close as possible to the condition of withdrawing zero current.

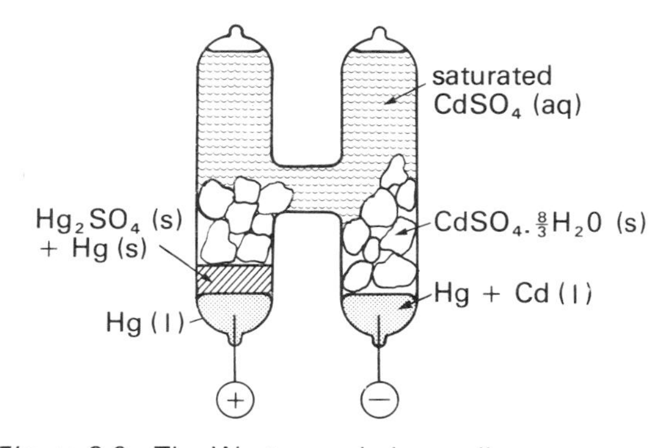

Figure 9.9 The Weston cadmium cell

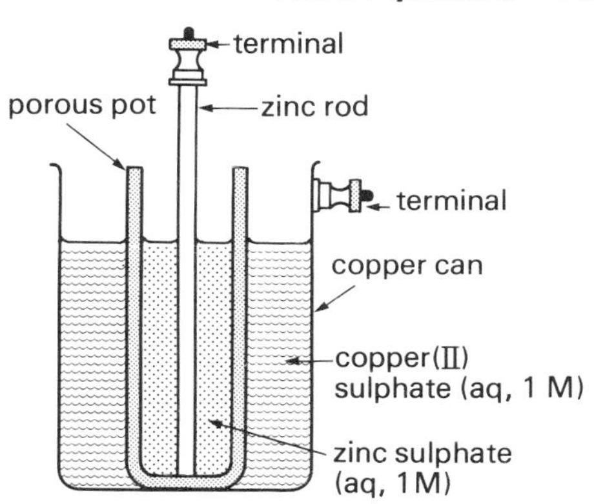

Figure 9.10 The Daniell cell

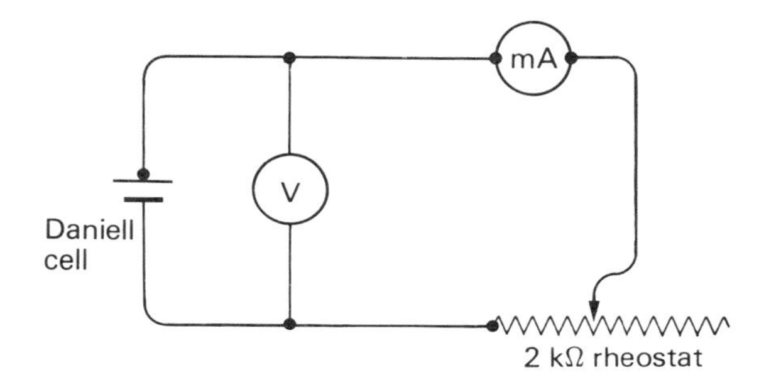

Figure 9.11 Circuit to investigate the effect of current demand on the potential of Daniell cell

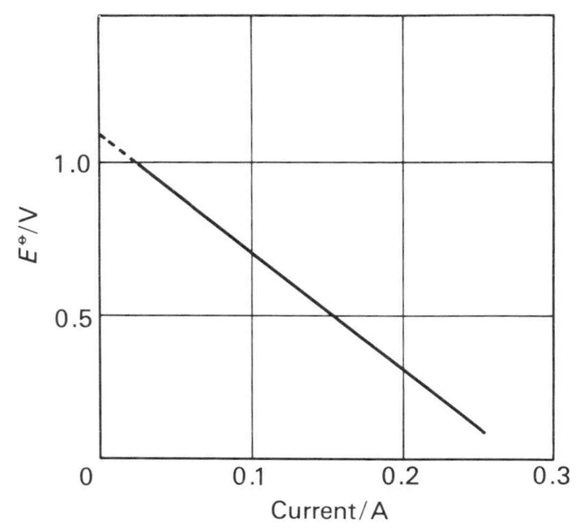

Figure 9.12 Effect of current demand on the potential of a Daniell cell

Methods of measuring cell potentials

(*a*) *High resistance voltmeters* can have resistances of 20 kΩ and are thus satisfactory for most purposes.

(*b*) *Valve or FET voltmeters* are d.c. amplifiers with either an electrometer valve (resistance 10^{14} Ω) or a Field Effect Transistor (resistance 10^{9} Ω) at input stage.

(*c*) *Potentiometers* consist of a uniform resistance wire (which may, however, be coiled or arranged in a helix) with a working cell across its ends. A sliding contact enables variable fractions of the total voltage to be balanced against the potential of the cell under test. The experimental arrangement is shown in Figure 9.13. The sliding contact is adjusted until the galvonometer reading is zero. The potentials at A and B are then equal, and this potential is that of the cell under test. It will be observed that the measurement is made under conditions in which no current is withdrawn from the test cell. The potentiometer must be calibrated against a suitable voltage standard.

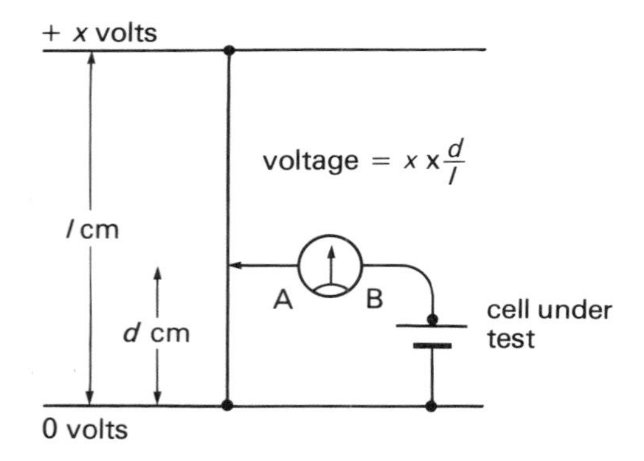

Figure 9.13 The potentiometer principle

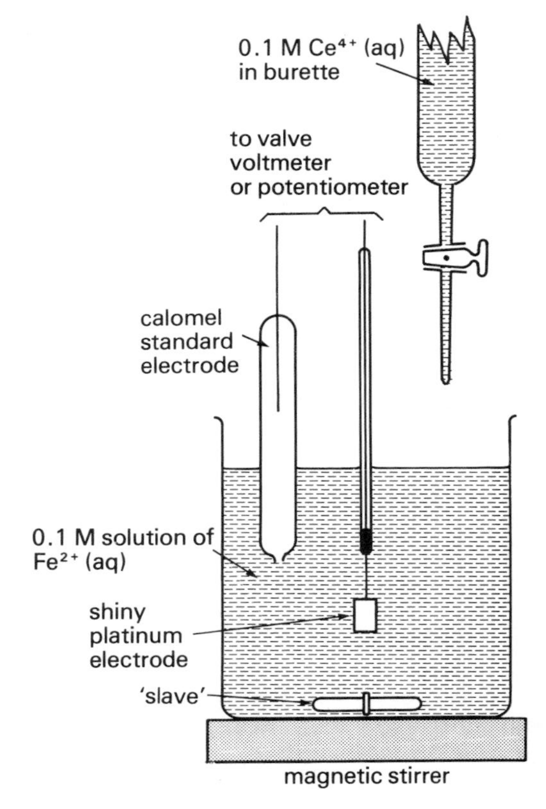

Figure 9.14 Apparatus for the potentiometric titration of iron(II) by cerium(IV)

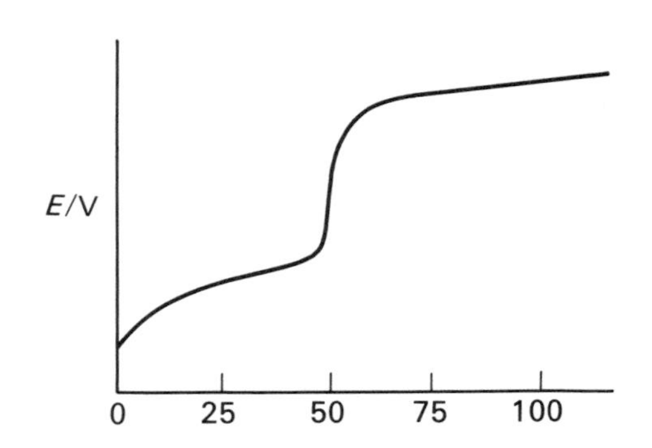

Figure 9.15 Potentiometric titration of iron(II) with cerium(IV)

Potentiometric titrations

Some titrations can be monitored using potentiometric measurements on a cell incorporating the reaction. An example of such a reaction is the titration of iron(II) with cerium(IV):

$$Fe^{2+}(aq) + Ce^{4+}(aq) \rightarrow Fe^{3+}(aq) + Ce^{3+}(aq)$$

This can be conducted in the apparatus shown in Figure 9.14. The electrode potentials for the iron(II)/iron(III) and cerium(IV)/cerium(III) systems are:

$$E_{Fe} = E^{\ominus}_{Fe} + \frac{RT}{F} \ln \frac{[Fe^{3+}(aq)]}{[Fe^{2+}(aq)]}$$

$$E_{Ce} = E^{\ominus}_{Ce} + \frac{RT}{F} \ln \frac{[Ce^{4+}(aq)]}{[Ce^{3+}(aq)]}$$

Figure 9.15 shows how the potential of the cell changes as cerium(IV) is added to 50 cm^3 of 0.1 M iron(II). At the equivalence point, when 50 cm^3 of 0.1 M cerium(IV) have been added, there is a sharp escalation of potential which indicates the end-point. Potentiometric titrations of this type are quite versatile and are particularly useful when dealing with coloured solutions in which an indicator cannot be employed. The pH titrations discussed later in this chapter are further examples of potentiometric titrations.

The pH concept

As discussed previously, acids are materials rich in hydrogen ions and alkalis are rich in hydroxide ions. Water, which ionizes slightly according to the equation:

$$H_2O(l) \rightleftharpoons H^+(aq) + OH^-(aq)$$

or in terms of the Lowry–Brønsted theory:

$$2\,H_2O(l) \rightleftharpoons H_3O^+(aq) + OH^-(aq)$$

is said to be neutral, because there are equal numbers of hydrogen and hydroxide ions. The Danish chemist Sørensen defined the *pH scale* so that the relative acidity/alkalinity of solutions commonly encountered could be compared conveniently. His definition was:

$$pH = -\lg [H^+(aq)]$$

The negative sign ensures that the pH value assigned in the vast majority of cases is positive. The small 'p' is derived from the German *potenz* meaning power, and refers to the power to which the hydrogen ion concentration is raised in order that this logarithmic relationship can be applied.

Calculating the pH values of some solutions

It should be appreciated that the pH concept can only be directly applied to water and to aqueous solutions.

1. *For strong acids* In 0.01 M hydrochloric acid, for example,

$$[H^+(aq)] = 10^{-2}$$

therefore,

$$pH = 2$$

2. *For weak acids* For these calculations we require the equilibrium constant for the dissociation of the acid into ions, K_c. These equilibrium constants are often known as *acid dissociation constants* and given the symbol K_a. These values tend to be small, and it is convenient to define the term by analogy with pH:

$$pK_a = -\lg K_a$$

Acidity constants for some common weak acids are given in Table 9.3; these have generally been determined using the conductivity method discussed in Chapter 8.

Table 9.3 Acidity constants for some weak acids at 298 K

Acid	*Formula*	*K_a/mol dm^{-3}*	*pK_a*
Methanoic (formic)	HCO_2H	1.8×10^{-4}	3.8
Ethanoic (acetic)	CH_3CO_2H	1.8×10^{-5}	4.8
Propanoic	$C_2H_5CO_2H$	1.3×10^{-5}	4.9
Chloroethanoic	CH_2ClCO_2H	1.4×10^{-3}	2.9
Dichloroethanoic	$CHCl_2CO_2H$	5.0×10^{-2}	1.3
Phosphoric(V)	H_3PO_4	7.5×10^{-3}	2.1
Dihydrogenphosphate	$H_2PO_4^-$	6.2×10^{-8}	7.2
Hydrogenphosphate	HPO_4^{2-}	2.2×10^{-13}	12.7
Hydrocyanic	HCN	4.9×10^{-10}	9.3
Hydrogen sulphide	H_2S	9.1×10^{-8}	7.0
Hydrogen sulphide ion	HS^-	1.1×10^{-12}	12.0
Phenol	C_6H_5OH	$1.3 = 10^{-10}$	9.9

To calculate the pH of 0.1 M ethanoic acid, for example, using the value of K_a from Table 9.3:

$$K_a = 1.8 \times 10^{-5}$$

$$= \frac{[CH_3CO_2^-(aq)]_{eqm}[H^+(aq)]_{eqm}}{[CH_3CO_2H(aq)]_{eqm}}$$

In a solution of ethanoic acid,

$$[H^+(aq)]_{eqm} = [CH_3CO_2^-(aq)]_{eqm}$$

$$\frac{[H^+(aq)]^2_{eqm}}{[CH_3CO_2H(aq)]_{eqm}} = 1.8 \times 10^{-5}$$

As ethanoic acid is so slightly dissociated, it can reasonably be assumed that the concentration of undissociated acid remains at its original value, i.e. $[CH_3CO_2H(aq)]_{eqm} = 0.1$. Substituting this in the above expression yields:

$$\frac{[H^+(aq)]^2_{eqm}}{0.1} = 1.8 \times 10^{-5}$$

$$[H^+(aq)]_{eqm} = \sqrt{(1.8 \times 10^{-6})} = 1.34 \times 10^{-3}$$

$$pH = -\lg(1.34 \times 10^{-3}) = 2.9$$

N.B.—When K_a for the acid concerned is too large for the concentration of the acid to be unaltered by dissociation, it is necessary to apply Ostwald's dilution law (derived in Chapter 8) to calculate the degree of dissociation. This is multiplied by the concentration of acid to give the hydrogen ion concentration, from which the pH can be calculated.

3. *For water*

$$H_2O(l) \rightleftharpoons H^+(aq) + OH^-(aq)$$

$$K_c = \frac{[H^+(aq)]_{eqm}[OH^-(aq)]_{eqm}}{[H_2O(l)]_{eqm}}$$

The value of $[H_2O(l)]_{eqm}$ is extremely large in all aqueous solutions (approximately 56 mol dm^{-3}) and so is not appreciably altered by changes in the extent to which dissociation takes place. This constant value can therefore be combined with K_c to give a new constant, K_w, known as the *ionic product of water.*

$$K_w = [H^+(aq)]_{eqm}[OH^-(aq)]_{eqm}$$

At room temperature K_w has the value 10^{-14} mol^2 dm^{-6}. In pure water

$$[H^+(aq)]_{eqm} = [OH^-(aq)]_{eqm}.$$

Hence

$$[H^+(aq)]^2_{eqm} = 10^{-14}$$

$$[H^+(aq)]_{eqm} = 10^{-7}$$

This gives the familiar value of 7 for the pH of water.

4. *For strong alkalis* In 0.001 M sodium hydroxide solution, $[OH^-(aq)]_{eqm} = 10^{-3}$

$$[H^+(aq)]_{eqm} = \frac{K_w}{[OH^-(aq)]_{eqm}} = \frac{10^{-14}}{10^{-3}} = 10^{-11}$$

$$pH = -\lg 10^{-11} = 11$$

5. *For weak alkalis* Values of basicity constants are required. These correspond to reactions of the type:

$$NH_3(aq) + H_2O(l) \rightleftharpoons NH_4^+(aq) + OH^-(aq)$$

$$K_b = \frac{[NH_4^+(aq)]_{eqm}[OH^-(aq)]_{eqm}}{[NH_3(aq)]_{eqm}}$$

N.B.—A term for the concentration of water is not necessary, for the same reason that it does not appear in the ionic product for water.

Table 9.4 Basicity constants for some weak bases at 298 K

Base	*Formula*	K_b/*mol dm*$^{-3}$	pK_b
Ammonia	NH_3	1.8×10^{-5}	4.8
Methylamine	CH_3NH_2	4.5×10^{-4}	3.3
Ethylamine	$C_2H_5NH_2$	6.5×10^{-4}	3.2
Phenylamine (aniline)	$C_6H_5NH_2$	4.3×10^{-10}	9.4
Pyridine	C_5H_5N	4.5×10^{-9}	8.3

These constants are normally determined from conductivity measurements.

To calculate the pH of 0.01 M of methylamine:

$$K_b = 4.5 \times 10^{-4}$$

$$= \frac{[CH_3NH_3^+(aq)]_{eqm}[OH^-(aq)]_{eqm}}{[CH_3NH_2(aq)]_{eqm}}$$

Assuming that, because dissociation is slight, $[CH_3NH_3(aq)]_{eqm} = 10^{-2}$ and because $[CH_3NH_3^+(aq)]_{eqm} = [OH^-(aq)]_{eqm}$

$$\frac{[OH^-(aq)]^2_{eqm}}{10^{-2}} = 4.5 \times 10^{-4}$$

$$[OH^-(aq)]_{eqm} = \sqrt{(4.5 \times 10^{-6})} = 2.12 \times 10^{-3}$$

$$[H^+(aq)]_{eqm} = \frac{K_w}{[OH^-(aq)]_{eqm}} = \frac{10^{-14}}{2.12 \times 10^{-3}}$$

$$= 4.7 \times 10^{-12}$$

$$pH = -\lg(4.7 \times 10^{-12}) = 11.3$$

6. *For buffer solutions* These are solutions that can withstand the addition of substantial quantities of either acid or alkali added to them with little change in their pH. They consist either of a weak acid and one of its salts in the same solution or of a weak base and one of its salts in the same solution. Before discussing buffer solutions themselves, it is instructive to consider how 'unbuffered' pure water is. Consider 1 dm^3 of pure water with pH 7. If one drop of 1 M hydrochloric acid (approximately 0.1 cm^3) is added:

$$[H^+(aq)] = 1 \times \frac{0.1}{1000} = 10^{-4}$$

The pH thus changes from 7 to 4, the full significance of which is apparent when it is appreciated that a logarithmic scale is involved. Living systems are particularly sensitive to even small changes in conditions and buffer solutions are therefore extremely important biologically. Most mammalian body fluids contain buffer systems.

The action of buffer solutions is best understood in terms of the equation derived below. Consider the weak acid HAc that dissociates to $H^+(aq)$ and $Ac^-(aq)$

$$K_a = \frac{[H^+(aq)]_{eqm}\,[Ac^-(aq)]_{eqm}}{[HAc(aq)]_{eqm}}$$

This can be rearranged:

$$[H^+(aq)]_{eqm} = K_a \times \frac{[HAc(aq)]_{eqm}}{[Ac^-(aq)]_{eqm}}$$

$$pH = pK_a + \lg\frac{[Ac^-(aq)]_{eqm}}{[HAc(aq)]_{eqm}}$$

If we wish to calculate the pH of a solution that is 0.1 M with respect to both ethanoic acid (pK_a = 4.8) and sodium ethanoate,

$$pH = 4.8 + \lg\frac{10^{-1}}{10^{-1}} = 4.8$$

It is worth observing that the pH of a buffer solution is equal to the pK_a for the acid, when it consists of an equal concentration of the acid and its salt.

Consider the addition of 1 drop of 1 M hydrochloric acid to 1 dm^3 of this buffer solution; the acid will react with some of the ethanoate ions:

$$CH_3CO_2^-(aq) + H^+(aq) \rightarrow CH_3CO_2H(aq)$$

As the drop contains 0.0001 moles of H^+, the concentration of ethanoate ions will drop to (0.1000–0.0001) = 0.0999 mol dm^{-3}. Similarly the concentration of ethanoic acid will rise to 0.1001 mol dm^{-3}. When these values are substituted in the above equation and the log ratio evaluated, it will be seen that for all practical purposes the pH of the solution will be unaffected. The addition of a quantity of alkali can also be shown to have a negligible effect on the pH.

The experimental determination of pH

Universal (pH) indicator paper and solution can be used to determine pH values but such measurements lack precision. Consider the possible experimental arrangement in Figure 9.16. This is an example of a *concentration cell*, so called because its two electrodes are identical except for the concentration of the electrolyte solution. It is because of this concentration difference that a potential is observed. The Nernst equation can be applied to electrode B in the form:

$$E = E^{\ominus} + 0.06 \lg[H^+(aq)]$$

Because $E^{\ominus}$ for the hydrogen electrode is zero and from the definition of pH:

$$E = -0.06\,pH$$

The assembly could therefore, in principle, be used in the measurement of pH. However, the cell would be cumbersome to set up, difficult to transport, and require a considerable volume of test liquid. Modern pH meters have alternative electrode systems to overcome these problems. The electrode A is replaced as a reference electrode by a length of silver wire coated with silver chloride and dipping into saturated potassium chloride solution. The second hydrogen electrode is replaced by a glass electrode. The theory of the glass electrode is complex and advanced texts often contain

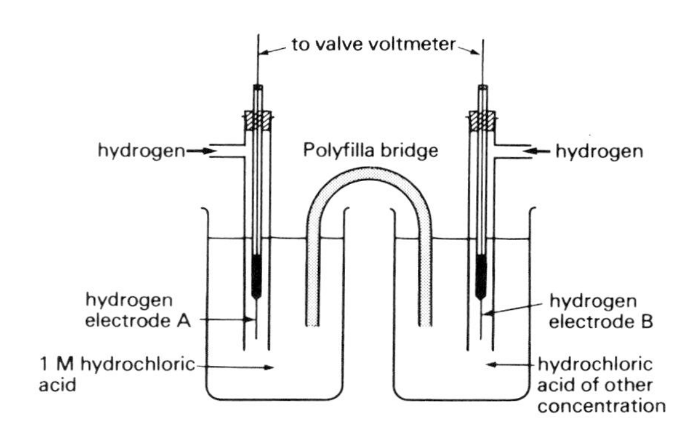

Figure 9.16 Assembly to measure pH with hydrogen electrodes

conflicting explanations of how it works. It is one of several 'ion selective electrodes' that are now available which enable the direct determination of a specific ion to be carried out potentiometrically. (Many combination electrodes are now available which contain both an ion sensitive and a reference electrode.) The glass electrode is sensitive to hydrogen ions. Figure 9.17 shows a reference and glass electrode system assembled to determine the unknown pH of a liquid. We saw how the potential of a Daniell cell is reduced as current is drawn from it; this electrode assembly is even more strongly affected in this manner, and for this reason a valve (or FET) voltmeter must be employed. Many such meters can be interconverted between reading potentials in volts and reading directly in pH units at the turn of a switch. Because the reference electrode does not develop zero potential, pH meters must be calibrated against a standard buffer solution. The international pH standard is a 0.05 M solution of potassium hydrogen benzenedicarboxylate (potassium hydrogenphthalate), which is taken to have a pH of 4.00 at 298 K. The 0.06 term in the pH expression is only accurate at 298 K; in order to allow measurements to be made at temperatures significantly different from this, the pH meter may contain a temperature compensation control. On some instruments an external thermometer is incorporated in the electrode assembly to allow automatic compensation for temperature differences. A commercial combined FET voltmeter and pH meter is shown in Figure 9.18.

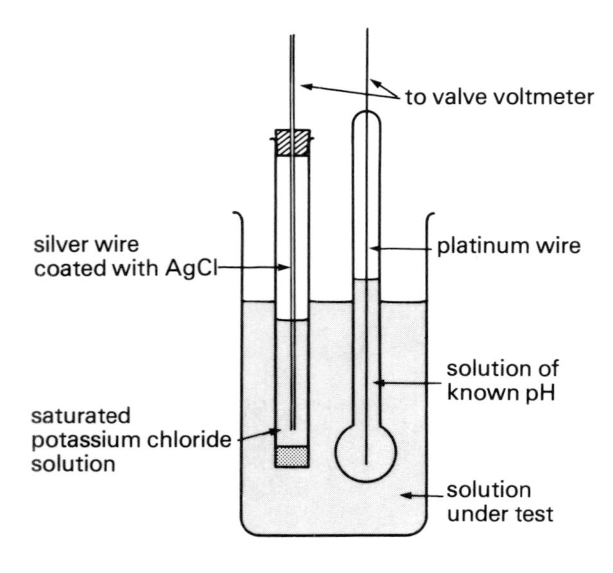

Figure 9.17 Modern reference and glass electrodes for the measurement of pH

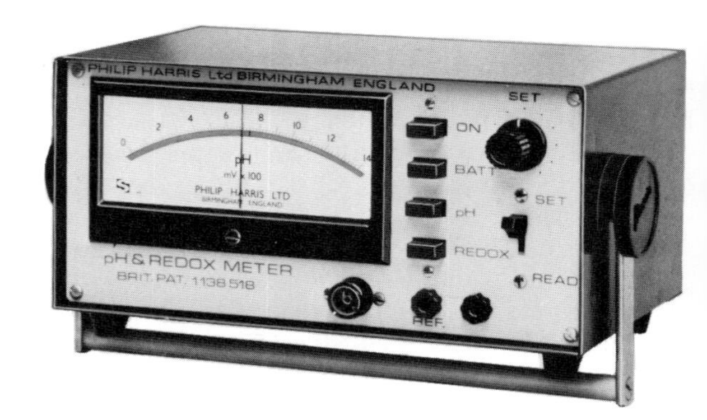

Figure 9.18 A combined FET voltmeter and pH meter

pH changes during acid/base titrations

These can be investigated experimentally by pipetting 25 cm^3 portions of 1 M acid into a conical flask and stirring magnetically. A pH meter electrode assembly is dipped into it and the pH is observed as successive volumes of 1 M alkali are added from a burette. It is most convenient to carry out an initial titration in which alkali is added in 1 cm^3 portions and then to perform a subsequent titration in which 0.1 cm^3 (1 drop) portions are added over any region when the pH changed rapidly. Every combination of strong and weak acid (e.g. hydrochloric and ethanoic acids) and strong and weak alkali (e.g. sodium hydroxide and ammonia) can be titrated in turn. Graphs of the type shown in Figure 9.19 are obtained. Data from which these graphs are drawn can also be obtained by calculation.

1. *Strong acid/strong base* In this situation we will apply the necessary theory to explain the enormous change in pH that occurs immediately around the end-point. 1 drop *before* the end-point there is 25 cm^3 of 1 M hydrochloric acid and 24.9 cm^3 of 1 M sodium hydroxide. Assuming the excess 0.1 cm^3 of acid to be in a total volume of 50 cm^3:

$$[H^+(aq)] = 1 \times \frac{0.1}{50} = 0.002$$

$$pH = 2.7$$

By a similar argument, 1 drop *after* the end-point:

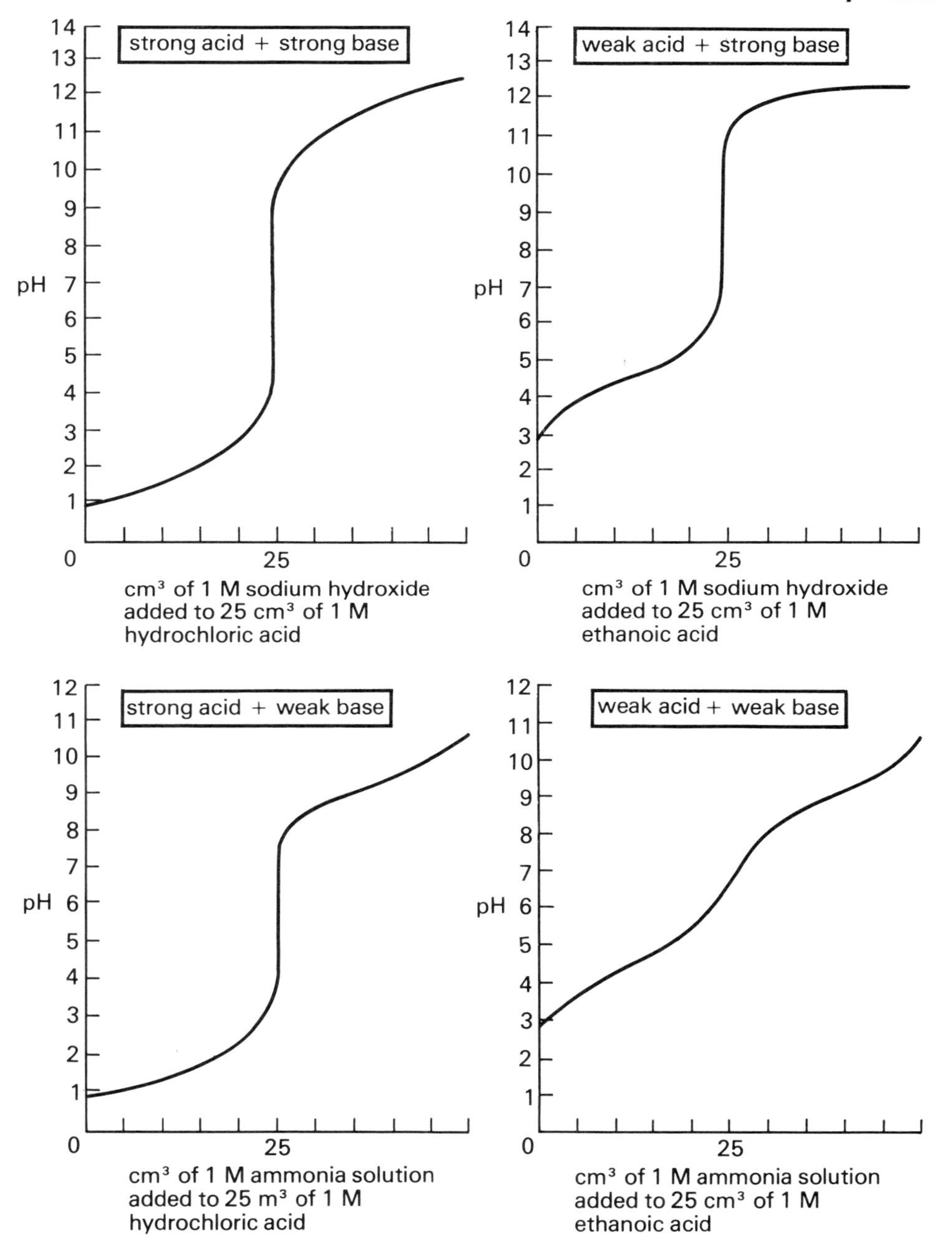

Figure 9.19 pH changes during acid/base titrations

$$[OH^-(aq)] = 0.002 = 2 \times 10^{-3}$$

$$[H^+(aq)] = \frac{10^{-14}}{2 \times 10^{-3}} = 5 \times 10^{-12}$$

$$pH = 11.3$$

All the other points on the curve can be calculated in a similar manner.

2. *Weak acid/strong base* v cm^3 of 1 M sodium hydroxide are added to 25 cm^3 of 1 M ethanoic acid. For values of v from 0 to 24.9:

$$[H^+(aq)]_{eqm} = \frac{K_a \times [CH_3CO_2H(aq)]_{eqm}}{[CH_3CO_2^-(aq)]_{eqm}}$$

$$= K_a \times \frac{(25 - v)}{(25 + v)} \times \frac{(25 + v)}{v}$$

$$pH = -\lg\left\{1.8 \times 10^{-5} \times \frac{(25 - v)}{v}\right\}$$

$$= -\lg\left\{(45/v - 1.8) \times 10^{-5}\right\}$$

For values of v greater than 25.0:

$$[H^+(aq)]_{eqm} = \frac{K_w}{[OH^-(aq)]_{eqm}}$$

$$pH = -\lg\left\{10^{-14} \times \frac{(v + 25)}{(v - 25)}\right\}$$

From these equations a computer can be programmed to calculate the values of pH at volume intervals of 0.1 cm^3.

Similar calculations can be carried out for the cases of strong acid/weak base and weak acid/weak base.

Acid/base indicators

Several natural and artificial dyes are known that have different colours depending on whether they are added to acidic or alkaline solutions. Their action can be investigated in more detail if drops of such indicator solutions are added to a series of buffer solutions covering the pH range 3 to 10. It is then observed that the pH range over which indicators change colour is not always the same. In general, there is a pH interval of approximately two units over which the indicator changes colour; this is known as its *working range*.

Ostwald's theory of indicators

According to this theory, acid/base indicators are either weak acids or weak bases. Consider an indicator that functions as a weak acid:

$$\underset{\text{(colour 1)}}{HIn(aq)} \rightleftharpoons H^+(aq) + \underset{\text{(colour 2)}}{In^-(aq)}$$

The addition of acids swings the equilibrium in the direction of HIn which has colour 1, whereas the addition of alkali results in the formation of In^- with colour 2. At the mid-point of the working range, the concentrations of undissociated acid and its anion should be equal. This enables the acid dissociation constant for such indicators to be determined from their working range. The mid-point of the working range of methyl orange, for example, is 3.8.

$$K_a = \frac{[H^+(aq)]_{eqm}\,[In^-(aq)]_{eqm}}{[HIn(aq)]_{eqm}}$$

At the mid-point of the working range, $[HIn(aq)]_{eqm} = [In^-(aq)]_{eqm}$.

$[H^+(aq)]_{eqm}$ (calculated from the pH) = 3.07 $\times 10^{-4}$, and therefore

$$K_a = 3.07 \times 10^{-4}$$

One of the chemically simplest substances that can act as an acid/base indicator is 4-nitrophenol. This is thought to exist in solution in the following series of equilibrium forms:

Table 9.5 Working ranges of some acid-base indicators

Indicator	*Acid colour*	*Alkaline colour*	*Working range*
Methyl violet	yellow	blue	0.0–1.6
Thymol blue (1st change)	red	yellow	1.2–2.8
Methyl orange	red	yellow	3.2–4.4
Bromocresol green	yellow	blue	3.8–5.4
Methyl red	yellow	red	4.8–6.0
Bromothymol blue	yellow	blue	6.0–7.6
Phenol red	yellow	red	6.8–8.4
Thymol blue (2nd change)	yellow	blue	8.0–9.6
Phenolphthalein	colourless	red	8.2–10.0

NO_2 (colourless, with OH) ⇌ NO_2H (brown, with =O) ⇌ NO_2^- (brown, with =O) $+ H^+$

(colourless) (brown) (brown)

A solution of 4-nitrophenol is colourless in acid conditions, but brown in alkalis. This is consistent with the equilibrium situation proposed above.

Selection of the correct indicator

For an indicator to operate in a satisfactory manner its colour must change sharply at the end-point. For this to happen, the rapid rise in pH in the region of the end-point must pass through the working range of the indicator. The two most commonly used indicators in acid/base titrations are methyl orange (working range 3.2–4.4) and phenolphthalein (working range 8.2–10.0).

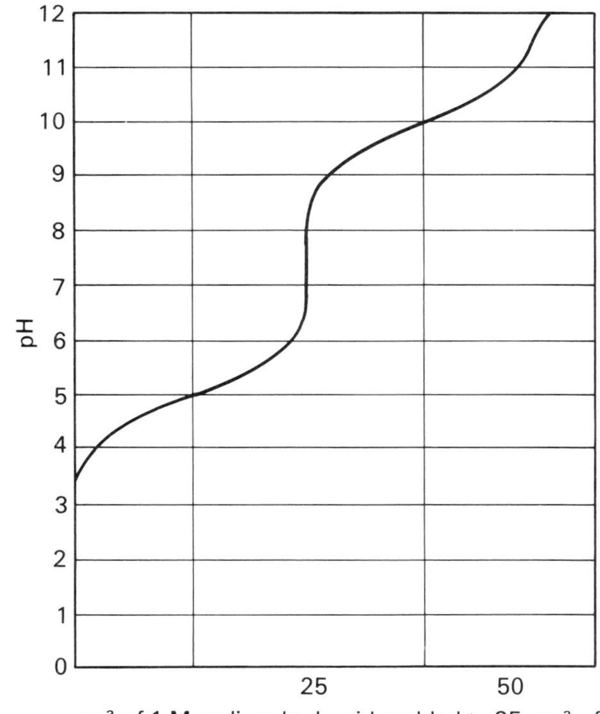

Figure 9.20 Titration curve for a diprotic (dibasic) acid

Examination of Figure 9.19 shows that either of these indicators fulfils this condition in the case of strong acid/strong base titrations. For strong acid/weak base titrations, however, only methyl orange is suitable and in the titration of weak acid/strong base phenolphthalein must be used. There is no satisfactory indicator for the titration of a weak acid with a weak base.

Quantitative treatment of salt hydrolysis

As discussed in Chapter 8, the sodium salts of weak acids (such as ethanoic acid) give alkaline solutions in water, e.g.

$$CH_3CO_2^-(aq) + H_2O(l) \rightleftharpoons CH_3CO_2H(aq) + OH^-(aq)$$

The equilibrium constant, known as the hydrolysis constant, K_h, for this process can be written:

$$K_h = \frac{[CH_3CO_2H(aq)]_{eqm}\,[OH^-(aq)]_{eqm}}{[CH_3CO_2^-(aq)]_{eqm}}$$

A useful relationship is obtained if this is multiplied by the acid dissociation constant for the parent acid:

$$\begin{aligned} K_h \times K_a &= \frac{[CH_3CO_2H(aq)]_{eqm}\,[OH^-(aq)]_{eqm}}{[CH_3CO_2^-(aq)]_{eqm}} \times \frac{[H^+(aq)]_{eqm}\,[CH_3CO_2^-(aq)]_{eqm}}{[CH_3CO_2H(aq)]_{eqm}} \\ &= [H^+(aq)]_{eqm}\,[OH^-(aq)]_{eqm} \\ &= K_w \end{aligned}$$

From this relationship, K_a for a weak acid can be determined from the measurement of the pH of a solution of its sodium salt.

Diprotic (dibasic) acids

These possess two moles of ionizable hydrogen in each mole. Examples include both strong and weak acids such as sulphuric acid, ethanedioic (oxalic) acid, and hydrogen sulphide. The titration curves for such acids show two regions in which a rapid increase in pH follows the addition of relatively small quantities of alkali. An example of such a titration curve is shown in Figure 9.20. The first pH jump is associated with the change:

$$H_2X(aq) + OH^-(aq) \rightarrow HX^-(aq) + H_2O(l)$$

The second jump corresponds to the change:

$$HX^-(aq) + OH^-(aq) \rightarrow X^{2-}(aq) + H_2O(l)$$

Inspection of Figure 9.20 shows that the volume of alkali required in such a titration will depend upon the working range of the indicator employed.

We can write the equilibrium expressions for the ionization of a weak diprotic acid such as hydrogen sulphide:

$$H_2S(aq) \rightleftharpoons H^+(aq) + HS^-(aq) \rightleftharpoons 2H^+(aq) + S^{2-}(aq)$$

$$K_{a_1} = \frac{[H^+(aq)]_{eqm}\,[HS^-(aq)]_{eqm}}{[H_2S(aq)]_{eqm}}$$

$$K_{a_2} = \frac{[H^+(aq)]_{eqm}\,[S^{2-}(aq)]_{eqm}}{[HS^-(aq)]_{eqm}}$$

Multiply these expressions together:

$$K_{a_1} \times K_{a_2} = \frac{[H^+(aq)]_{eqm}\,[HS^-(aq)]_{eqm}}{[H_2S(aq)]_{eqm}} \times \frac{[H^+(aq)]_{eqm}\,[S^{2-}(aq)]_{eqm}}{[HS^-(aq)]_{eqm}}$$

$$= \frac{[H^+(aq)]^2_{eqm}\,[S^{2-}(aq)]_{eqm}}{[H_2S(aq)]_{eqm}}$$

i.e. $K_{a_1} \times K_{a_2} = K_{a(\mathrm{overall})}$

This can be used to find the overall equilibrium constant for other situations involving sequential equilibria.

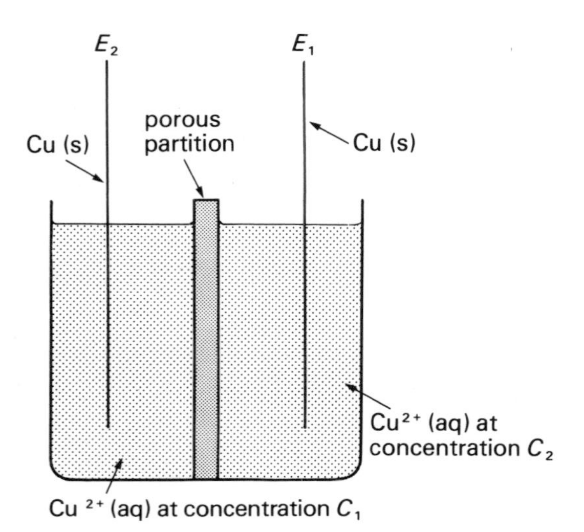

Figure 9.21 A copper/copper ion concentration cell

A more rigorous approach to the Nernst equation

Consider the copper/copper ion concentration cell illustrated in Figure 9.21. If 1 mole of copper is dissolved from one electrode and deposited on the other the charge passing through the cell is $2F$ coulombs. The potential difference between the two electrodes is $E_1 - E_2$; therefore, assuming $E_1 > E_2$, the electrical work (w) done during the transfer is given by the expression:

$$w = 2F(E_1 - E_2)$$

An alternative method of calculating the work done in this process is to consider it as an expansion against the concentration gradient doing work opposing the osmotic pressure difference. By analogy with the expansion of an ideal gas, we have seen in Chapter 4:

$$w = p\Delta V$$

for a small change in w,

$$\delta w = p\delta V$$

From the general gas equation, for 1 mole,

$$p = \frac{RT}{V}$$

$$\delta w = \frac{RT}{V}\delta V$$

Hence in the limit,

$$\mathrm{d}w = \frac{RT}{V}\mathrm{d}V$$

Integrating between volume limits V_1 and V_2 ($V_2 > V_1$)

$$w = \int_{V_1}^{V_2} \frac{RT}{V}\mathrm{d}V$$

$$= RT \ln \frac{V_2}{V_1}$$

As the concentrations are inversely related to the volumes

$$w = RT \ln \frac{c_1}{c_2}$$

Equating these expressions for the work done:

$$2F(E_1 - E_2) = RT \ln \frac{c_1}{c_2}$$

This can be rearranged in the form:

$$E_1 = E_2 + \frac{RT}{2F} \ln \frac{c_1}{c_2}$$

When c_2 is unity, then E_2 is $E^{\ominus}$,

$$E = E^{\ominus} + \frac{RT}{2F} \ln [Cu^{2+}(aq)]$$

The general case for a metal with ions M^{z+},

$$E = E^{\ominus} + \frac{RT}{zF} \ln [M^{z+}(aq)]$$

Activities

Because of inter-ionic attractions, the activity of an ion is not identical with its concentration. This effect is particularly pronounced in cases of solutions containing ions carrying a multiple charge, or as solutions become more concentrated and the ions are on average closer together. The activity of an ion at a particular concentration can be found by multiplying its concentration by the appropriate activity coefficient from Table 9.6.

Table 9.6 Some activity coefficients at 298 K

Concentration/ mol dm^{-3}	*$AgNO_3$*	*$CuSO_4$*	*HCl*
0.001	0.97	0.74	—
0.005	0.92	0.53	0.93
0.010	0.90	0.41	0.91
0.050	0.79	0.21	0.83
0.100	0.72	0.16	0.80
0.500	0.54	—	0.77
1.00	0.43	—	0.81

Thus activity of 0.1 M $Ag^+(aq)$ in silver nitrate $= 0.1 \times 0.72$

$$a_{Ag^+} = 0.072$$

Such values should be used instead of concentrations for accurate work.

Suggestions for further reading

ROBBINS, J. *Ions in Solution 2: An Introduction to Electrochemistry* (Oxford: Oxford University Press, 1972)

BROWNING, D. R. (ed.) *Electrometric Methods* (Maidenhead: McGraw-Hill, 1969)

ROBINSON, R. A. and STOKES, R. H. *Electrolyte Solutions* (London: Butterworth, 1970)

Problems

1. Use the data in Table 9.1 (page 126) to determine $E^{\ominus}$ for the cell:

   $$Ni(s) \mid Ni^{2+}(aq) \,\vdots\, Mn^{2+}(aq) \mid Mn(s)$$

   What reaction will take place when the cell is used to operate a low voltage light bulb? Explain why the potential of the cell gradually drops as it is operated.
2. Use the data in Table 9.1 to predict any reactions occurring when: (i) tin is placed in a 1 M solution of iron(III) chloride, (ii) 2 M solutions of tin(II) chloride and mercury(II) chloride are mixed in equivolume proportions, (iii) an acidified solution of 1 M potassium manganate(VII) is added to an equal volume of 1 M iron(II) sulphate.

   Write equations for the changes that would take place.
3. Use the data in Table 9.2 to (i) predict what will happen when equal volumes of 1 M calcium chloride and 1 M sodium sulphate are mixed, (ii) calculate the solubility in g dm^{-3} of lead(II) chloride in water at 298 K, (iii) calculate the pH of a saturated solution of magnesium hydroxide at 298 K.
4. Use the data in Table 9.3 to calculate the pH of (i) 0.01 M methanoic acid, (ii) a solution that is 0.01 M with respect to both methanoic acid and sodium methanoate, (iii) 0.1 M sodium methanoate.
5. Explain the origin of the potential difference which appears when a metal electrode is dipped into a solution of its own ions. Why does the magnitude of this potential depend on the concentration of the ions in the solution? Why is the electrode potential of a metal not simply related to the ionization potential of that metal?

   Describe how you would measure the electrode potential of a metal. Explain the reasons for your choice of apparatus.

   (Oxford and Cambridge 'A')
6. The equilibrium constant for the reaction between iron(III) ions and iodide ions can be determined either by setting up a suitable voltaic cell and varying the concentration of one of the ions, making measurements of e.m.f., or by calculations based upon standard e.m.f.s and the Nernst equation:

   $$E = E^{\ominus} + \frac{2.3RT}{zF} \lg \frac{[\text{oxidized form}]}{[\text{reduced form}]}$$

Describe in detail how **one** of these methods could be put into practice. (Nuffield 'A')

7. Discuss (*a*) the factors which influence the mobility of ions in solution, (*b*) a technique used for the measurement of electrolytic conductivity, (*c*) the use of conductivity data for the calculation of the solubility product of a sparingly soluble salt.

Discuss critically the use of the term *dissociation* as applied to molecules in the gaseous phase and to electrolytes in solution.

From the data given below calculate the limiting value of the degree of dissociation in aqueous solution of HCN (the value approached as the concentration of HCN tends to zero at 298 K).

[K_a(HCN) = 7.2×10^{-10} mol dm^{-3}; ionic product of water, $K_w = 10^{-14}$ mol^2 dm^{-6} at 298 K.] (Oxford and Cambridge 'S')

8. The standard electrode potentials, $E^{\ominus}$, of copper and silver are:

$$Cu^{2+}(aq) + 2e^- \rightarrow Cu(s);\ E^{\ominus} = +0.80\ V$$

$$Ag^{+}(aq) + e^- \rightarrow Ag(s);\ E^{\ominus} = +0.34\ V$$

If the following cell is set up

$$Ag(s) \mid Ag^{+}(aq) \vdots Cu^{2+}(aq) \mid Cu(s)$$

(*a*) what is the e.m.f. of the cell, (*b*) which metal forms the positive pole, (*c*) when the copper electrode is connected by a wire to the silver electrode, write an equation for the reaction that takes place and give the states (aq or s) of the reactants and products. (Oxford 'A')

9. A saturated solution of silver chromate, Ag_2CrO_4, has a concentration of 5×10^{-5} mol dm^{-3} at 20 °C. Outline an experiment by which you could verify this solubility, stating clearly the theoretical principles involved and *essential* practical details.

Using the data provided above, calculate the concentration of silver ions in a saturated solution of silver chromate at 20 °C. Similarly, calculate the concentration of silver ions in a saturated solution of silver chloride if the solubility product of silver chloride at 20 °C is 1.2×10^{-10} mol^2 dm^{-6}. Discuss, qualitatively, the relevance of your two answers to the use of silver chromate as an indicator of the end point in titrations of silver ions against chloride ions. (WJEC 'A')

10. What is an acid? What is meant by 'the strength of an acid'? Give an account of the principles underlying the methods by which the strengths of acids may be compared quantitatively. (London 'S')

11. (*a*) Outline the principles of the method you would use to measure the standard redox potential for the reaction

$$MnO_4^- + 8H^+ + 5e^- \rightarrow Mn^{2+} + 4H_2O$$

(*b*) The standard redox potentials for Ce^{4+}/Ce^{3+} (Ce = cerium) and Fe^{3+}/Fe^{2+} are 1.610 V and 0.771 V respectively. Deduce the direction of the reaction

$$Ce^{3+} + Fe^{3+} = Ce^{4+} + Fe^{2+}$$

and outline an experiment you could use to find the end-point when the reaction is carried out as a titration. (*N.B.*—Both Ce^{4+} and Fe^{3+} ions are yellow in aqueous solution.)

(*c*) What explanation can you offer for the fact that the standard electrode potentials of copper and zinc are +0.34 V and −0.76 V respectively, although the sums of the first two ionization energies for both metals are approximately 2640 kJ mol^{-1}? (Cambridge 'A')

12. Explain *concisely* what is meant by the terms *salt hydrolysis*, *common ion effect*, and *buffer solution*.

The dissociation constant of acetic acid is 1.8×10^{-5} mol dm^{-3} and the ionic product of water is 10^{-14} mol^2 dm^{-6}.

(*a*) Calculate the pH of 0.1 M acetic acid—you may assume the degree of dissociation is small.

(*b*) Calculate the pH of 0.1 M sodium acetate.

(*c*) Calculate the pH of a solution which is 0.1 M in sodium acetate and 0.1 M in acetic acid.

(*d*) What is the resulting pH when *one drop* of HCl (making negligible volume change) containing 10^{-3} mol of acid is added to 0.1 dm^3 of the above solution in (*c*)?

(Cambridge Entrance and Awards)

10. Rates of Reaction – Chemical Kinetics

An important aspect of a chemical reaction is obviously the rate at which it proceeds. Reactions have been studied that are so fast as to be virtually complete within 10^{-9} seconds; others are so slow as to occur only to a minimal extent even after many millions of years. In this chapter we shall be concerned with a wide range of methods that can be used to investigate the rates of reactions, what can be learned from such studies on rates about the mechanisms of reactions, and how such considerations are important in biochemistry and chemical engineering.

The catalytic decomposition of hydrogen peroxide

Hydrogen peroxide is sold by pharmacists in aqueous solution for use as an antiseptic. If such a solution is examined there is little evidence of its decomposition according to the equation:

$$2H_2O_2(aq) \rightarrow 2H_2O(l) + O_2(g)$$

Many substances are known, however, which greatly speed up the rate of this decomposition; manganese(IV) oxide is the most commonly em-employed. During the course of this reaction, the concentration of hydrogen peroxide remaining after various time intervals can be determined by withdrawing samples of the reaction mixture, adding this to a large excess of dilute sulphuric acid to 'freeze' the reaction, and titrating against a solution of potassium manganate(VII) (potassium permanganate). The reaction must be carried out using a buffer solution made by dissolving 3.1 g of boric(V) acid (orthoboric acid) in 25 cm^3 of 1 M sodium hydroxide and making up to 250 cm^3.

In such an experiment, 10 cm^3 of 6-volume hydrogen peroxide solution were pipetted into a conical flask containing 150 cm^3 of water and 50 cm^3 of the buffer solution and then clamped in a thermostat water bath at 20 °C. 10 cm^3 of 0.004 M potassium manganate(VII) solution were added which reacted with some of the hydrogen peroxide under the pH conditions present in the flask to generate the catalyst manganese(IV) oxide in finely divided form. The reaction was allowed to settle down to steady decomposition for five minutes when a 10 cm^3 sample was withdrawn and rapidly added to 50 cm^3 of dilute sulphuric acid in a second conical flask. This 'freezes' the reaction efficiently as the dilution involved substantially reduces the rate. The acidified sample was then titrated with 0.004 M potassium manganate(VII), the end-point being when the sample would no longer decolourize the purple manganate(VII) ions and the contents of the flask acquired a permanent pink colouration. Sampling and titration were repeated at approximately five minute intervals until the titre fell below 5 cm^3. The volume of potassium manganate(VII) solution used in each titre can be taken directly as a measure of the concentration of peroxide that remained at the time of sampling. The results (Table 10.1) are plotted in Figure 10.1.

Table 10.1 Concentration changes with time during the catalytic decomposition of hydrogen peroxide solution

Time/s	*$[H_2O_2]/cm^3$ 0.004 M manganate(VII)*
0	32.5
297	24.8
617	18.9
904	13.9
1215	10.5
1654	6.5
1939	4.5

It will be observed that rate of the decomposition, measured in terms of the gradient of the curve, decreases during the course of the experiment. This can be explained as due to the drop in concentration of hydrogen peroxide as reaction proceeds. A reasonable hypothesis is that the rate of the reaction is proportional to the concentration of hydrogen peroxide at that instant in time. We can test this hypothesis by taking gradients at a representative range of concentrations.

$$\text{Rate of reaction} = -\frac{d[H_2O_2]}{dt}$$

The results (Table 10.2) are plotted in Figure 10.2.

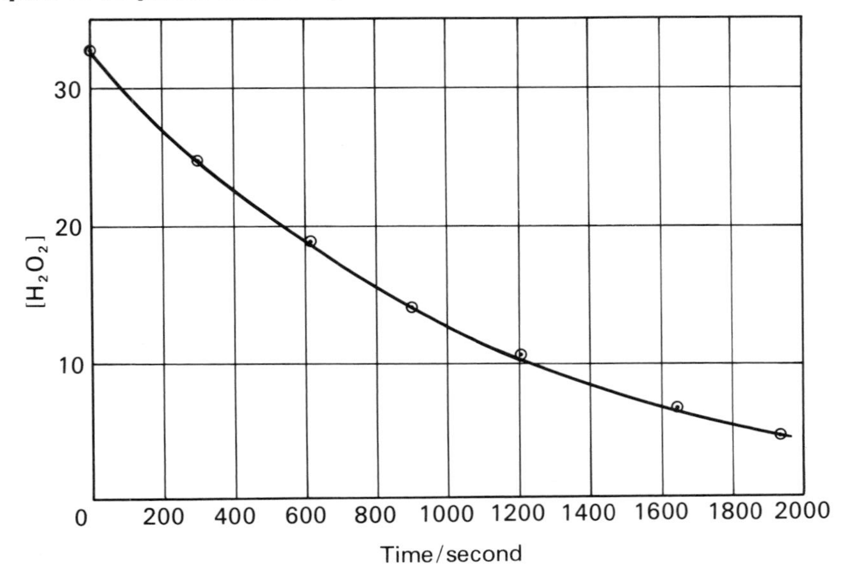

Figure 10.1 Concentration changes with time during the MnO_2 (s) catalysed decomposition of hydrogen peroxide solution

Figure 10.2 Effect of concentration on the rate of decomposition of hydrogen peroxide

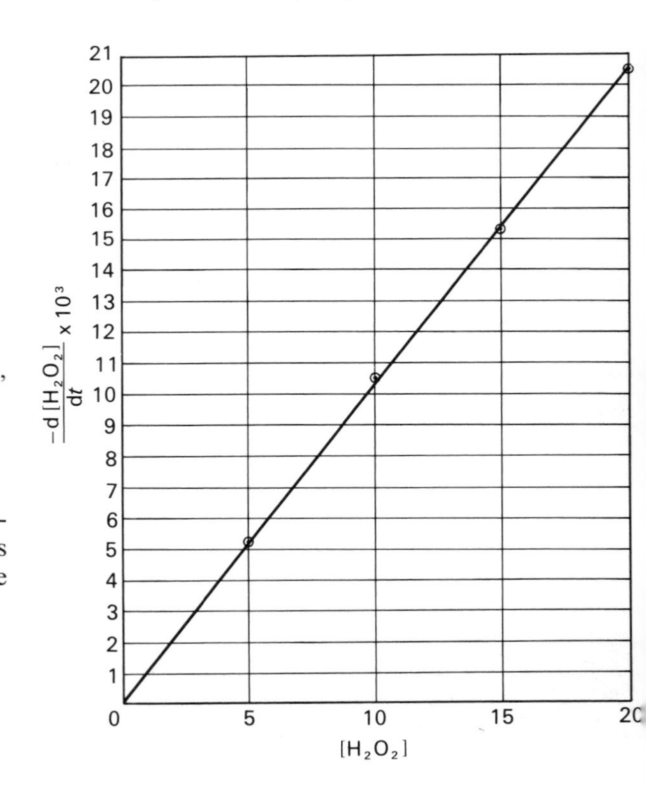

Table 10.2 Effect of concentration on rate of decomposition of hydrogen peroxide solution

$-\dfrac{d[H_2O_2]}{dt} \times 10^3$ (*rate of reaction*)	$[H_2O_2]$
20.6	20
15.4	15
10.5	10
5.2	5

The straight line, passing through the origin, justifies writing the relationship:

$$-\frac{d[H_2O_2]}{dt} = k[H_2O_2]$$

This is known as the *rate expression* for the reaction in which k, the proportionality constant, is called the *rate constant*. The rate expression can be rearranged in a form suitable for integration:

$$-\frac{d[H_2O_2]}{[H_2O_2]} = k\,dt$$

On integration this yields:

$$-\ln [H_2O_2] = kt + \text{constant}$$

The constant can be evaluated by considering the initial concentration of reactant, designated

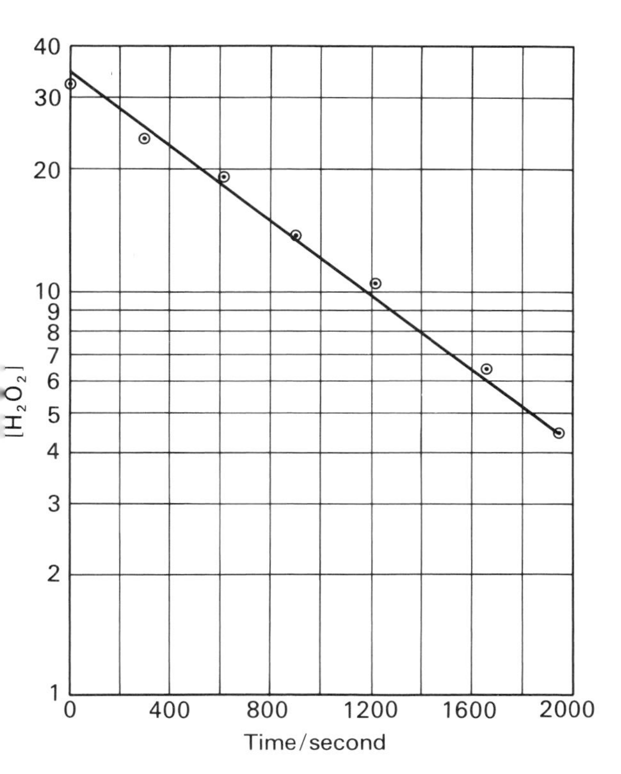

Figure 10.3 Logarithmic relationship between concentration and time for the catalytic decomposition of hydrogen peroxide solution

$[H_2O_2]_0$, which is present when $t = 0$. Thus $kt = 0$ and the constant is thus $-\ln [H_2O_2]_0$. Substituting this into the equation above yields:

$$-\ln [H_2O_2] = kt - \ln [H_2O_2]_0$$

which rearranges to:

$$\ln \frac{[H_2O_2]_0}{[H_2O_2]} = kt$$

This result can be confirmed by plotting the results in Table 10.1 with the concentration of hydrogen peroxide on a logarithmic scale, as shown in Figure 10.3.

An important relationship emerges when the time taken for half the original amount of hydrogen peroxide to react is considered. This time interval is known as the *half-life* of the reaction and given the symbol $t_{\frac{1}{2}}$.

$$\ln 2 = kt_{\frac{1}{2}}$$

$$t_{\frac{1}{2}} = \frac{\ln 2}{k} = \frac{0.693}{k}$$

It will be observed that for this reaction the time taken for half of any given quantity of reactant to disappear is independent of the initial quantity of reactant. From Figure 10.3 it can be deduced that the half-life for this reaction is 720 seconds. This value can be used to evaluate k, the rate constant:

$$k = \frac{0.693}{t_{\frac{1}{2}}} = \frac{0.693}{720\ \text{s}} = 9.6 \times 10^{-4}\ \text{s}^{-1}$$

Mechanism of the reaction

Proposing a mechanism for a reaction involves suggesting, at a molecular level, the route by which the reaction proceeds. Evidence concerning the validity of any proposed mechanism comes fundamentally from a study of the kinetics of the reaction concerned.

For the catalytic decomposition of hydrogen peroxide solutions discussed above, two possible mechanisms by which the reaction proceeds are:

(*a*) two molecules of hydrogen peroxide could collide to form a briefly existing double molecule which decomposes to give the products water and oxygen. The equation given below differs from the purely stoichiometric equation given previously. (*Stoichiometric* equations are book-keeping statements which relate the number of moles of reactants and products involved. *Mechanistic* equations describe the interaction between individual molecules and ions in the proposed mechanism of the reaction.)

$$2H_2O_2(aq) \rightarrow \underset{\text{(activated complex)}}{[H_4O_4(aq)]} \rightarrow 2H_2O(l) + O_2(g)$$

(*b*) A single molecule of hydrogen peroxide could undergo fission to give a water molecule plus an oxygen atom. A second reaction then occurs in which pairs of oxygen atoms combine to form molecules:

$$H_2O_2(aq) \rightarrow H_2O(l) + O(g)$$

then $$2O(g) \rightarrow O_2(g)$$

It is possible to decide between the two mechanisms by considering the results of the kinetics investigation. These show that the rate of the reaction is proportional to the concentration of hydrogen peroxide. If mechanism (*a*) is correct, the rate of the reaction would be proportional to the square of the hydrogen peroxide concentration, because doubling the concentration would quadruple the rate of collision between molecules. The known rate expression for the reaction is thus consistent with mechanism (*b*). The catalyst, however, plays a part in this—possibly the individual molecules undergoing fission are among those adsorbed on its surface.

The role of such catalysts is discussed in more detail later in this chapter.

Order of reaction

The catalytic decomposition of hydrogen peroxide is said to be a *1st-order reaction* as the rate is proportional to the concentration of hydrogen peroxide raised to the power of 1. The order of a reaction must always be related to the experimentally determined rate expression; it cannot be inferred from the stoichiometric equation. The following examples amplify the concept of order of reaction.

1. In aqueous solutions, chlorate(I) (hypochlorite) ions disproportionate into chlorate(V) (chlorate) and chloride ions. The stoichiometric equation is:

$$3ClO^-(aq) \rightarrow ClO_3^-(aq) + 2Cl^-(aq)$$

The experimentally determined rate expression is:

$$-\frac{d[ClO^-(aq)]}{dt} = k[ClO^-(aq)]^2$$

The reaction is said to be *2nd-order*.

2. Ethanal (acetaldehyde) decomposes in the vapour phase to methane and carbon monoxide. The rate expression is:

$$CH_3CHO(g) \rightarrow CH_4(g) + CO(g)$$

$$-\frac{d[CH_3CHO(g)]}{dt} = k[CH_3CHO(g)]^2$$

The reaction is 2nd-order.

3. Hydrogen peroxide solution oxidizes hydriodic acid to iodine.

$$H_2O_2(aq) + 2HI(aq) \rightarrow 2H_2O(l) + I_2(aq)$$

The rate expression is:

$$-\frac{d[H_2O_2(aq)]}{dt} = k[H_2O_2(aq)][HI(aq)]$$

The reaction is 1st-order with respect to hydrogen peroxide and 1st-order with respect to hydriodic acid. It can be described as 2nd-order overall.

4. Carbon monoxide combines with chlorine to give carbonyl chloride (phosgene) in the vapour phase.

$$CO(g) + Cl_2(g) \rightarrow COCl_2(g)$$

The rate expression is:

$$-\frac{d[CO(g)]}{dt} = k[CO(g)][Cl_2(g)]^{\frac{3}{2}}$$

Fractional orders are sometimes observed.

The hydrolysis of benzenediazonium ions

Benzenediazonium ions are formed by the reaction between phenylamine (aniline) and nitric(III) (nitrous) acid in aqueous solutions close to its freezing point. They undergo hydrolysis according to the equation:

$$⌬N_2^+(aq) + H_2O(l) \rightarrow$$
$$⌬OH(aq) + H^+(aq) + N_2(g)$$

The reaction can be conveniently monitored by collecting the nitrogen in a gas syringe and taking corresponding readings of volume and time. The volume of nitrogen is zero when the benzenediazonium ion is at its initial concentration, and at a maximum when the benzenediazonium ions have all reacted. This is said to be the infinity reading and is given the symbol V_∞. From this it can be seen that:

concentration of benzenediazonium ion at time $t \propto (V_\infty - V_t)$

where V_t is the volume of nitrogen at time t. This difference in volumes can therefore be used as a measure of the concentration of benzenediazonium ion.

In an experimental investigation of this reaction, using the apparatus shown in Figure 10.4, 0.78 g of phenylamine were dissolved in 2.5 cm^3 of concentrated hydrochloric acid and cooled by surrounding with ice. A similarly cooled solution of 0.57 g of sodium nitrate(III) (nitrite) in 10 cm^3 of water was gradually added and the solution made up to 100 cm^3 with water. This solution was placed in the apparatus and, after allowing 10 minutes for the reaction to reach steady conditions, gas was collected in the syringe and a series of volume and time readings were obtained. The results are shown graphically in Figures 10.5 and 10.6. The straight line obtained in the logarithmic plot shows the reaction to be 1st-order. (The integration of 2nd-order rate expressions does not lead to a logarithmic relationship between concentration and time.)

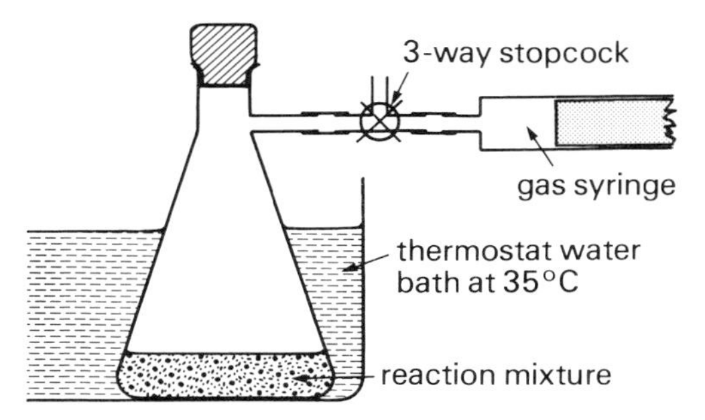

Figure 10.4 Apparatus to investigate the hydrolysis of benzenediazonium ions

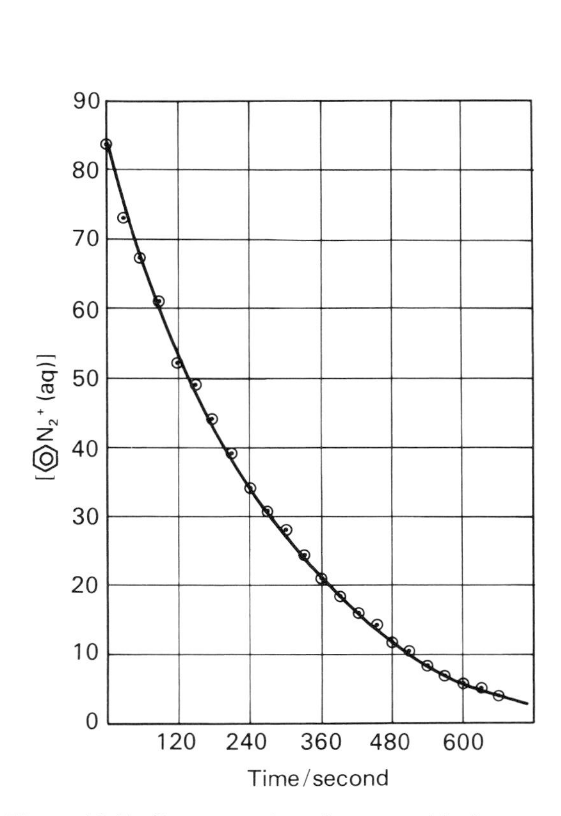

Figure 10.5 Concentration changes with time for the hydrolysis of benzenediazonium ions

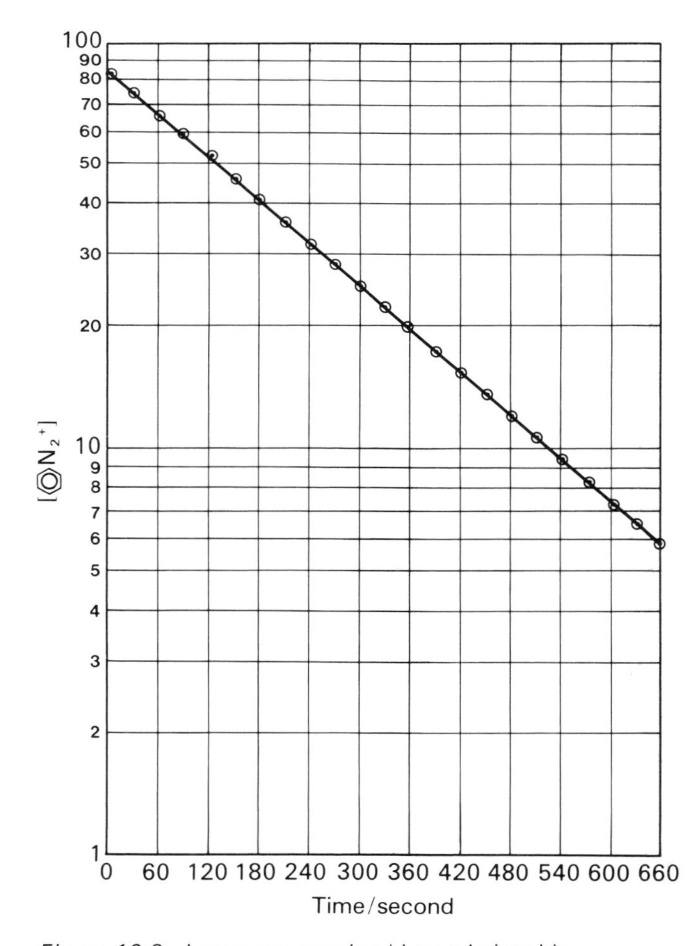

Figure 10.6 Log concentration/time relationship for the hydrolysis of benzenediazonium ions

Integration of 2nd-order rate expressions

Consider a reaction for which the rate expression is:

$$-\frac{d[A]}{dt} = k[A]^2$$

This can be rearranged to give:

$$-\frac{d[A]}{[A]^2} = kdt$$

On integration this yields:

$$\frac{1}{[A]} = kt + \text{constant}$$

When $t = 0$, it will be seen that the constant $= \frac{1}{[A]_0}$, where $[A]_0$ is the initial concentration of A. On substitution into the equation this yields:

$$\frac{1}{[A]} = kt + \frac{1}{[A]_0}$$

When $t = t_{\frac{1}{2}}$, $[A] = \frac{[A]_0}{2}$. Thus:

$$\frac{1}{[A]_0/2} = kt_{\frac{1}{2}} + \frac{1}{[A]_0}$$

$$t_{\frac{1}{2}} = \frac{1}{[A]_0 k}$$

This shows that no logarithmic relationship exists between concentration and time for reactions of this type, nor is the half-life independent of the initial concentration.

Techniques for monitoring reaction kinetics

The iodination of propanone (acetone)

Propanone reacts in aqueous solution with iodine according to the equation:

$$CH_3COCH_3(aq) + I_2(aq) \rightarrow CH_2ICOCH_3(aq) + H^+(aq) + I^-(aq)$$

The reaction can be investigated kinetically by making up suitable mixtures and timing how long it takes for the brown colour due to iodine to disappear as this reactant is used up. The reaction is catalysed by acid, and so the concentration of acid must be one of the variables investigated. Table 10.3 shows the results of such an experiment.

Inspection of the table shows the rate expression to be

$$-\frac{d[I_2(aq)]}{dt} = k[CH_3COCH_3][I_2]^0[H^+]$$

It is surprising to observe that the concentration of iodine does not affect the rate, when the concentration of hydrogen ions, which do not appear as reactants in the stoichiometric equation, is important. This is consistent with the following mechanism:

$$(CH_3)_2C{=}O + H^+ \xrightarrow{\text{fast}} (CH_3)_2COH^+ \xrightarrow[\text{(rate determining step)}]{\text{slow}} CH_2{=}C(OH)CH_3 + H^+$$

'enol' form

Table 10.3 Iodination of propanone

	Experiment 1	2	3	4
Volume of 2 M hydrochloric acid/cm³	5	2.5	5	5
Volume of 2 M propanone/cm³	2	2	1	2
Volume of water/cm³	0	2.5	1	0.5
Volume of 0.01 M iodine/cm³	1	1	1	0.5
Time for brown colour to be discharged/second	75	150	152	36
$-\frac{\Delta[I_2(aq)]}{\Delta} \times 10^5$ (rate)	1.7			

$$
\begin{array}{lllll}
 & CH_2 & CH_2I & & CH_2I \\
I\text{—}I\cdots\| & & | & & | \\
 & COH \xrightarrow{\text{fast}} & {}^{+}COH + I^- & \xrightarrow{\text{fast}} & C{=}O + H^+ + I^- \\
 & | & | & & | \\
 & CH_3 & CH_3 & & CH_3
\end{array}
$$

As iodine enters only after the slow rate-determining step, the rate of reaction is independent of its concentration. For a full account of this and similar reaction mechanisms, the student should consult a suitable textbook of organic chemistry.

Oxidation of copper metal by acidified dichromate(VI) solutions

A widely used method of monitoring the rate of a chemical reaction is to follow changes in the absorption spectrum of the reaction mixture caused by reactants being used up and products formed. Although this method has been most useful in the infra-red and ultra-violet portions of the spectrum, we shall discuss the oxidation of copper metal to copper ions by dichromate(VI) ions, which are reduced to chromium(III).

$$3Cu(s) + Cr_2O_7^{2-}(aq) + 14H^+(aq) \rightarrow 3Cu^{2+}(aq) + 2Cr^{3+}(aq) + 7H_2O(l)$$

This reaction can be followed in the visible region of the spectrum as the orange colour of dichromate(VI) changes to the blue-green colour of the mixture of copper(II) and chromium(III). Details of suitable instruments to follow this change are described in Chapter 3 which also includes the visible absorption spectra of copper(II) and chromium(III).

In such an experiment, a piece of copper foil of 4 cm^2 area was placed into 250 cm^3 of a solution which was 1 M with respect to sulphuric acid and M/60 with respect to potassium dichromate(VI). The mixture was stirred magnetically and small samples withdrawn after regular time intervals on which determinations of the visible absorption spectrum were made. The results are shown in Figure 10.7 It will be seen how the spectrum gradually changes to that expected for a mixture of copper(II) and chromium(III) ions.

The inversion of sucrose (cane-sugar)

In the presence of an acid as catalyst, sucrose in aqueous solution undergoes hydrolysis to glucose and fructose:

$$\underset{\text{sucrose}}{C_{12}H_{22}O_{11}(aq)} + H_2O(l) \rightarrow \underset{\text{glucose}}{C_6H_{12}O_6(aq)} + \underset{\text{fructose}}{C_6H_{12}O_6(aq)}$$

Sucrose, glucose, and fructose all exhibit optical activity (see Chapter 5). Sucrose is dextro-rotatory, while fructose rotates the plane of polarized light more powerfully to the left than glucose rotates it to the right. As the reaction proceeds, therefore, the initial angle of rotation to the right diminishes until, when the reaction is complete, the mixture is overall laevo-rotatory. It is from this inversion of the direction of rotation that the description of the reaction comes. The construction of a polarimeter suitable for monitoring the progress of the reaction is described in Chapter 5.

The reaction is 1st-order with respect to sucrose and 1st-order with respect to water; however, because the reaction is carried out in solutions in which the concentration of water is very large compared with the concentration of sucrose, the concentration of water stays virtually constant and the rate of reaction is found to depend solely on the sucrose concentration. The true rate expression is:

$$-\frac{d[C_{12}H_{22}O_{11}(aq)]}{dt} = k[C_{12}H_{22}O_{11}(aq)][H_2O(l)]$$

Under the experimental conditions, however, $[H_2O(l)]$ is virtually constant and the observed rate expression is:

$$-\frac{d[C_{12}H_{22}O_{11}(aq)]}{dt} = k'[C_{12}H_{22}O_{11}(aq)]$$

in which

$$k' = k[H_2O(l)]$$

In situations such as this where a higher order reaction appears to exhibit 1st-order kinetics because one reactant is present at a far higher concentration than the other, the reaction is said to be *pseudo 1st-order*.

If α_0 is the initial angle of rotation, α_t the angle of rotation at time t, and α_∞ the angle of rotation at the completion of the reaction (i.e. after infinite time), the initial concentration of sucrose, $[A]_0$, measured in terms of angles of rotation is $(\alpha_0 - \alpha_\infty)$ and the concentration at time t, $[A]$, in similar terms is $(\alpha_t - \alpha_\infty)$. The general form of the integrated rate equation for a 1st-order reaction, derived earlier when considering the decomposition of hydrogen peroxide, is:

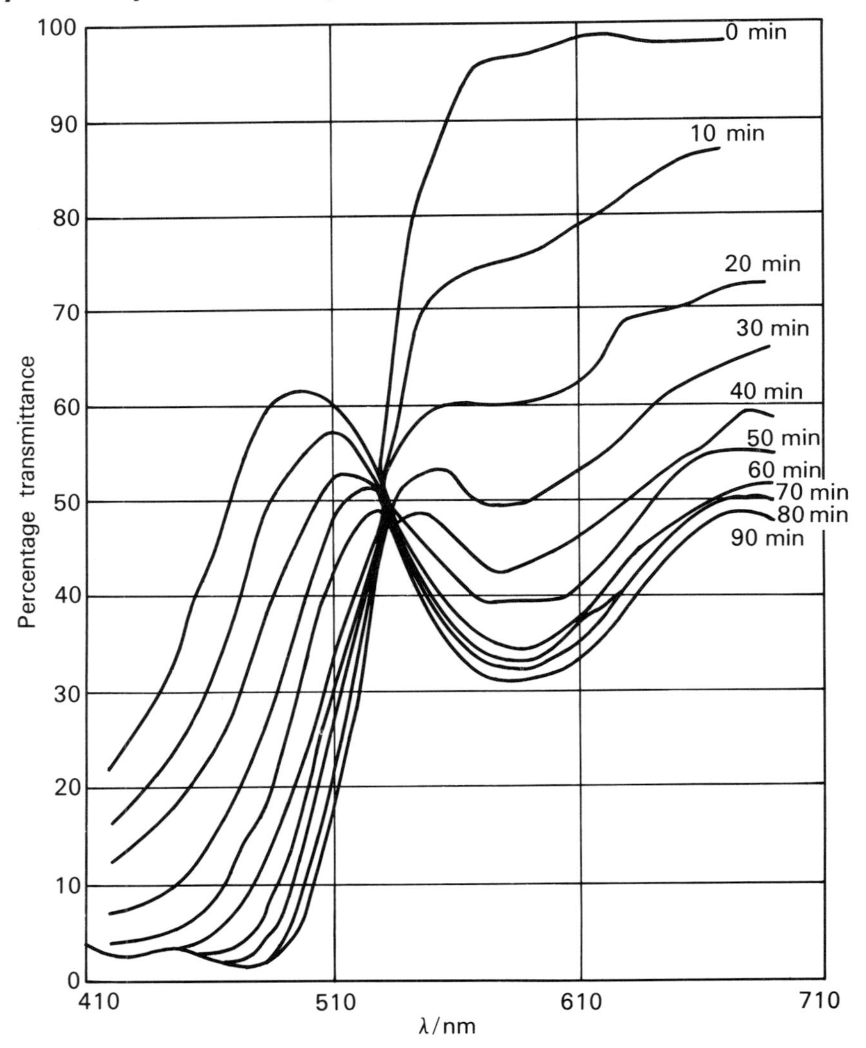

Figure 10.7 Changes in the visible absorption spectrum as copper is oxidized by acidified dichromate(VI) solution

$$\ln \frac{[A]_0}{[A]} = kt$$

This rearranges to:

$$k = \frac{1}{t} \ln \frac{[A]_0}{[A]} = \frac{1}{t} \ln \frac{(\alpha_0 - \alpha_\infty)}{(\alpha_t - \alpha_\infty)}$$

This relationship can be used to calculate values for the rate constant from data observed using a polarimeter. In one experiment, the results in Table 10.4 were obtained.

Table 10.4 Polarimetric data on the inversion of a 20% solution of sucrose in the presence of 0.5 M 2-hydroxypropanoic acid at 298 K

time/10⁵ sec	α	$k'/10^{-7}$ sec^{-1}
0	34° 5′	—
0.861	31° 1′	9.07
2.59	25° 0′	9.10
4.24	20° 2′	8.97
6.82	14° 0′	8.85
8.50	10° 0′	9.17
11.9	5° 1′	9.18
18.0	−1° 7′	8.90
∞	−10° 8′	—

The hydrolysis of bromobutane and its isomers

$$C_4H_9Br(aq) + H_2O(aq) \rightarrow C_4H_9OH(aq) + H^+(aq) + Br^-(aq)$$

As the progress of this reaction is accompanied by a large increase in the concentration of ions in the solution, the reaction is conveniently monitored by following changes in conductance with time. (Conductivity meters and electrodes are discussed in Chapter 8.)

In such an experiment, 150 cm^3 of a mixture of 80% ethanol and 20% water* were placed in a beaker and stirred magnetically. The conductance was determined and this value taken as a 'background' reading to be subtracted from later experimental values. 1 cm^3 of 2-bromo-2-methylpropane were added and a stop-clock started. The conductance was measured at suitable time intervals. The final value of conductance was taken after the assembly had been standing overnight; this is γ_∞. The concentration of 2-bromo-2-methylpropane at time t can therefore be taken as $\gamma_\infty - \gamma_t$. These measurements were made at 18 °C. The results are shown in Figure 10.8.

Inspection of Figure 10.8 shows that the reaction appears to obey 1st-order kinetics and has a half-life of 28.5 minutes. From this we can calculate the rate constant, k:

$$k = \frac{0.693}{28.5_{\text{ min}}} = 2.43 \times 10^{-2}\ \text{min}^{-1}$$

* The mixed solvent is necessary as the bromobutanes are not sufficiently soluble in water.

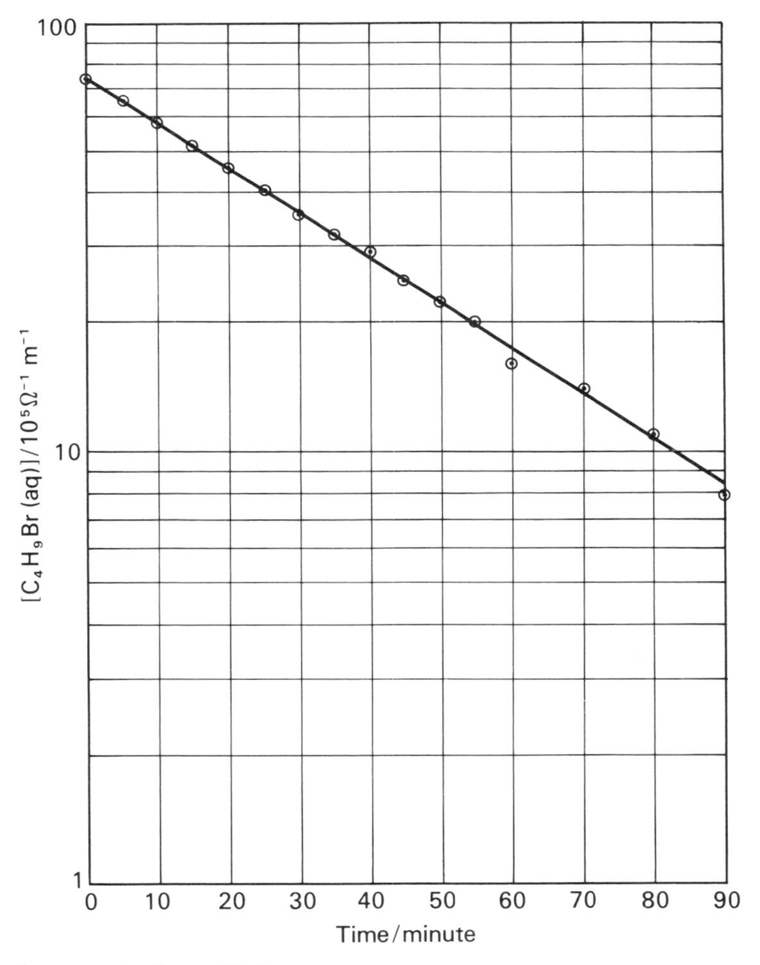

Figure 10.8 Change in concentration with time on the hydrolysis of 2-bromo-2-methylpropane

When the experiment is repeated using isomers of 2-bromo-2-methylpropane, considerable differences in the rates of hydrolysis, measured in terms of the rate constant, are observed. The fastest rate of hydrolysis is with 2-bromo-2-methylpropane and the slowest with 1-bromobutane. The next slowest is 2-bromobutane, followed by 1-bromo-2-methylpropane. In each case the reactions seem to follow a 1st-order law; however, it is possible that any or all could be pseudo 1st-order because of the excess of water present. Further experiments must be carried out if the situation is to be clarified.

Mechanism of the reactions Two possible mechanisms can be advanced:

(1) The reaction involves a single stage process whereby a molecule of C_4H_9Br is attacked by a hydroxide ion from the water:

$$C_4H_9Br + OH^- \rightarrow C_4H_9OH + Br^-$$

(2) There are two stages in the reaction; firstly C_4H_9Br ionizes:

$$C_4H_9Br \rightarrow C_4H_9{}^+ + Br^- \quad \text{(slow)}$$

A second step then occurs in which $C_4H_9{}^+$ combines with a hydroxide ion:

$$C_4H_9{}^+ + OH^- \rightarrow C_4H_9OH \quad \text{(fast)}$$

The kinetic information so far obtained is consistent with either of these interpretations. If mechanism (2) is the route by which the reaction occurs then the reaction will be zero-order with respect to hydroxide ion as this enters into the reaction after the slow (rate-determining) step; if the reaction proceeds by mechanism (1), however, it will be 1st-order with respect to hydroxide ion also. This can be investigated using solutions containing hydroxide ion of differing concentration as the hydrolysing medium. When this is done, the reaction is found to be zero-order with respect to hydroxide ion in the case of 2-bromo-2-methylpropane, and 1st-order in the case of 1-bromobutane. The two other isomers have fractional orders with respect to hydroxide ion which indicates that they proceed by a mixture of both possible mechanisms.

We can explain the choice of routes in terms of the structures of C_4H_9Br isomers.

(1)

$$\begin{array}{ccccc} CH_3 & & CH_3 & & CH_3 \\ | & & | & & | \\ CH_2 & & CH_2 & & CH_2 \\ | & \rightarrow & | & \rightarrow & | \\ CH_2 & & CH_2 & & CH_2 \\ | & & | & & | \\ HO^- + C(H)(H)Br & & HO\cdots C(H)(H)\cdots Br & & HO{-}C(H)(H) + Br^- \end{array}$$

This route is not possible in the case of 2-bromo-2-methylpropane as the carbon atom involved in attack by the hydroxide ion is shielded by the methyl group attached to it. This is best understood by building space-filling models of the two structures.

(2)

$$\begin{array}{ccccc} & CH_3 & & CH_3 & \\ & | & & | & \\ CH_3{-} & C{-}Br & \rightarrow CH_3{-} & C^+ & + \quad Br^- \\ & | & & | & \\ & CH_3 & & CH_3 & \end{array}$$

followed by:

$$\begin{array}{ccccc} & CH_3 & & CH_3 & \\ & | & & | & \\ CH_3{-} & C^+ & + OH^- \rightarrow CH_3{-} & C{-}OH & \\ & | & & | & \\ & CH_3 & & CH_3 & \end{array}$$

It may be wondered why this second, faster route is not taken in the case of 1-bromobutane. This is because the $C_4H_9{}^+$ ion that would have to be formed in this case would have a positive charge at its end which could not be so readily covered by pulling electrons from surrounding sections of the molecule as when the positive charge is more centrally situated.

Effect of solvent on reaction rate Solvents of high polarity promote reactions involving ionization by reducing the attractive forces between the ions of opposite charge. The hydrolysis of 2-bromo-2-methylpropane is thus less rapid in ethanol/water mixtures than it would be in pure water.

Molecularity

It is important to distinguish between the terms *order of reaction* (which essentially derives from a kinetic study of the reaction) and *molecularity*. The molecularity of a reaction is the number of molecules (or ions) that must be present for a reaction to take place at some step in the mechanism. A unimolecular reaction of the type:

$$A \rightarrow B$$

could occur if only one molecule of A was present Bimolecular reactions such as:

$$C + D \rightarrow E$$

and
$$2F \rightarrow G$$

can occur only when two molecules of the right species collide. The concept of molecularity is most usefully applied to the slow, rate-determining step of a complex reaction.

Theories of reaction rates

The collision theory

Many bimolecular reactions in the gaseous phase, such as:

$$A(g) + B(g) \rightarrow \text{products}$$

have rate constants (at any given temperature) of similar magnitude. The rates of such reactions are obviously related to the rate at which collisions occur between the reacting species. Calculations can be made of the total number of collisions occurring using the kinetic theory of gases. This can be compared with the number of effective collisions (i.e. those leading to product formation) obtained from measurements of the rate of reaction. These indicate that a very small percentage (in the order of 10^{-11} per cent) of all collisions are effective at 750 K. For many reactions of this type, a rise of temperature of 10 K approximately doubles the rate of reaction. Such a temperature rise, therefore, doubles the rate of *effective* collisions; however, the *total* number of collisions is increased to a very much smaller extent.

This effect can be explained if it is assumed that a reaction producing products takes place only when collisions occur between molecules having energy greater than a certain minimum value. This minimum energy, known as the *activation energy*, E_a, is the energy required to break bonds in the reactants before bonds in the products can be formed. The energy changes during the course of an exothermic and an endothermic reaction are shown in Figure 10.9. The distribution of molecular energies within samples of a gas at two temperatures, T_1 and T_2 are shown in Figure 10.10. (This distribution is discussed more fully in

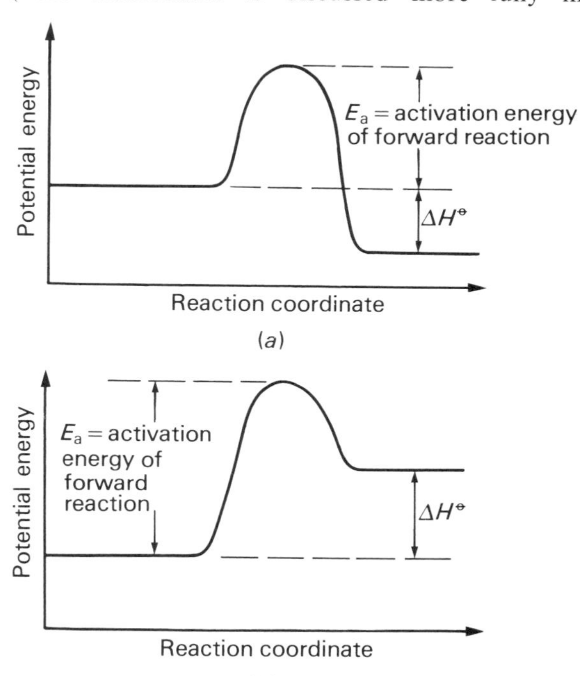

Figure 10.9 Energy profiles for (*a*) an exothermic and (*b*) and endothermic reaction

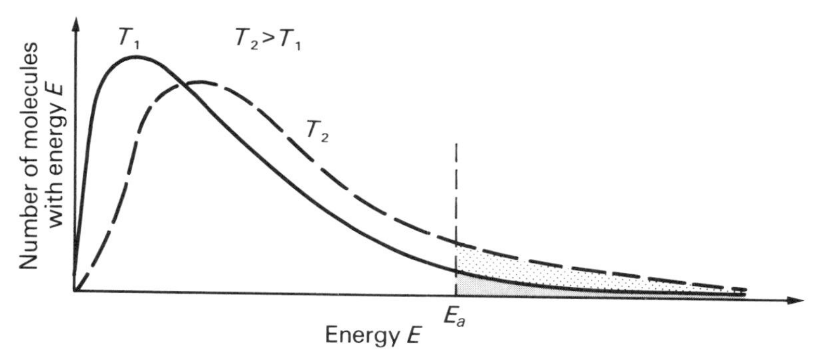

Figure 10.10 Distribution of molecular energies in a gas at two temperatures

Chapter 1.) The molecules having at least the activation energy, E_a, is proportional to the shaded area under the appropriate curve. Examination of this figure shows how it is possible for the number of molecules possessing activation energy to double for a moderate increase in temperature.

The transition state theory

In this theory attention is focused on the intermediate, known as the activated complex, that represents the state of the reacting molecules at the peak of the activation energy curve. This situation is depicted in Figure 10.11 for the hydrolysis of 1-bromobutane discussed earlier.

In some cases the energy profile for the reaction takes the form shown in Figure 10.12 and it is sometimes possible to isolate the intermediate in a metastable form.

A quantitative approach to activation energy

As discussed in Chapter 1, it can be shown that

$$\text{Fraction of molecules with energy } E \text{, or greater} = \exp(-E/RT)$$

Svante Arrhenius in 1889 realized that the rate of reaction, measured in terms of the magnitude of the rate constant, is proportional to the fraction of molecules having the activation energy or greater. He proposed the relationship, known as the *Arrhenius equation*:

$$k = A \exp(-E_a/RT)$$

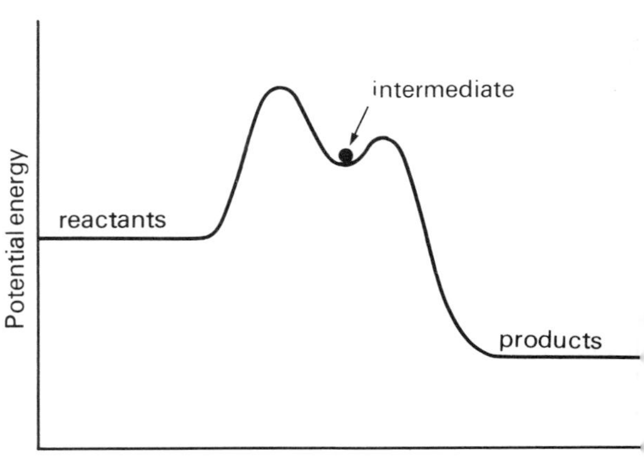

Figure 10.12 Energy profile for a reaction with a metastable intermediate

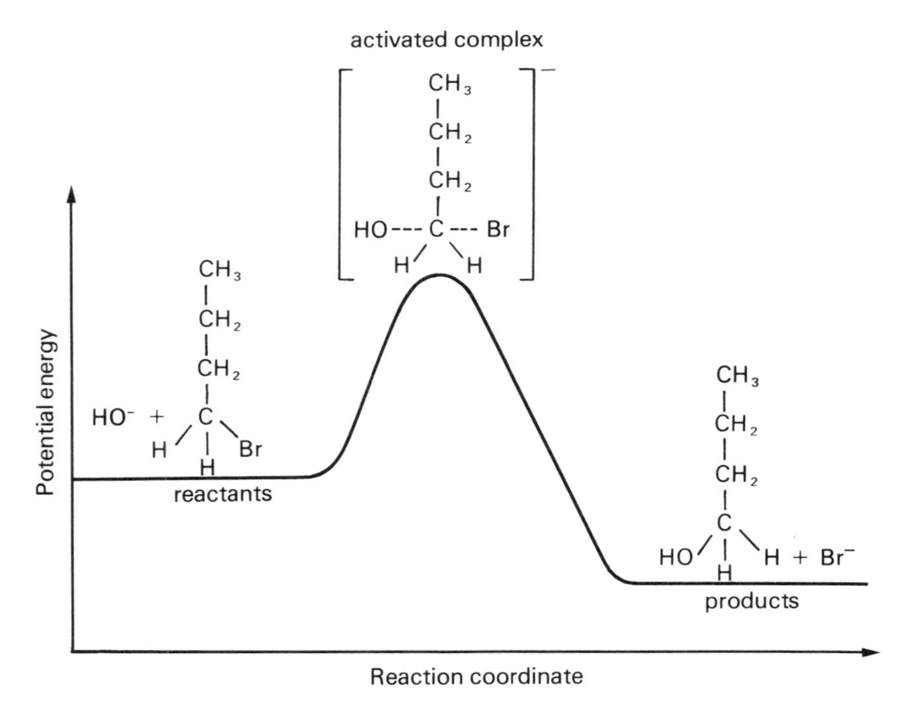

Figure 10.11 Transition state theory interpretation of the hydrolysis of 1-bromobutane

where A, the proportionality constant, is known as the collision factor. The collision factor can be regarded as made up of two factors, Z the chance of two reacting molecules meeting within unit time and p the so-called steric factor. The steric factor allows for the fact that although a collision may occur between two molecules possessing activation energy, the collision may not be effective because the geometry of the collision makes this impossible. The Arrhenius equation can be written in the form

$$k = pZ \exp(-E_a/RT)$$

Taking natural logarithms of this equation yields

$$\ln k = -\frac{E_a}{RT} + \ln pZ$$

The activation energy of a reaction and the collision factor can thus be determined if values for the rate constant are known at a series of temperatures.

The Harcourt–Esson reaction

In this reaction, hydrogen peroxide oxidizes iodide ions to iodine in acid solution.

$$H_2O_2(aq) + 2H^+(aq) + 2I^-(aq) \rightarrow 2H_2O(l) + I_2(aq)$$

The reaction is slow enough to be conveniently timed. This can be done by adding thiosulphate ions to the mixture to react with the iodine as it forms turning it back into iodide ions.

$$I_2(aq) + 2S_2O_3{}^{2}(aq) \rightarrow 2I^-(aq) + S_4O_6{}^{2-}(aq)$$

Starch is also added to the reaction mixture so that when all the thiosulphate has been used up, it reacts with the free iodine liberated and turns blue. This is an example of an 'iodine-clock' reaction. The time taken from mixing to the sudden appearance of a blue colour in the reaction mixture is thus the time required for the reaction to proceed to a specific extent. Three solutions were prepared and used to carry out such an experiment.

(1) Solution A: 5 g of potassium iodide were dissolved in 150 cm^3 of 2 M sulphuric acid and made up to 1 dm^3 with distilled water.

(2) Solution B: A '2-volume' solution of hydrogen peroxide was prepared by dilution of a stock solution.

(3) Solution C: 2 g of starch were made into a paste with water and added with stirring to 500 cm^3 of boiling water. This was cooled and to it was added 500 cm^3 of 0.1 M sodium thiosulphate.

25 cm^3 each of B and C were placed in a conical flask. 25 cm^3 of A was placed in a boiling tube. These were clamped in a thermostat bath at 20 °C until their contents reached thermal equilibrium. The boiling tube was emptied into the conical flask and the liquids well mixed; simultaneously a stop-clock was started. The time taken from mixing to the formation of a blue colour was noted. The experiment was repeated with the thermostat at 10 °C higher temperature intervals up to a maximum of 50 °C.

The results of the experiment are shown graphically in Figure 10.13. From the Arrhenius equation:

$$\ln k = \frac{-E_a}{RT} + \ln pZ$$

a graph of $\ln k$ against $1/T$ should give a straight line of gradient $-E_a/R$ and intercept $\ln pZ$. From Figure 10.13:

$$E_a = \text{gradient} \times R$$
$$= \frac{2.26}{0.30} \text{K}^{-1} \times 8.31 \times 10^3 \text{ J K mol}^{-1}$$
$$= 62.6 \text{ kJ mol}^{-1}$$

In this example it is more convenient to plot $\ln t$ than $\ln k$. This merely has the effect of changing the direction of the gradient but keeps its magnitude identical.

Table 10.5 Activation energies of selected reactions

Reaction	*Conditions*	*E_a/kJ mol^{-1}*
$2H_2O_2(aq) \rightarrow 2H_2O(l) + O_2(g)$	Uncatalysed	79
	MnO_2 catalysed	34
	Peroxidase catalysed	23
$2HI(g) \rightarrow H_2(g) + I_2(g)$	Uncatalysed	184
	Au catalysed	105
	Pt catalysed	59

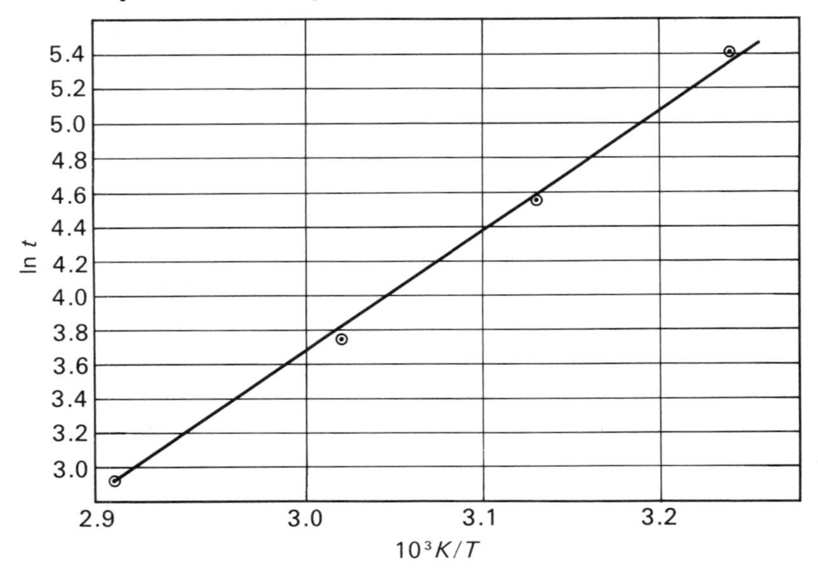

Figure 10.13 Arrhenius equation plot for the Harcourt–Esson reaction

Catalysis

Catalysts are substances which increase the rate of chemical reactions without being permanently consumed in the reaction. When the catalyst is in the same phase as the reactants it is described as *homogeneous catalysis*; when the reaction occurs at an interface between catalyst and reactants this is known as *heterogeneous catalysis*. Inspection of the data in Table 10.5 shows that catalysts work by providing an alternative pathway for the reaction with a lower activation energy. When a reactant is capable of undergoing two different competing reactions, selection of a particular catalyst can favour one of these at the expense of the other. Ethanol, for example, undergoes decomposition between 550 and 600 K to produce either ethene and water or ethanal (acetaldehyde) and hydrogen. The principal products with copper and aluminium oxide as heterogeneous catalysts are:

$$C_2H_5OH(g) \xrightarrow{Cu} CH_3CHO(g) + H_2(g)$$

$$C_2H_5OH(g) \xrightarrow{Al_2O_3} C_2H_4(g) + H_2O(g)$$

In the commercial production of sulphuric acid, sulphur dioxide is combined with oxygen to form sulphur trioxide, using vanadium(V) oxide as a catalyst. The overall reaction is:

$$2SO_2(g) + O_2(g) \rightarrow 2SO_3(g)$$

The reaction is extremely slow without the presence of a catalyst, and it is replaced by two faster reactions involving the vanadium(V) oxide:

$$SO_2(g) + V_2O_5(s) \rightarrow SO_3(g) + V_2O_4(s)$$
$$O_2(g) + 2V_2O_4(s) \rightarrow 2V_2O_5(s)$$

Compounds containing carbon-to-carbon double bonds are often reacted with hydrogen to form saturated compounds. An example of such a reaction is the production of ethane from ethene:

$$C_2H_4(g) + H_2(g) \rightarrow C_2H_6(g)$$

Such reactions can be carried out in the presence of finely divided nickel as a catalyst. Figure 10.14 shows the mechanism of this reaction.

I: as the hydrogen molecules approach the metal surface they are adsorbed, forming weak links with adjacent atoms.

II: as the bonding between the metal and hydrogen atoms strengthens, the bond between the hydrogen atoms weakens.

III: On the approach of the ethene molecule weak links are established between the adsorbed hydrogen atoms and the carbon atoms; there is a simultaneous weakening of the carbon-to-carbon double bond.

IV: the bond between the metal atoms and the hydrogen is broken as the new carbon–hydrogen bonds strengthen and the product molecule separates from the catalyst surface.

An important commercial example of such a catalytic hydrogenation reaction is the production of margarine from vegetable oils.

In Figure 10.15 the delivery of catalyst to a fixed bed reactor on a 'sweetening' unit at an oil refinery is shown. These units remove corrosive and evil-smelling substances from jet fuels, petrols and kerosene. These undesirable compounds are oxidized with controlled quantities of air in the presence of a copper(II) chloride catalyst.

The role of hydrogen ions as an homogeneous catalyst in the iodination of propanone has been discussed earlier in this chapter.

Most biochemical reactions are controlled by catalysts known as enzymes. This is discussed later in this chapter.

Autocatalysis

Some reactions are known which produce their own catalyst. An example of such autocatalysis occurs in the oxidation of ethanedioc (oxalic) acid in acid conditions by manganate(VII) ions:

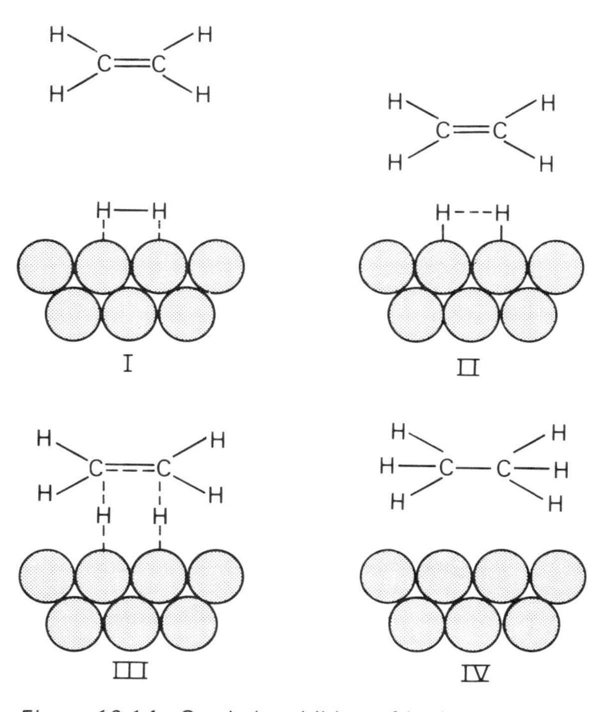

Figure 10.14 Catalytic addition of hydrogen across the double bond in ethene

Figure 10.15 Delivering catalyst to a 'sweetening' unit at an oil refinery

$$2MnO_4^-(aq) + 16H^+(aq) + 5C_2O_4^{2-}(aq) \rightarrow 2Mn^{2+}(aq) + 8H_2O(l) + 10CO_2(g)$$

The only specie involved that is significantly coloured is the manganate(VII) ion and so the reaction can be conveniently monitored by allowing it to take place in a tube contained in a colorimeter. The working of colorimeters is described in Chapter 3.

In such an experiment a colorimeter tube filled with water was placed in the colorimeter and the instrument adjusted to give maximum meter reading (m_0). 1 cm^3 of 0.02 M potassium manganate(VII) solution was placed in a matching tube and 10 cm^3 of 2 M sulphuric acid added. As there is no ethanedioc acid present, no reaction occurs and the colorimeter reading with this tube can be used to obtain the zero time reading. In a third matching tube, 1 cm^3 of 0.02 M potassium manganate(VII) solution was placed and a solution 1.2 M with respect to ethanedioc acid and 2 M with respect to sulphuric acid was quickly added and a stop-clock started simultaneously. This was speedily placed in the colorimeter and readings taken at suitable time intervals.

The concentration of manganate(VII) ions at any time is related to the meter reading at that time, m_t, by the relationship:

$$\lg \frac{m_o}{m_t} \propto [MnO_4^-(aq)]$$

The results are shown in Figure 10.16. It will be seen that after the reaction has proceeded for some time, there is an unexpected acceleration of the rate. This is because the reaction is catalysed by the manganese(II) ions as they are formed.

Light promoted reactions

An equimolar mixture of hydrogen and chlorine reacts extremely slowly at room temperature to give hydrogen chloride:

$$H_2(g) + Cl_2(g) \rightarrow 2HCl(g)$$

The reaction is accelerated and occurs with explosive violence when the mixture is subjected to a burst of light rich in ultra-violet radiation. Nernst suggested that this results from a chain reaction involving single atoms, known as *free radicals*. A free radical is an atom, or group of atoms, containing an odd number of electrons. The reaction is initiated by the absorption of a quantum of light of suitable frequency by a chlorine molecule, resulting in dissociation:

$$Cl_2(g) \xrightarrow{h\nu} Cl\cdot(g)$$

A propagation step then occurs, many millions of times over for each quantum initially absorbed:

$$Cl\cdot(g) + H_2(g) \rightarrow HCl(g) + H\cdot(g)$$
$$H\cdot(g) + Cl_2(g) \rightarrow HCl(g) + Cl\cdot(g)$$

This is eventually terminated when two chlorine atoms react to form a molecule:

$$2Cl\cdot(g) \rightarrow Cl_2(g)$$

An important industrial example of such a reaction is that between chlorine and methane to produce a variety of chlorosubstituted methanes. For details of this and similar processes, a suitable textbook of organic chemistry should be consulted.

Some advanced methods of measuring reaction rates

Gas phase reactions

In 1926 C. N. Hinshelwood and T. E. Green used an apparatus of the type shown in Figure 10.17 to investigate the reaction between hydrogen and nitrogen oxide at elevated temperatures. It is necessary that the glass tubing connections are as fine as possible to ensure that the dead-space is

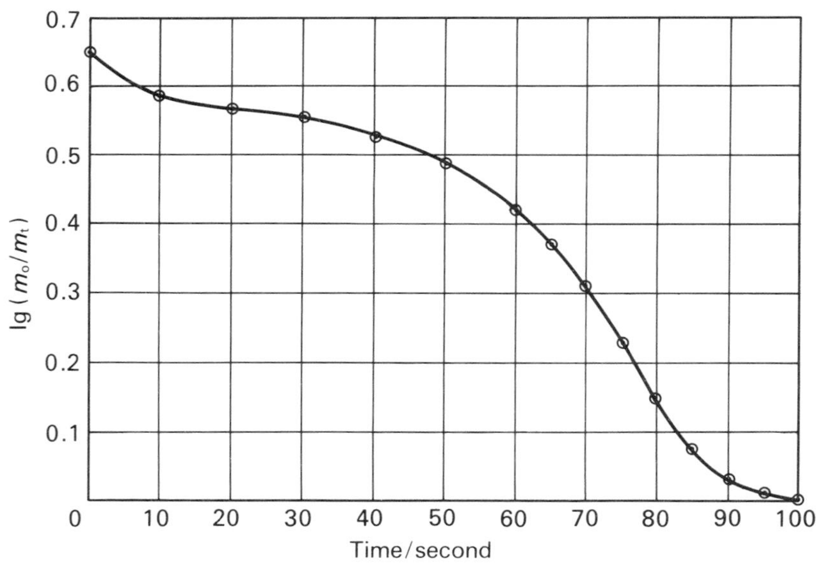

Figure 10.16 Change in concentration of manganate(VII) ions with time during its reaction with acidified ethanedioic acid

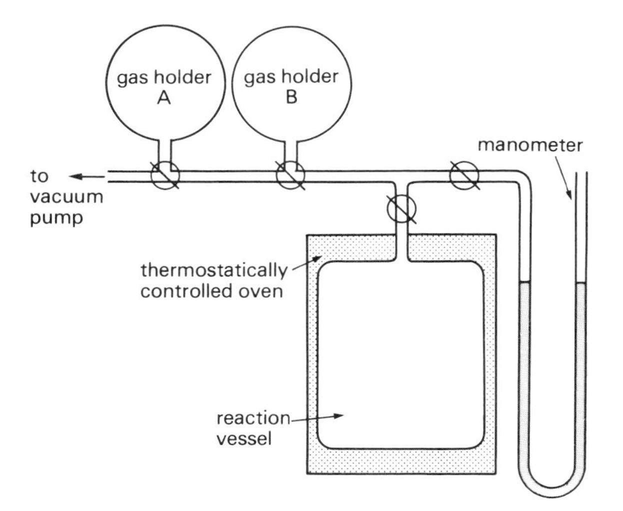

Figure 10.17 Apparatus for investigating the kinetics of a reaction between two gases at elevated temperatures

small compared with that of the reaction vessel, the reaction then proceeds close to conditions of constant volume. The apparatus is completely evacuated and one reactant admitted to gas holder A. Its pressure is measured via the manometer. The procedure is repeated for the other reactant admitted into gas holder B. The two reactants are simultaneously admitted, using the pressure gradient, to the reaction vessel. From the geometry of the apparatus, their initial partial pressures can be deduced. The partial pressure of each gas is proportional to its concentration:

$$p_X = \frac{n}{V} RT = [X]\,RT$$

The reaction occurring is

$$2NO(g) + 2H_2(g) \rightarrow N_2(g) + 2H_2O(g)$$

As the reaction proceeds, there is a reduction in the total number of moles of gas and thus, at constant volume, a total drop in pressure is registered on the manometer. From this and the initial partial pressures and, knowing the stoichiometry of the reaction, the change in partial pressures with time can be calculated. By taking the appropriate gradient of the graph of partial pressure against time, the initial rate of reaction can be obtained for each run of the experiment (Table 10.6).

It will be observed that in runs 1–3 in Table 10.6, there is the same initial partial pressure of hydrogen and in runs 4–6 the same initial partial pressure of nitrogen oxide.

Table 10.6 Kinetic data for the reaction between hydrogen and nitrogen oxide at 1100 K

Run	*Partial pressure/mmHg*		*Initial rate of reaction/10^{-2} mmHg s^{-1}*
	Hydrogen	*Nitrogen oxide*	
1	400	152	25
2	400	300	103
3	400	359	150
4	300	400	174
5	205	400	110
6	147	400	79

The rate expression will be:

$$\text{rate} = k\,[NO(g)]^m\,[H_2(g)]^n.$$

In runs 1–3 $[H_2(g)]^n$ is constant, therefore:

$$\text{rate} = k'\,[NO(g)]^m$$

On taking logs,

$$\lg \text{rate} = m \lg [NO(g)] + \lg k'$$

Thus a graph of lg rate against lg [NO(g)] should take the form of a straight line of gradient m and intercept k'. This plot is shown in Figure 10.18.

$$\text{gradient} = m = \frac{0.50}{0.24} \approx 2$$

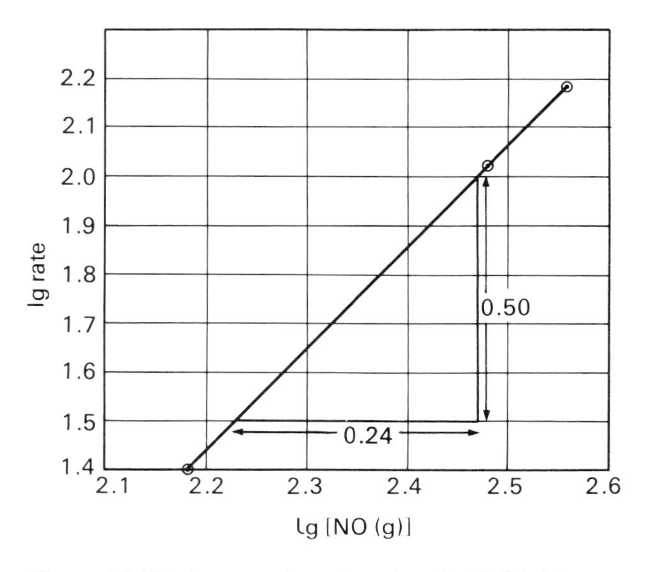

Figure 10.18 Lg rate plotted against lg [NO(g)] for the reaction between hydrogen and nitrogen oxide

In runs 4–6, $[NO(g)]^m$ is constant, therefore

$$\text{rate} = k'' [H_2(g)]^n$$

On taking logs,

$$\lg \text{rate} = n \lg [H_2(g)] + \lg k''$$

A plot of lg rate against lg $[H_2(g)]$ yields a straight line of unit slope. The rate expression is thus seen to be:

$$\text{rate} = k [NO(g)]^2 [H_2(g)]$$

The continuous flow method

Hartridge and Roughton in 1923 devised a method of rapidly mixing reacting solutions by forcing them under pressure from separate containers into a mixing chamber. From this they flowed along an observation tube at one point on which the absorption of light by a species formed in the reaction could be detected via a photocell. Their apparatus is shown in Figure 10.19. This technique enabled them to study the kinetics of some very fast reactions with half-lives as short as 1 millisecond. An example of a reaction that can be investigated in this way is that between solutions containing iron(III) and thiocyanate ions:

$$Fe^{3+}(aq) + NCS^-(aq) \rightarrow FeNCS^{2+}(aq)$$

The $FeNCS^{2+}(aq)$ ion is a deep blood-red colour. The flow-rate can be varied by altering the applied pressure and so can the position of observation along the tube.

Relaxation methods

This method of investigating fast reactions was first used by Manfred Eigen in 1954. In this technique a system in equilibrium is rapidly disturbed by altering of one of the external conditions such as temperature, pressure or electrical field strength. The progress of the system as it relaxes into the new conditions can then be followed by absorption spectrophotometry. In one of the most widely used versions of this technique, known as the temperature jump method, a temperature rise of a few degrees is achieved within microseconds by discharging a high voltage condenser through the solution. An example of a reaction studied in this way is the formation of a complex ion between magnesium(II) and adenosine-5′-triphosphate (ATP^{4-}):

$$Mg^{2+}(aq) + ATP^{4-}(aq) \rightarrow MgATP^{2-}(aq)$$

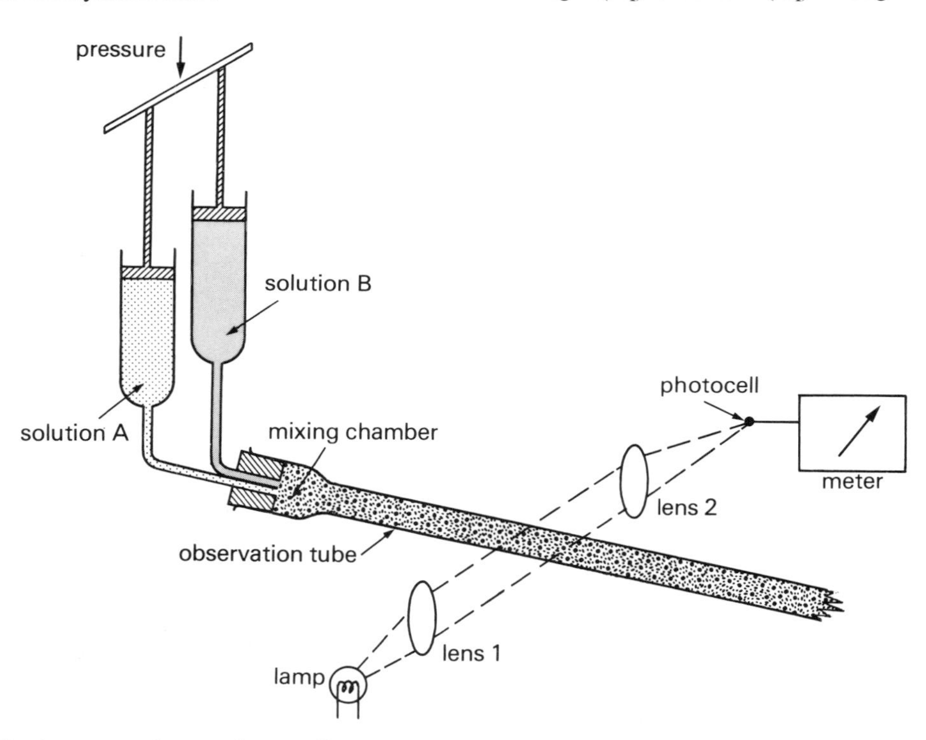

Figure 10.19 Apparatus for continuous flow investigations

Flash photolysis

The apparatus for this technique, introduced by Porter and Norrish in 1953, is shown in Figure 10.20. The trigger unit is used to fire a photoflash. Light from this initiates free radical formation in the reaction mixture held in the reaction vessel. The photocell is also triggered by the flash which activates the delay unit which, after a predetermined interval, causes a second flash from the spectroscopic flash unit to pass through the reaction vessel. The absorption spectrum from this second flash by species in the reaction vessel is observed on the spectrograph. A series of flashes at fixed intervals is then sent through the reaction vessel and in each case the absorption spectrum is recorded. Figure 10.21 shows the absorption spectrum at various times after flash photolysis of a mixture of oxygen and chlorine. From the change in intensity with time in various parts of the spectrum, the decay of the species responsible for that absorption can be monitored. In our example the decaying species is the ClO· free radical.

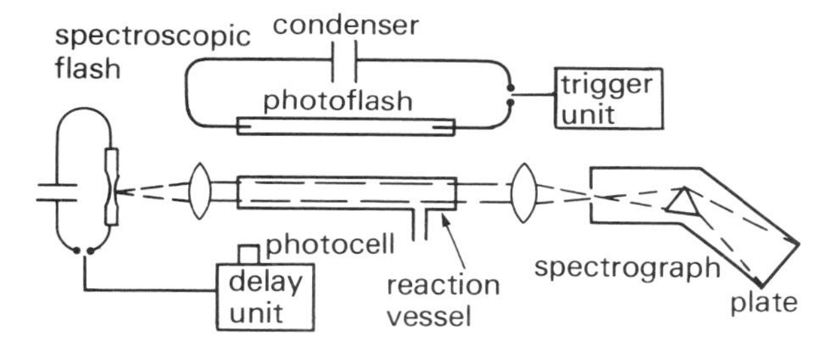

Figure 10.20 Apparatus for flash photolysis

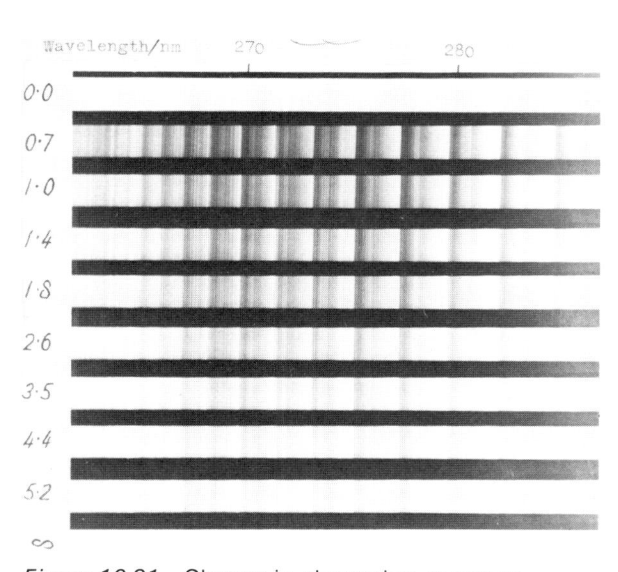

Figure 10.21 Change in absorption spectrum with time of the ClO radical after flash photolysis of an oxygen/chlorine mixture

Enzymes

Enzymes are catalysts for the biochemical reactions that occur in plants and animals. Enzyme-controlled reactions do not obey the Arrhenius rate equation. This is because they are proteinaceous in character and, with rising temperature, the structure of the protein is steadily broken down until the enzyme is completely inactive above 60 °C. Enzymes operate with maximum efficiency at 37 °C, the normal body temperature of warm-blooded animals.

The Michaelis–Menten theory of enzyme kinetics

This theory is based on the following postulates.

(*a*) The enzyme and the material involved in the biochemical process, known as the substrate, react to form a 'complex' by a reaction which is 1st-order with respect to both enzyme and substrate. The complex then either dissociates to reform enzyme and substrate, or gives products:

$$E + S \underset{k_2}{\overset{k_1}{\rightleftharpoons}} ES \xrightarrow{k_3} E + \text{products}$$

Let $[E]$ be the *total* enzyme concentration;
$[S]$ be the substrate concentration;
$[ES]$ be the complex concentration.
thus $[E] - [ES]$ is the concentration of *free* enzyme.

(*b*) When the rate of reaction is at a maximum, there is virtually no free enzyme and thus $[E] = [ES]$.

(*c*) The rate of the reaction between enzyme and substrate to form the complex is very much more rapid than the dissociation of the complex to give enzyme and products.

From (*a*) the rate of formation of the complex is:

$$\frac{\mathrm{d}[ES]}{\mathrm{d}t} = k_1 \{[E] - [ES]\} \times [S]$$

and the rate of dissociation of the complex is:

$$-\frac{\mathrm{d}[ES]}{\mathrm{d}t} = k_2 [ES] + k_3 [ES]$$

In the steady state, the rate of formation of the complex equals the rate at which it dissociates, therefore

$$k_1 \{[E] - [ES]\} \times [S] = k_2 [ES] + k_3 [ES]$$

This rearranges to give:

$$\frac{\{[E] - [ES]\} \times [S]}{[ES]} = \frac{k_2 + k_3}{k_1} = K_m$$

The constant, K_m, defined in this way is known as Michaelis' constant. Thus

$$K_m = \frac{[\text{free enzyme}]\,[\text{substrate}]}{[\text{complex}]}$$

Michaelis' constant can thus be seen to be an equilibrium constant, the magnitude of which is a measure of the affinity of the substrate for the enzyme.

Further rearrangement yields:

$$[ES] = \frac{[E]\,[S]}{[S] + K_m} \qquad (1)$$

The overall velocity, v, from postulates (*a*) and (*c*) must be

$$v = k_3 [ES] \qquad (2)$$

It follows from postulate (*b*) that when the velocity of the reaction is maximum, v_{max}, $[E] = [ES]$

$$v_{max} = k_3 [E] \qquad (3)$$

Combining equations (1) and (2) yields:

$$v = \frac{k_3 [E]\,[S]}{[S] + K_m}$$

Combining this with equation (3) yields

$$v = \frac{v_{max} [S]}{[S] + K_m}$$

This rearranges to give:

$$\frac{1}{v} = \frac{K_m}{v_{max}} \left\{\frac{1}{[S]}\right\} + \frac{1}{v_{max}} \qquad (4)$$

which takes the form y = mx + c. Thus if a plot is made of $1/v$ against $1/[S]$, a straight line graph is obtained of gradient K_m/v_{max} and of intercept $1/v_{max}$. If the values of these constants are evaluated, they can be combined to give K_m.

An alternative method of deducing K_m utilizes the fact that it equals the substrate concentration which gives half the maximum velocity.

When $v = v_{max}/2$, let $[S] = [S]^*$

which on substitution into (4), yields:

$$\frac{2}{v_{max}} = \frac{K_m}{v_{max}} \times \frac{1}{[S]^*} + \frac{1}{v_{max}}$$

This simplifies to

$$K_m = [S]^*$$

Kinetics in chemical engineering

The work of the chemical engineer involves scaling up processes that have been tested on a laboratory scale so that they can be carried out in industrial sized plant. One type of reactor used for carrying out reactions in chemical factories is the continuous flow reactor. In these the reactants are fed in continuously, stirred within the reactor and products and unchanged reactants continually run out. It generally pays to allow less than full conversion of reactants into products because after a certain percentage conversion, the additional time the material is held within the reactor is not compensated for by the better yield.

Consider a 1st-order reaction of the type

$$A \rightarrow \text{products}$$

A will be fed into a continuous reactor in which perfect mixing occurs. If the volume of the reactor is V dm^3 and the volumetric flow rate is u dm^3 min^{-1}, then the mean retention time, t, of the material in the reactor is:

$$t = \frac{V}{u}$$

For an initial concentration $[A]_0$ of A entering the reactor, the moles of A entering in time t is $[A]_0ut$. Similarly if $[A]$ is the concentration of A leaving the reactor, the moles of A exiting in time t is $[A]ut$. That is,

moles reacting in time t = rate × $Vt = k[A]Vt$

or

moles entering the reactor	=	moles leaving the reactor	+	moles which have reacted
$[A]_0ut$	=	$[A]ut$	+	$k[A]Vt$

Dividing by ut yields:

$$[A]_0 = [A] + k[A]\frac{V}{u}$$

This rearranges to

$$\frac{V}{u} = \frac{[A]_0 - [A]}{k[A]} = t$$

This design equation can be used to calculate the efficiency of conversion of A into products within a reactor of known volume under a variety of conditions of volumetric flow rate, thus allowing the optimum conditions to be deduced.

Problems

1. Dinitrogen pentoxide decomposes in solution in tetrachloromethane according to the equation:

$$2N_2O_5(CCl_4) \rightarrow 2N_2O_4(CCl_4) + O_2(g)$$

The following data was obtained for the reaction at 45 °C:

Time/s	0	184	319	526
$[N_2O_5(CCl_4)]$	2.33	2.08	1.9	1.67
Time/s	867	1198	1877	
$[N_2O_5(CCl_4)]$	1.36	1.11	0.72	

Determine (*a*) the order of the reaction and (*b*) the rate constant.

2. A solution of hydrogen peroxide was decomposed in the presence of the enzyme peroxidase at 20 °C. At suitable time intervals, samples of the reaction mixture were withdrawn, added to excess dilute sulphuric acid and titrated against a solution of potassium manganate(VII). The results obtained were:

Time/min	0	5	10
Vol. $MnO_4^-(aq)/cm^3$	46.1	37.1	29.8
Time/min	20	30	50
Vol. $MnO_4^-(aq)/cm_3$	19.6	12.3	5.0

Show that the reaction is 1st-order and determine (*a*) the half-life and (*b*) the rate constant. Make a sketch graph showing how the value of k will vary for this reaction between 20 and 60 °C.

3. In an experiment to investigate the oxidation of bromide ions by bromate(V) ions in acid solution, the following mixtures were made and their relative rates established:

Experiment	*Volume/cm³* 1 M $BrO_3^-(aq)$	1 M $Br^-(aq)$	1 M $H^+(aq)$	$H_2O(l)$	*Relative rate*
A	5	25	30	40	1
B	5	25	60	10	4
C	10	25	30	35	4
D	5	50	30	15	2

The stoichiometry of the reaction is:

$$BrO_3^-(aq) + 5Br^-(aq) + 6H^+(aq) \rightarrow 3Br_2(aq) + 3H_2O(l)$$

The rate expression is:

$$\text{rate} = k\,[BrO_3^-(aq)]^x\,[Br^-(aq)]^y\,[H^+(aq)]^z.$$

What values of x, y, and z are consistent with the kinetics data?

4. For the reaction (occurring in 80 per cent ethanol, 20 per cent water):

$$(CH_2)_6\langle^{Cl}_{CH_3} + H_2O \rightarrow (CH_2)_6\langle^{OH}_{CH_3} + H^+ + Cl^-$$

The rate constants at various temperatures are:

Temperature/°C	0	25	35	45
$k/10^{-4}\ s^{-1}$	0.106	3.19	9.86	29.2

Calculate the activation energy of the reaction.

5. Account for four of the following observations.
(*a*) When a gaseous mixture of hydrogen and deuterium is kept over a nickel surface molecules of HD are formed.
(*b*) A mixture of 2 volumes of hydrogen and 1 volume of oxygen explodes when ignited by means of a spark.
(*c*) The time taken for a 1st-order reaction to be half completed is independent of the initial concentration of the reactant.
(*d*) The order of a reaction cannot be inferred from the stoichiometric equation for the reaction.
(*e*) When a quartz bulb containing nitrogen dioxide (NO_2) is exposed to ultra-violet light the colour fades and the pressure in the bulb rises. When the light is removed the dark brown colour returns and the pressure falls to its original value.

(Oxford and Cambridge 'S')

6. Hydrogen peroxide and iodide ions react in acid solution to form iodine and water. The rate of the reaction is given by:

$$\text{Rate} = k[H_2O_2]^a[I^-]^b$$

Describe, without giving practical details, how you would determine *a* and *b*, given sulphuric acid, aqueous potassium iodide and hydrogen peroxide of suitable concentration, a solution containing sodium thiosulphate and starch and a stopwatch.

(Associated Examining Board 'S')

7. The following results are those from an experiment in which equal volumes of equimolar solutions of a bromoalkane of formula C_4H_9Br and of potassium hydroxide were mixed and then 20 cm^3 samples taken at intervals, the reaction quenched in an excess of ice-cold water, and titrated against a standard acid solution.

Time/10^3 sec	0	0.45	0.9	1.8	2.7	3.6
Titre/cm^3 acid	20	11.5	8.0	5.0	3.55	2.8

Time/10^3 sec	4.5	5.4	6.3	7.2	8.1
Titre/cm^3 acid	2.35	2.0	1.75	1.55	1.4

(*a*) Use these results to find the overall order of the reaction.

(*b*) From your result in (*a*), deduce the most probable mechanism for the hydrolysis, explaining your deduction.

(*c*) What is the most probable formula for the bromoalkane? Explain your reasoning carefully. (Nuffield 'S')

8. Write an essay on the catalysis of reactions. Your answer should include an explanation of how catalysts work. (Nuffield 'A')

9. The rate of the reaction

$$2A + B = C + D$$

is given by the equation

$$\text{rate} = k\,[A]^2\,[B]$$

where k is the rate constant.

State (*a*) the units in which the rate constant would be expressed if concentrations are given in mol dm^{-3}, (*b*) the effect on the rate when the concentrations of A and B are simultaneously doubled.

Give a graphical representation of the distribution of velocities of gas molecules at different temperatures and, in the light of this graph, discuss the effect of temperature on reaction rate.

Distinguish between *homogeneous* and *heterogeneous* catalysis. Illustrate your answer by reference to a suitable reaction in each case.

(WJEC 'S')

10. Define the term *rate of reaction* and describe the factors that affect it.

Distinguish between *molecularity* and *reaction order*.

The pyrolysis of acetaldehyde can be represented by the stoichiometric equation

$$CH_3CHO \rightarrow CH_4 + CO$$

The following pressures were measured as a function of time in a reaction vessel containing acetaldehyde vapour at 800 K.

Time/second	0	105	242	480	840	1440
Pressure/kPa	48.4	58.2	66.2	74.2	80.9	86.2

Find the order of and the rate constant for the reaction, clearly stating the units of the latter quantity. (Cambridge Entrance and Awards)

11. Chemical Thermodynamics

The First Law

Up to and during the eighteenth century the distinction between matter and energy was not fully understood. In his *Traité des substances élémentaires*, published in 1789, Antoine Laurent Lavoisier included heat and light in his table of elements. Until well into the nineteenth century many scientists regarded heat as a fluid, known as 'caloric'. Hot bodies were thought to contain high levels of caloric and when they were connected to a colder body by a good thermal conductor, the fluid was thought to flow from the hotter to the colder. In 1798, while engaged in cannon boring at the Munich Arsenal, Benjamin Thompson, who later became Count Rumford, was impressed by the large quantity of heat produced. He was able to show, within the limits of experimental accuracy, that the same quantity of heat always resulted from the same expenditure of mechanical energy. James Joule, in a series of experiments started in 1840, showed that the expenditure of given quantities of mechanical or electrical energy resulted in the production of an equivalent quantity of heat. Such experiments established the modern view of heat as a form of energy and gave rise to the *First Law of Thermodynamics* which states: *energy can neither be created nor destroyed*. When energy in one form is converted into another, the quantity of energy produced is exactly equivalent to that which disappears. This law is discussed in detail in Chapter 4. In SI this equivalence of various energy forms is implicit in the use of one unit, the *joule*, for all types of energy.

The study of thermodynamics involves the quantitative relationships that are observed when heat and other forms of energy are interconverted. It developed substantially in the second half of the nineteenth century as scientists and engineers strove to understand the factors governing the efficiencies of steam and other heat engines. It soon became apparent that the principles involved were of far wider application than this. Indeed there is probably no branch of science on which they do not have some bearing. The historical approach to chemical thermodynamics via the study of heat engines is very time consuming and involves mathematics with which you are unlikely to be familiar. It serves our present purpose best, therefore, to take some short cuts to arrive at useful ideas to help us in our present chemical studies.

Spontaneous and non-spontaneous processes

A *spontaneous process* is one that occurs without energy from outside the changing system being supplied. (*N.B.*—The word *spontaneous* should not be confused with *instantaneous*.) A stone falls spontaneously under gravity, an ideal gas diffuses spontaneously into a vacuum and a mixture of two moles of hydrogen and one of oxygen combine spontaneously to form water. The reverse of these processes does not take place spontaneously, energy from outside must be supplied if they are to be reversed. Processes are generally spontaneous in one direction but non-spontaneous in the other. What criteria can be applied to predict the direction of spontaneous change? Consideration of mechanical systems, such as the falling stone, suggests that such systems change spontaneously in the direction that leads to them having minimum potential energy. A more complex situation is shown in Figure 11.1 which shows a rectangular block of wood changing from metastable to stable equilibrium. During this process the centre of gravity rises, with consequent increase in potential energy, until it passes the vertical and from then onwards the potential energy decreases, finishing at a lower level than previously. The potential energy changes are shown in Figure 11.2. The difference in potential energy between A and C is analogous to the activation energies of chemical reactions discussed in Chapter 10. Is it possible to find a similar energy criterion to predict the direction of chemical changes? The equation for the reaction between hydrogen and oxygen is

$$2H_2(g) + O_2(g) \rightarrow 2H_2O(l); \quad \Delta H^{\ominus} = -572 \text{ kJ mol}^{-1}$$

Because of its high activation energy, this is an extremely slow reaction at room temperature. With a platinum catalyst, however, it rapidly proceeds to virtual completion in the forward direction. It does not proceed spontaneously in the reverse direction but can be made to do so by the

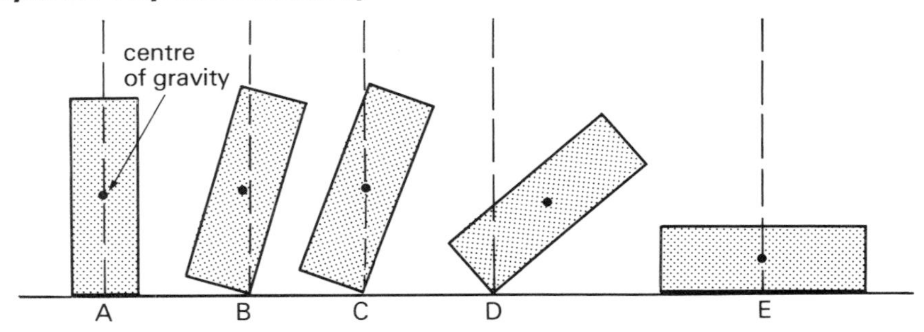

Figure 11.1 A rectangular block of wood changing from metastable to stable equilibrium

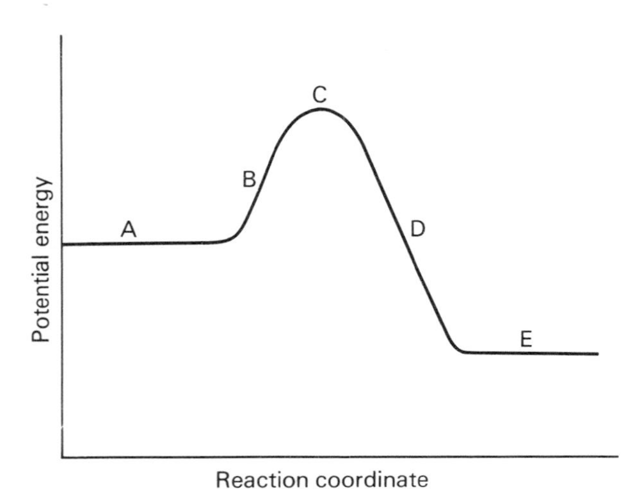

Figure 11.2 Changes in its potential energy as a rectangular wooden block changes from metastable to stable equilibrium

injection of energy from outside as in electrolysis. Is the change to lower enthalpy (i.e. a negative value for $\Delta H^{\ominus}$) the energy criterion we are seeking? Many reactions are known which are spontaneous *and* endothermic of which the two described below are spectacular examples.

1. About 25 cm^3 of thionyl chloride are stirred in a beaker with a thermometer. Measures of iron(III) chloride 6-water are added and a highly endothermic reaction occurs:

$$6SOCl_2(l) + FeCl_3 . 6H_2O(s) \rightarrow FeCl_3(s) + 6SO_2(g) + 12HCl(g)$$

$$\Delta H^{\ominus} = +1271 \text{ kJ mol}^{-1}$$

2. A beaker is placed on a sheet of glass with a small drop of water between their surfaces. 10 g of barium hydroxide 8-water and 5 g of ammonium thiocyanate are added and stirred together. An endothermic reaction takes place which freezes the beaker to the glass sheet:

$$Ba(OH)_2 . 8H_2O(s) + 2NH_4NCS(s) \rightarrow Ba(NCS)_2(aq) + 2NH_3(g) + 10H_2O(l)$$

$$\Delta H^{\ominus} = +786 \text{ kJ mol}^{-1}$$

The sign of the enthalpy change is therefore not a reliable indicator of the direction of spontaneous chemical change. In Chapter 9, the use of standard electrode potentials ($E^{\ominus}$) to predict the direction of chemical change is discussed. This, of course, can only be applied in the case of redox reactions. $E^{\ominus}$ values are not energy changes but they can be combined with another factor to give an energy change.

Intensity and capacity factors

An intensity factor is independent of the mass of the system for which it is quoted. A capacity factor depends on the mass of the system. Some examples are shown in Table 11.1.

Table 11.1 Intensity and capacity factors

Intensity factor	*Capacity factor*	*Energy*
Temperature	Thermal capacity	Heat
Voltage	Charge	Electrical
Velocity^2	Mass/2	Kinetic

It will be seen that energies are the products of the appropriate intensity and capacity factors. To calculate the electrical energy generated when a redox reaction occurs, E for the cell must be multiplied by the charge in coulombs that passes. For the cell discussed in Chapter 9:

$$Cu(s) \mid Cu^{2+}(aq) \vdots Ag^{+}(aq) \mid Ag(s)$$
$$E^{\ominus} = +0.46 \text{ V}$$

Application of the *ABCD* rule predicts that the reaction will proceed spontaneously in the direction:

$$Cu(s) + 2Ag^{+}(aq) \rightarrow Cu^{2+}(aq) + 2Ag(s)$$

When the reaction occurs in the mole quantities specified in the equation, the charge transferred is two faradays.

$$\text{Energy transferred} = 0.46 \times 2 \times 96\,500 \text{ J mol}^{-1} = 88 \text{ kJ mol}^{-1}$$

The electrical energy change for such a reaction, carried out under the standard conditions described, is known as the *Standard Gibbs free energy** change and given the symbol $\Delta G^{\ominus}$. For redox reactions $\Delta G^{\ominus}$ is related to the standard electrode potential by the equation:

$$\Delta G^{\ominus} = -zFE^{\ominus}$$

(z and F have their usual meanings)

It is necessary to introduce the minus sign to reconcile the convention regarding cell notation with the fact that as a cell operates it *loses* energy to the surroundings. The determination of values of $\Delta G^{\ominus}$ for reactions that cannot be carried out in electrolytic cells is discussed later in this chapter.

* Named after Willard Gibbs (1839–1903), a celebrated American Physical Chemist.

Gibbs free energy and chemical equilibrium

We have seen that for a chemical reaction to take place spontaneously, there must be a negative value of ΔG in the direction of change. Figure 11.3 gives the relationship between composition of the reaction mixture and free energy for three reactions of the type:

$$A \rightarrow B$$

The following observations can be made about these reactions.

I: This reaction goes virtually to completion, the free energy has a minimum value when there is almost 100% of B.

II: This reaction occurs only to a negligible extent, the minimum value of free energy occurs when hardly any of A has reacted.

III: This is an equilibrium situation, the minimum value of free energy occurs when appreciable quantities of both A and B are present.

A direct quantitative relationship exists between the Standard Gibbs free energy change for a reaction and its equilibrium constant. Consider the Daniell cell reaction:

$$Zn(s) + Cu^{2+}(aq) \rightarrow Zn^{2+}(aq) + Cu(s)$$

The corresponding cell diagram is:

$$Zn(s) \mid Zn^{2+}(aq) \vdots Cu^{2+}(aq) \mid Cu(s)$$

Applying the Nernst equation (Chapter 9) to each electrode gives:

Figure 11.3 Gibbs Free Energy and composition for three reactions of the type A ⟶ B

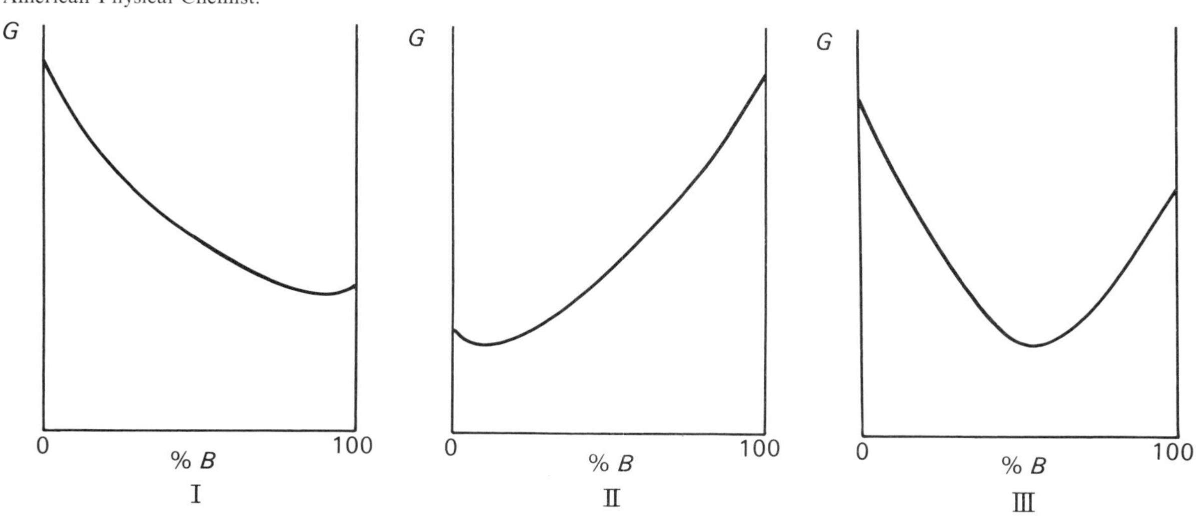

$$E_{Cu} = E^{\ominus}_{Cu} + \frac{RT}{zF} \ln [Cu^{2+}(aq)]$$

and

$$E_{Zn} = E^{\ominus}_{Zn} + \frac{RT}{zF} \ln [Zn^{2+}(aq)]$$

Subtracting the equation for the zinc electrode from that for copper gives the following equation relating to the cell overall:

$$E_{cell} = E_{Cu} - E_{Zn}$$
$$= E^{\ominus}_{Cu} - E^{\ominus}_{Zn} + \frac{RT}{zF} \ln \frac{[Cu^{2+}(aq)]}{[Zn^{2+}(aq)]}$$

or

$$E_{cell} = E^{\ominus}_{cell} + \frac{RT}{zF} \ln \frac{[Cu^{2+}(aq)]}{[Zn^{2+}(aq)]}$$

When the two electrodes of the cell have concentrations of $Cu^{2+}(aq)$ and $Zn^{2+}(aq)$ that are in equilibrium, then E_{cell} is zero, hence:

$$-E^{\ominus}_{cell} = \frac{RT}{zF} \ln \frac{[Cu^{2+}(aq)]_{eqm}}{[Zn^{2+}(aq)]_{eqm}}$$
$$-zFE^{\ominus}_{cell} = RT \ln \frac{1}{K_c}$$
$$\Delta G^{\ominus} = -RT \ln K_c$$

This relationship holds true for reactions in solution. For gaseous equilibria the equivalent relationship is $\Delta G^{\ominus} = -RT \ln K_p$. This equation shows that as all reactions have a $\Delta G^{\ominus}$ value, all reactions should be regarded as equilibria. However, reactions having values of $\Delta G^{\ominus}$ greater than $+40$ kJ mol^{-1} have such small values for their equilibrium constant that they can be regarded as occurring only to a negligible extent. Similarly reactions having $\Delta G^{\ominus}$ values more negative than 40 kJ mol^{-1} have such large equilibrium constants that they can be regarded as going virtually to completion.

Equilibrium constants from cell potentials

$$Fe^{2+}(aq) + Ag^{+}(aq) \rightleftharpoons Fe^{3+}(aq) + Ag(s)$$

The cell corresponding to the equilibrium above is:

$$Pt \mid [Fe^{2+}(aq), Fe^{3+}(aq)] \vdots Ag^{+}(aq) \mid Ag(s)$$

for which $E^{\ominus}_{cell}$, determined from Table 9.1 is $+0.06$ V.

$$\Delta G^{\ominus} = -zFE^{\ominus} = -RT \ln K_c$$
$$\ln K_c = \frac{zFE^{\ominus}}{RT}$$

$$\ln K_c = \frac{1 \times 96\,500 \times 0.06}{8.3 \times 298}$$
$$= 2.34$$
$$K_c = \ln^{-1} 2.34 = 10.4$$

This value can be verified by a wide range of additional experimental evidence.

(*a*) Equilibrium mixtures of known initial concentrations can be analysed volumetrically, either by determining the $Fe^{2+}(aq)$ with manganate(VII) ions in acid conditions or by determining $Ag^{+}(aq)$ with thiocyanate ions.
(*b*) Using a suitable radioactive isotope of iron or silver, reaction mixtures can be made up in which the concentration of the ion so 'tagged' can be determined by radiochemical counting procedures.
(*c*) A series of cells of the type shown in Figure 11.4 can be constructed and their potentials measured. A graph of cell potential against lg $[Ag^{+}(aq)]$ is plotted and from this the concentration of silver ions in equilibrium with the iron(II)/iron(III) solution can be deduced.

$$K_c = \frac{[Fe^{3+}(aq)]_{eqm}}{[Fe^{2+}(aq)]_{eqm}[Ag^{+}(aq)]_{eqm}}$$

when

$$[Fe^{3+}(aq)]_{eqm} = [Fe^{2+}(aq)]_{eqm}$$
$$K_c = \frac{1}{[Ag^{+}(aq)]_{eqm}}$$

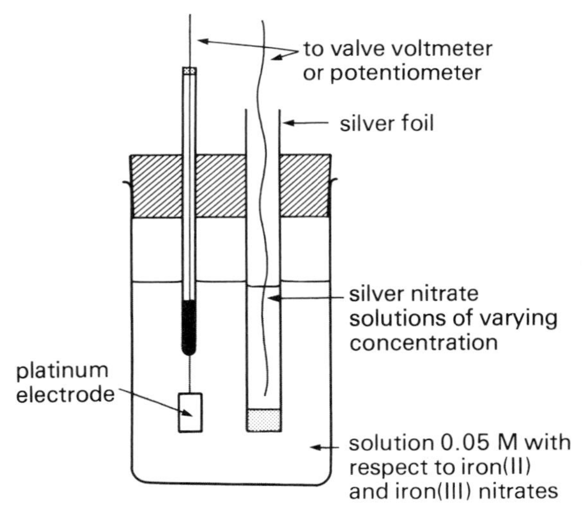

Figure 11.4 Electrochemical determination of an equilibrium constant

N.B.—In all these methods it is essential that the silver and iron ions used are balanced by the nitrate ions. This is because other anions replace water molecules as ligands around the metal ions and change the potential. Because iron(II) and nitrate is unstable, such solutions must be prepared immediately prior to use.

The extraction of metals

Very few metals occur in the earth's crust in the uncombined state. Metals always exhibit positive oxidation numbers in their compounds and their extraction therefore involves reduction. Carbon is a particularly economical reducing agent to use for this purpose.

Figure 11.5 shows the Gibbs free energy changes for some reactions involving oxide formation at a variety of temperatures. Below about 1000 K, carbon dioxide is the stable oxide of carbon, above this temperature carbon monoxide is stable. We can use this data to investigate the feasibility of reducing a metal oxide to the metal at various temperatures. The reduction of zinc oxide will be used as an example (*a*) at 1000 K and (*b*) at 1500 K.

(*a*) *at* 1000 *K*:

$$C + O_2 \rightarrow CO_2; \quad \Delta G^{\ominus} = -390 \text{ kJ mol}^{-1} \quad (1)$$

$$2Zn + O_2 \rightarrow 2ZnO; \quad \Delta G^{\ominus} = -490 \text{ kJ mol}^{-1} \quad (2)$$

Subtracting (2) from (1) and rearranging gives:

$$2ZnO + C \rightarrow 2Zn + CO_2; \quad \Delta G^{\ominus} = +100 \text{ kJ mol}^{-1}$$

The reaction is not feasible at this temperature.

(*b*) *at* 1500 *K*:

$$2C + O_2 \rightarrow 2CO; \quad \Delta G^{\ominus} = -470 \text{ kJ mol}^{-1} \quad (3)$$

$$2Zn + O_2 \rightarrow 2ZnO; \quad \Delta G^{\ominus} = -300 \text{ kJ mol}^{-1} \quad (4)$$

Subtracting (4) from (3) and rearranging gives:

$$2ZnO + 2C \rightarrow 2Zn + 2CO; \quad \Delta G^{\ominus} = -170 \text{ kJ mol}^{-1}$$

At this temperature the reduction of zinc oxide with carbon is thermodynamically feasible.

Similar reasoning can be applied in the case of other metals. In each case this involves subtracting $\Delta G^{\ominus}$ for the reaction involving the formation of the metal oxide from $\Delta G^{\ominus}$ for the formation of the appropriate oxide of carbon. $\Delta G^{\ominus}$ for the reduction process is therefore zero at the point of intersection of the lines that correspond to these changes. Examination of the data in Figure 11.5, known after its originator as an Ellingham diagram, reveals that uneconomically high temperatures would be required to extract such reactive metals as calcium and magnesium from their oxides using carbon. Such metals are normally extracted electrolytically.

Fuel Cells

Fuel cells are devices that involve the oxidation of substances used as conventional fuels at the anode of an electrolytic cell. Consider the following data for the oxidation of hydrogen:

$$2H_2(g) + O_2(g) \rightarrow 2H_2O(l); \quad \Delta H^{\ominus} = -572 \text{ kJ mol}^{-1} \quad \Delta G^{\ominus} = -474 \text{ kJ mol}^{-1}$$

It will be seen that if the reaction could be carried out in an electrolytic cell, about 80 per cent of the total energy would be available directly as electricity. Using hydrogen as a conventional fuel in a heat engine and using this to drive a dynamo would realize only about 15 per cent of the total energy as electricity. An additional advantage of the fuel cell is that waste products are not exhausted into the atmosphere to add to the pollution menace.

The first fuel cell was described by Sir William Grove in 1839. His apparatus is shown in Figure 11.6 in which there are four hydrogen/oxygen fuel cells connected in series. Grove found that each cell gave about 1 volt. As discussed in Chapter 9, the voltage of electrolytic cells deteriorates as a greater current is withdrawn from them. Fuel cells suffer particularly in this way as Grove was the first to observe. The reactions at the electrodes are:

Oxygen electrode: $O_2(g) + 2H_2O(l) + 4e^- \rightarrow 4OH^-(aq)$

Hydrogen electrode: $2H_2(g) + 4OH^-(aq) \rightarrow 4H_2O(l) + 4e^-$

A fuel cell suitable for laboratory investigations is shown in Figure 11.7. A fuel cell, using methanol and capable of producing 4 kW, is shown operating an electric hammer in Figure 11.8. Research is still in progress to develop commercial fuel cells

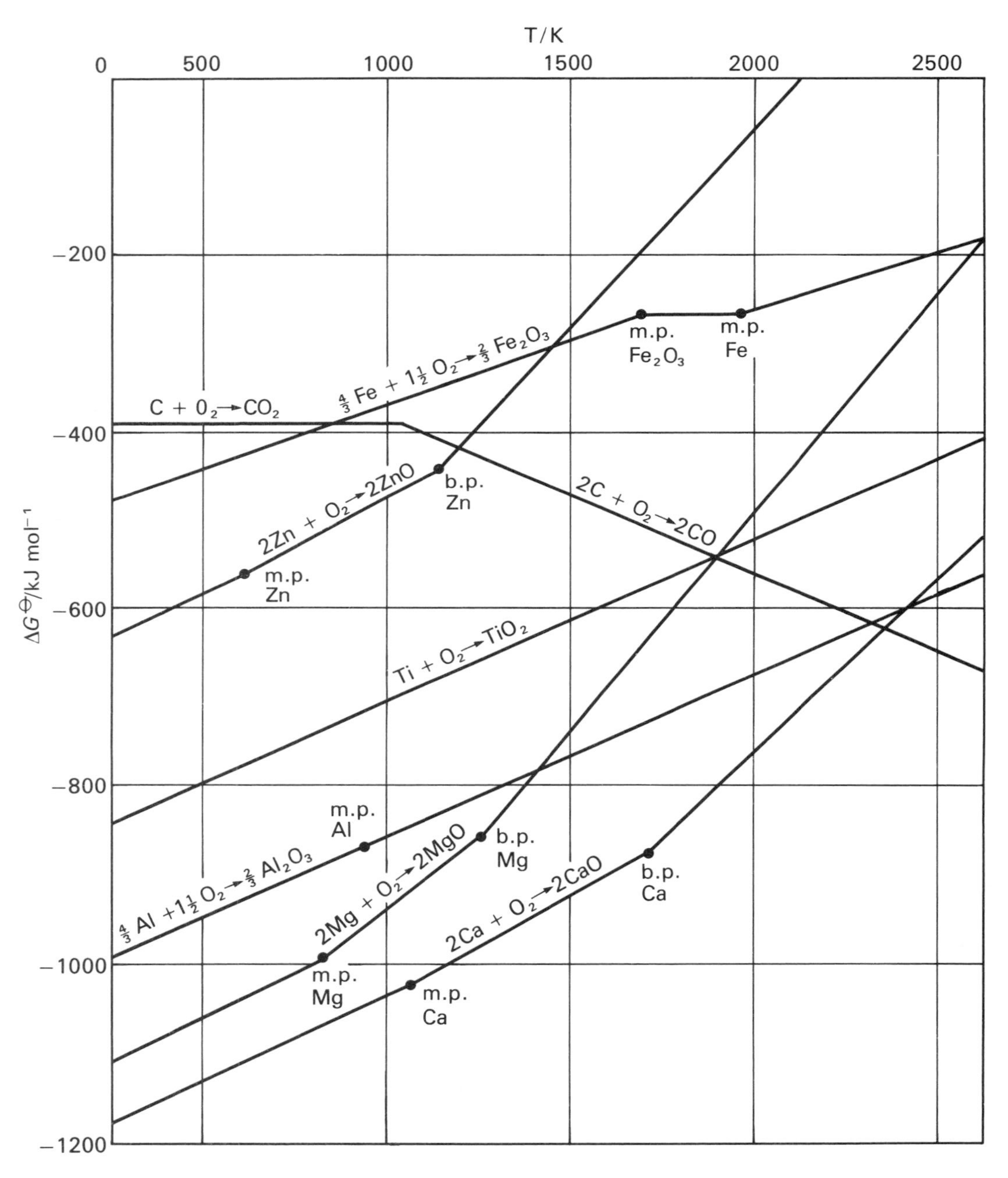

Figure 11.5 Ellingham diagram for carbon and selected metals

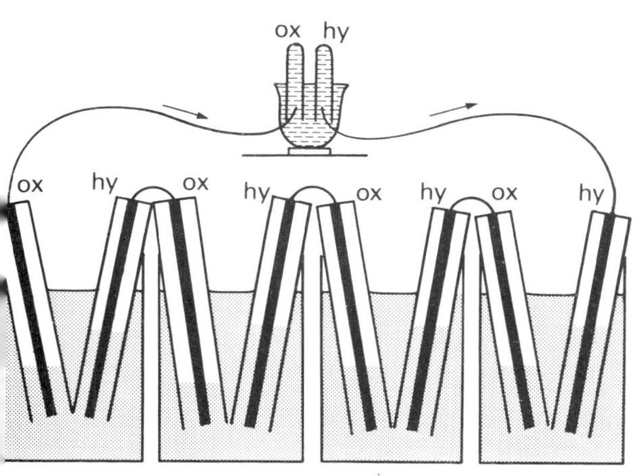

Figure 11.6 Sir William Grove's fuel cell of 1839

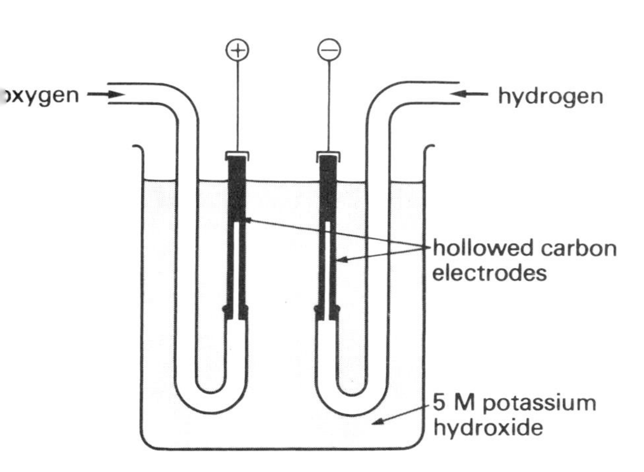

Figure 11.7 A fuel cell suitable for laboratory investigations

Figure 11.8 A fuel cell operating an electric hammer

that do not suffer from severe voltage deterioration with current demand and are not too bulky. Fuel cells have been successfully used in artificial satellites to power communications apparatus in which relatively small currents are required over long time intervals.

The relationship between $\Delta H^\ominus$ and $\Delta G^\ominus$

As we have seen, $\Delta H^\ominus$ and $\Delta G^\ominus$ for a reaction generally differ in magnitude. Spontaneous endothermic reactions must have negative values for $\Delta G^\ominus$ and so they must also differ in sign.

$$\Delta G^\ominus \neq \Delta H^\ominus$$

therefore,

$$\Delta G^\ominus = \Delta H^\ominus + \text{`X'}$$

The 'X' factor is the difference between $\Delta G^\ominus$ and $\Delta H^\ominus$. For a spontaneous endothermic change, the 'X' factor must be sufficiently negative to counterbalance the unfavourable value of $\Delta H^\ominus$ giving an overall negative value to $\Delta G^\ominus$. In order that the units are consistent throughout the equation, the 'X' factor must be an energy term and hence the product of an intensity and capacity factor. The concept of entropy first arose during the investigations into the workings of heat engines referred to previously. One way of regarding entropy is as a measure of the extent to which a system is disordered. Crystalline solids have a high degree of order in the arrangement of their component particles. When these melt and become liquids there is a greater degree of disorder. On boiling, the particles of the gas formed move rapidly in all directions at random and exhibit a high degree of disorder. It is a common experience that spontaneous changes in the physical world tend to result in increased disorder or entropy. Sugar dissolves spontaneously in tea, rocks erode, metals corrode and buildings crumble. Because a temperature rise is an increase in the mean kinetic energy of the particles comprising a system, the extent of disorder generally increases with increasing temperature. Let us examine the consequences of postulating that in the energy term 'X', temperature is the intensity factor and the change in entropy (given the symbol $\Delta S^\ominus$) is the capacity factor. The equation then becomes:

$$\Delta G^\ominus = \Delta H^\ominus - T\Delta S^\ominus$$

(The negative sign is necessary because we have postulated that an *increase* in $\Delta S^\ominus$ is associated with a more negative value of $\Delta G^\ominus$.) Reviewing the equations for the endothermic reactions discussed previously, we see that when they proceed in the forward direction there is a considerable entropy increase.

For reactions in which the same number of moles of similar particles are involved as reactants and products we should expect $\Delta S^\ominus$ to be smaller than when this is not so. In such cases there should therefore be a closer correspondence between $\Delta H^\ominus$ and $\Delta G^\ominus$. The following results are for two reactions for which $\Delta H^\ominus$ can be measured calorimetrically and $\Delta G^\ominus$ can be measured electrochemically:

$$Zn(s) + Cu^{2+}(aq) \rightarrow Zn^{2+}(aq) + Cu(s);$$
$$\Delta H^\ominus = -216 \text{ kJ mol}^{-1}$$
$$\Delta G^\ominus = -212 \text{ kJ mol}^{-1}$$

$$Cu(s) + 2Ag^{+}(aq) \rightarrow Cu^{2+}(aq) + 2Ag(s);$$
$$\Delta H^\ominus = -147 \text{ kJ mol}^{-1}$$
$$\Delta G^\ominus = -\ 89 \text{ kJ mol}^{-1}$$

For two gaseous reactions $\Delta H^\ominus$ and $\Delta G^\ominus$ are as follows:

$$H_2(g) + Cl_2(g) \rightarrow 2HCl(g);$$
$$\Delta H^\ominus = -184 \text{ kJ mol}^{-1}$$
$$\Delta G^\ominus = -190 \text{ kJ mol}^{-1}$$

$$N_2(g) + 3H_2(g) \rightarrow 2NH_3(g);$$
$$\Delta H^\ominus = -91 \text{ kJ mol}^{-1}$$
$$\Delta G^\ominus \doteq -33 \text{ kJ mol}^{-1}$$

In those examples where $\Delta G^\ominus$ is less negative than $\Delta H^\ominus$ there is a reduction in the entropy of the system as the reaction proceeds. This might be though to contradict the *Second Law of Thermodynamics* which can be stated in the form '*During a spontaneous change, the total entropy always tends to increase*'. As $\Delta H^\ominus$ for these reactions is negative, however, the exothermic reaction will warm molecules outside the system in contact with the reaction vessel and cause an increase in entropy there. Thus the *total* entropy change can still be in agreement with the Second Law.

Effect of temperature on $\Delta H^\ominus$, $\Delta G^\ominus$ and ΔS°

Calorimetric investigations show that $\Delta H^\ominus$ for reactions is largely constant even when measured over a wide range of temperatures. $\Delta G^\ominus$, however, must be temperature dependent as cell potentials vary considerably with temperature and these are directly related to $\Delta G^\ominus$ by the equation:

$$\Delta G^\ominus = -zFE^\ominus$$

From the relationship:

$$\Delta G^\ominus = \Delta H^\ominus - T\Delta S^\ominus$$

$\Delta S^\ominus$ could be largely independent of temperature. We can test this by determining the potential of a suitable cell at a series of temperatures. A suitable cell is:

$$Cu(s) \mid Cu^{2+}(aq) \,\vdots\, Ag^{+}(aq) \mid Ag(s)$$

The apparatus is shown in Figure 11.9. From $E^\ominus$ measurements, $\Delta G^\ominus$ at a series of temperatures can be calculated. If $\Delta H^\ominus$ and $\Delta S^\ominus$ are constant with temperature, a graph of $\Delta G^\ominus$ against T/K should give a straight line of intercept $\Delta H^\ominus$ and slope $-\Delta S^\ominus$. The results of such an experiment are shown in Figure 11.10.

$$\Delta H^\ominus = \text{intercept} = -148 \text{ kJ mol}^{-1}$$

This is in excellent agreement with values determined calorimetrically (page 47).

$$\Delta S^\ominus = -\text{slope} = -195 \text{ J mol}^{-1} \text{ K}^{-1}$$

This is one method by which standard entropy changes can be determined; other methods are discussed later in the chapter.

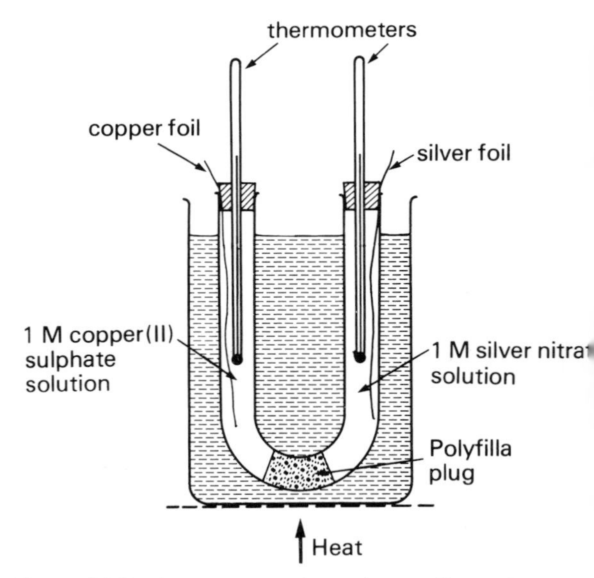

Figure 11.9 Apparatus to investigate effect of temperature on $\Delta G^\ominus$ for a cell reaction

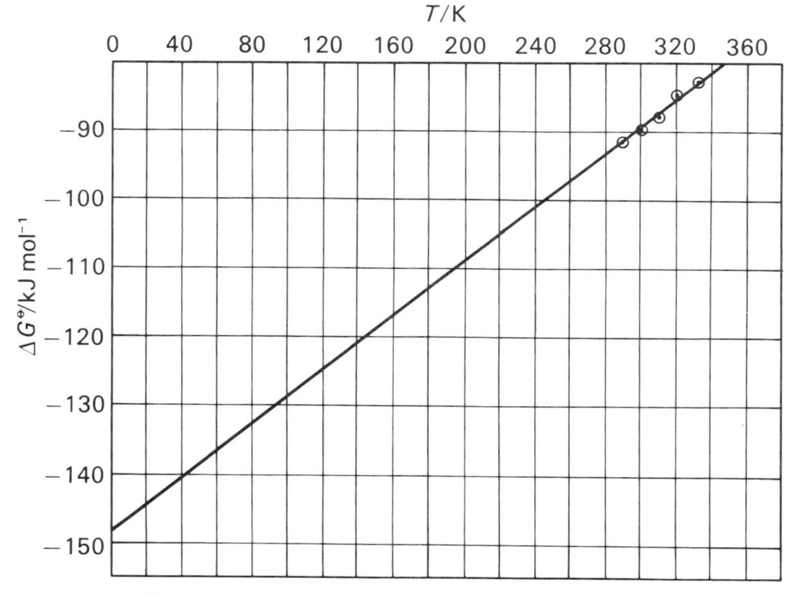

Figure 11.10 Variation of $\Delta G^{\ominus}$ with temperature for a cell reaction

Two reactions considered qualitatively

(1) Calcium carbonate decomposes according to the following equation:

$$CaCO_3(s) \rightarrow CaO(s) + CO_2(g); \quad \Delta H^{\ominus} = +178 \text{ kJ mol}^{-1}$$

The entropial energy term ($T\Delta S^{\ominus}$) must be sufficiently negative to counteract the positive values of $\Delta H^{\ominus}$ and thus make $\Delta G^{\ominus}$ negative:

$$\Delta G^{\ominus} = \Delta H^{\ominus} - T\Delta S^{\ominus}$$

As $\Delta S^{\ominus}$ will be positive, this is possible provided T is large enough. This agrees with the observation that calcium carbonate undergoes decomposition at elevated temperatures.

(2) Magnesium burns spontaneously in oxygen to form its oxide:

$$Mg(s) + \tfrac{1}{2}O_2(g) \rightarrow MgO(s); \quad \Delta H^{\ominus} = -602 \text{ kJ mol}^{-1}$$

There is a negative entropy change and therefore the reaction is driven by the negative value of $\Delta H^{\ominus}$.

Entropy and phase transitions

Consider a beaker of water containing ice and water in equilibrium at 0 °C:

$$H_2O(s) \rightleftharpoons H_2O(l)$$

If an infinitesimally small quantity of heat is added to the system a small quantity of ice absorbs latent heat of fusion and melts. If a similar quantity of heat is withdrawn a little water gives up its latent heat of fusion and solidifies. When a reaction is carried out involving only infinitesimal changes, and therefore virtually at equilibrium, it is said to happen under conditions of thermodynamic reversibility. For the ice and water to be in equilibrium they must have the same free energy, therefore ΔG for the phase change must be zero. (*N.B.*—it is ΔG which is zero and *not* $\Delta G^{\ominus}$, just as for an electrolytic cell in which the electrodes are in equilibrium, E is zero but $E^{\ominus}$ is not.) For this equilibrium situation:

$$\Delta G = 0$$
$$= \Delta H - T\Delta S$$

or

$$\Delta S = \frac{\Delta H}{T}$$

This relationship can be regarded as a definition of entropy. The finite changes involved in such phase changes can be regarded as the sum of the requisite number of infinitesimal changes that contribute to it. Thus the entropy of fusion of a solid changing to a liquid, $\Delta S_{fus} = \Delta H_{fus}/T_m$ where T_m is the melting point. Similarly the entropy of vaporization of a liquid, $\Delta S_{vap} = \Delta H_{vap}/T_b$. Both fusion and evaporation can be seen as processes that occur in a series of steps each one of which is conducted under thermodynamically reversible conditions.

It was discovered empirically by Trouton in 1884 that the fraction $\Delta H_{vap}/T_b$ had the same value for many liquids, an observation known as *Trouton's rule*. An equivalent statement of the law in thermodynamic terms is: '*many liquids have the same value for molar entropy of vaporization*'. Figure 11.11 shows molar latent heats of vaporization plotted against boiling point for a selection of liquids. The liquids with values which lie on the line obey Trouton's rule closely. This can be interpreted in terms of these liquids having the same molar entropy of vaporization because, having small or zero molecular dipole moments, there is similar entropy associated with each liquid. Water is anomalous because the high degree of hydrogen bonding between molecules in the liquid reduces the entropy in that state. The entropy change when vapour is formed is therefore correspondingly greater. The alcohols show a similar effect but to a lesser extent. The low value of molar entropy of vaporization of ethanoic acid can be explained in terms of the dimeric molecule formed in the vapour which reduces the entropy in that

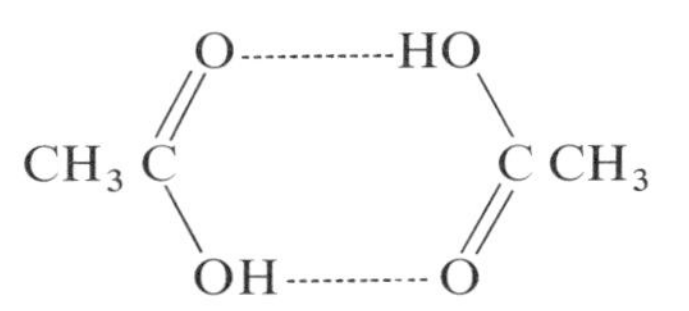

state: Hydrogen bonding is discussed in detail in Chapter 5.

The molar entropies of fusion of solids can be related to their crystal structure. The transition from solid to liquid for solids of the same crystalline structure is associated with similar values of molar entropy of fusion. This effect is illustrated in Figure 11.12.

Variation of equilibrium constant with temperature

$$\Delta G^{\ominus} = -RT \ln K$$
$$\Delta G^{\ominus} = \Delta H^{\ominus} - T\Delta S^{\ominus}$$

The two relationships above may be combined to give:

$$-RT \ln K = \Delta H^{\ominus} - T\Delta S^{\ominus}$$

which rearranges to

$$\ln K = -\frac{\Delta H^{\ominus}}{RT} + \frac{\Delta S^{\ominus}}{R}$$

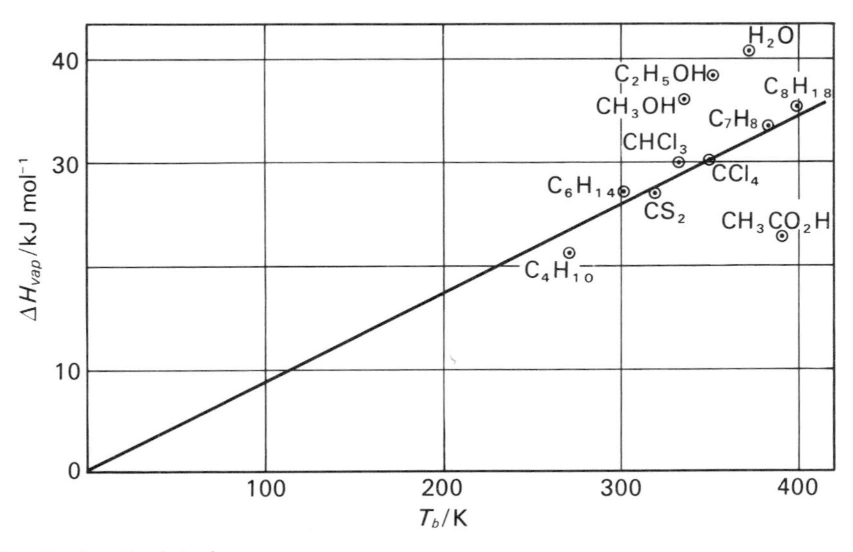

Figure 11.11 Trouton's rule data for selected liquids

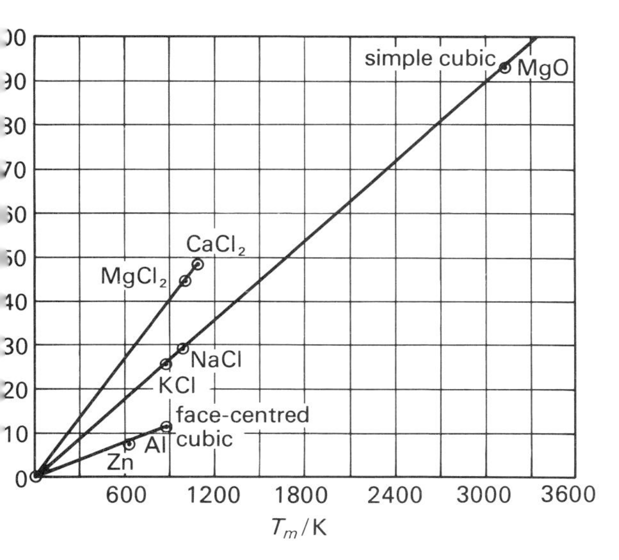

Figure 11.12 Molar entropy of fusion for some crystal systems

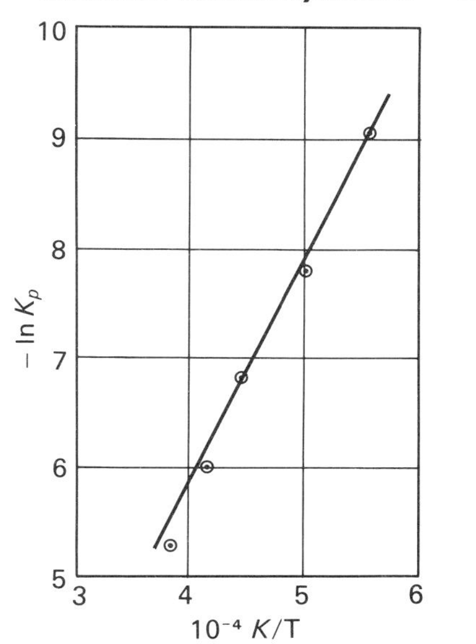

Figure 11.13 Equilibrium/temperature data for a reaction

This new equation can be used to study the variation of equilibrium constant with temperature. Appropriate data are given in Table 11.2 for the reaction:

$$N_2(g) + O_2(g) \rightleftharpoons 2NO(g)$$

Table 11.2

T/K	$10^{-4}K/T$	$10^{-4}K_p$	$\ln K_p$
1800	5.556	1.21	−9.02
2000	5.000	4.08	−7.80
2200	4.545	11.0	−6.81
2400	4.167	25.1	−5.99
2600	3.846	50.3	−5.29

Applying the linear law, y = mx + c, to the equation above shows that for a graph of ln K against the reciprocal of the temperature (in kelvin), there will be an intercept of $\Delta S^{\ominus}/R$ and a slope of $-\Delta H^{\ominus}/R$. This data has been plotted in Figure 11.13 and this gives +181 kJ mol^{-1} for $\Delta H^{\ominus}$. Because the bond energies of all the molecules involved in this reaction can be determined spectroscopically, an energy cycle for the formation of nitrogen oxide can be drawn. (For the spectroscopic determination of bond energies, diatomic molecules only may be studied as discussed in Chapter 3.) Figure 11.14 shows the energy cycle and the value of +185 kJ mol^{-1} for $\Delta H^{\ominus}$ is in good agreement with that obtained from equilibrium data.

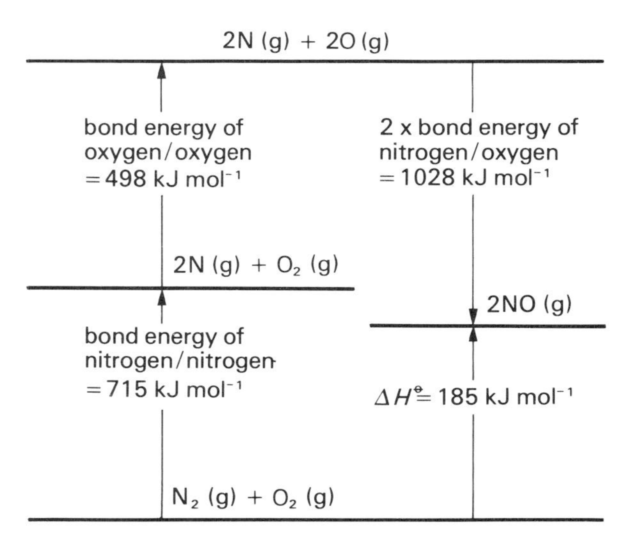

Figure 11.14 Energy cycle for the formation of nitrogen oxide

Determination of entropies

1. When $\Delta H^{\ominus}$ and $\Delta G^{\ominus}$ can both be readily measured directly, as for many reactions that can take place in an electrolytic cell, the entropy change can be calculated from the equation:

$$\Delta G^{\ominus} = \Delta H^{\ominus} - T\Delta S^{\ominus}$$

e.g. for the reaction

$$Cu(s) + 2Ag^{+}(aq) \rightarrow Cu^{2+}(aq) + 2Ag(s)$$

as we have seen

$$\Delta H^{\ominus} = -147 \text{ kJ mol}^{-1} \text{ at } 298 \text{ K}$$
$$\Delta G^{\ominus} = -\ \ 89 \text{ kJ mol}^{-1} \text{ at } 298 \text{ K}$$

substituting these values in the equation yields

$$(-89 \text{ kJ mol}^{-1}) = (-147 \text{ kJ mol}^{-1}) - (\Delta S^{\ominus} \times 298 \text{ K})$$

$$\Delta S^{\ominus} = \frac{(-89 - -147) \text{ kJ mol}^{-1}}{298 \text{ K}}$$
$$= -0.195 \text{ kJ mol}^{-1} \text{ K}^{-1}$$
$$= -195 \text{ J mol}^{-1} \text{ K}^{-1}$$

2. For phase transitions or transitions between polymorphic forms, if the temperature of transition and the enthalpy change for the transition are known, we can use the relationship:

$$\Delta S = \frac{\Delta H_{\text{reversible}}}{T}$$

e.g. for cyclohexane

$$\Delta H_{vap} = 30.1 \text{ kJ mol}^{-1}$$
$$T_b = 354 \text{ K}$$

which on substitution yields:

$$\Delta S_{vap} = \frac{30.1 \text{ kJ mol}^{-1}}{354 \text{ K}}$$
$$= 0.085 \text{ kJ mol}^{-1} \text{ K}^{-1}$$
$$= 85 \text{ J mol}^{-1} \text{ K}^{-1}$$

3. A system cannot have lower entropy than when it exists as a perfect crystal at the absolute zero of temperature. This is recognized in the *Third Law of Thermodynamics* which states: '*the entropy of perfect crystals can be taken as zero at the absolute zero of temperature*'. There is thus a natural baseline for entropy which does not exist in the case of enthalpy and free energy. This law enables calculations of absolute entropies, often called third law entropies, to be made.

$$\Delta S = \frac{\Delta H_{rev}}{T}$$

If the temperature of a substance is raised without any phase change occurring, the entropy increase as each infinitesimal increment of heat is added at constant pressure is:

$$dS = \frac{dH}{T}$$

or

$$\frac{dS}{dH} = \frac{1}{T}$$

$$\frac{dS}{dT} \cdot \frac{dT}{dH} = \frac{1}{T} \text{ (function of a function)}$$

$$\frac{dS}{dT} = \frac{1}{T} \cdot \frac{dH}{dT}$$

C_p, the molar heat capacity of a substance, is the rate at which a mole of the substance can gain enthalpy per kelvin, i.e.

$$C_p = \frac{dH}{dT}$$

Substituting this in the equation above yields

$$\frac{dS}{dT} = \frac{1}{T} C_p$$

If S_0 and S_{298} are the entropies of the substance at 0 K and 298 K respectively, this equation can be written in the form

$$S_{298} - S_0 = \int_0^{298} \frac{C_p}{T} dT$$

but from the third law of thermodynamics, $S_0 = 0$;

therefore
$$S_{298} = \int_0^{298} \frac{C_p}{T} dT$$

To determine S_{298}, therefore, it is necessary to plot C_p/T against T for the substance and find the area under the curve between 0 and 298 K. A plot of C_p against T for nickel metal is shown in Figure 11.15. Most of the values were obtained by standard calorimetrical methods, however, those close to 0 K are obtained by extrapolation. Values of C_p/T against T have been plotted to give Figure 11.16. The area under the curve can be found by cutting out and weighing or by counting squares. These give the value

$$S_{298}[Ni(s)] = 30 \text{ J mol}^{-1} \text{ K}^{-1}$$

It is such third law entropies that are generally printed in tables of thermodynamic data.

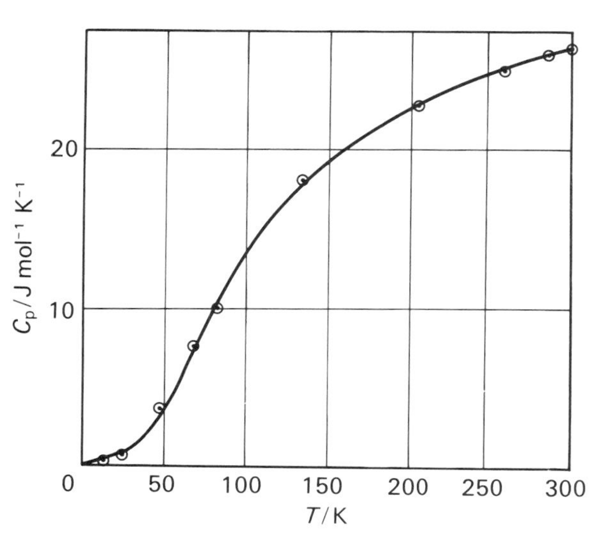

Figure 11.15 Variation of C_p with temperature for nickel

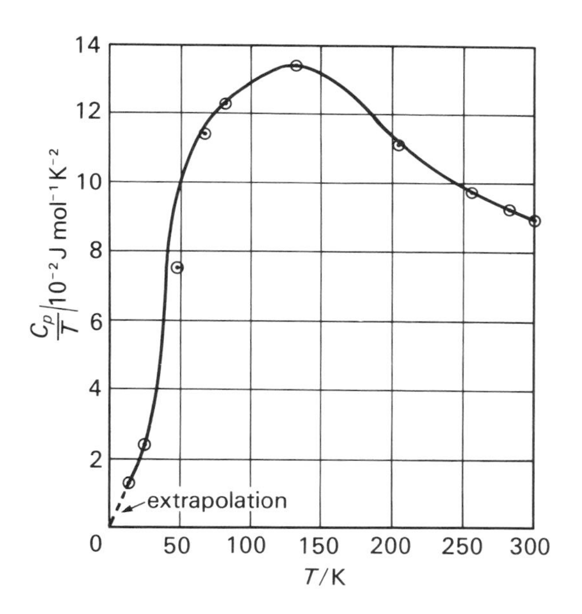

Figure 11.16 C_p/T against T for nickel

To find the entropy change for a reaction, the third law entropies taken from tables are used as in the example below:

$$S_{298}/\text{J mol}^{-1}\ \text{K}^{-1}: \quad \underset{30}{Ni(s)} + \underset{223}{Cl_2(g)} \rightarrow \underset{107}{NiCl_2(s)}$$

$$\Delta S = -146\ \text{J mol}^{-1}\ \text{K}^{-1}$$

$\Delta H_f^{\ominus}$ [$NiCl_2$(s)] can be determined directly by calorimetry; no such determination of $\Delta G_f^{\ominus}$ [$NiCl_2$(s)] is possible. The value of ΔS calculated above can be used to determine $\Delta G_{298}^{\ominus}$:

$$\Delta G_{298}^{\ominus} = \Delta H^{\ominus} - T\Delta S^{\ominus}$$

Using a suitable value of $\Delta H^{\ominus}$ from tables:

$$\Delta G_{298}^{\ominus} = -316 - \left(-\frac{298 \times 146}{1000}\right) \text{kJ mol}^{-1}$$
$$= -272\ \text{kJ mol}^{-1}$$

($\Delta G^{\ominus}$ at other temperatures can be calculated assuming that $\Delta H^{\ominus}$ and $\Delta S^{\ominus}$ do not change significantly with temperature.)

This value may be used to calculate K_p for the reaction at 298 K.

$$\Delta G_{298}^{\ominus} = -RT \ln K_p$$
$$\ln K_p = -\frac{\Delta G_{298}^{\ominus}}{RT}$$
$$= \frac{272}{8.30 \times 10^{-3} \times 298}$$
$$= 110$$
$$K_p = \ln^{-1} 110 = 5.7 \times 10^{47}$$

This large value of K_p shows that the reaction goes virtually to completion at 298 K.

Standard entropies when phase changes are involved

When the substance under consideration is in a different state at 298 K than it is at 0 K, it is necessary to make a summation of the methods appropriate to each stage in the process. This is illustrated by the calculation in Table 11.3 leading to the standard entropy, at 298 K, of chlorine gas.

Table 11.3

Temperature range/K	*Method*	*ΔS/J mol^{-1} K^{-1}*
0–15	Extrapolation	1.38
15–172(m.p.)	C_p/T. dT	69.33
172	$\Delta H_{fus}/T$	37.23
172–239(b.p.)	C_p/T. dT	21.88
239	$\Delta H_{vap}/T$	85.40
239–298	C_p/T. dT	7.78
	S_{298} =	223.0

Solubility of ionic solids

The solubility of a salt in water is equivalent to the combination of the following two processes:

$$A^+B^-(s) \rightarrow A^+(g) + B^-(g)$$

$\Delta H^\ominus$ is the lattice energy

$$A^+(g) + B^-(g) + (aq) \rightarrow A^+(aq) + B^-(aq)$$

$\Delta H^\ominus$ is the hydration enthalpy

Thus

$$\Delta H^\ominus \text{ (solution)} = \Delta H^\ominus \text{ (lattice energy)} + \Delta H^\ominus \text{ (hydration)}$$

$\Delta H^\ominus$ (lattice energy) is always positive and $\Delta H^\ominus$ (hydration) is always negative, therefore $\Delta H^\ominus$ (solution) can be either positive or negative. The entropy change is always positive as an ordered crystal lattice is replaced by ions moving randomly within the solution.

$$\Delta G^\ominus = \Delta H^\ominus - T\Delta S^\ominus$$

The more negative the value of $\Delta G^\ominus$ in the above equation, the more soluble will be the salt. It is common experience that, unless there is a change in hydration, the solubilities of salts increase with temperature. This is explicable in terms of the corresponding increase in the $T\Delta S^\ominus$ term in the equation above.

Thermodynamic interpretation of ebullioscopy

The use of elevation of the boiling point of solvents by dissolved solutes to determine molar masses is discussed in Chapter 7. The ebullioscopy constant, K_b, can be calculated using a thermodynamically derived equation. Consider the phase change:

$$Y(l) \rightleftharpoons Y(g)$$

for which $K_p = p_Y$

$$\ln K_p = \ln p_Y = -\frac{\Delta H^\ominus_{vap}}{RT} + \frac{\Delta S^\ominus_{vap}}{R}$$

When p_1 is the vapour pressure at temperature T_1 and p_2 and T_2 are similarily related, this can be written

$$\ln \frac{p_1}{p_2} = -\frac{\Delta H^\ominus_{vap}}{R}\left(\frac{1}{T_1} - \frac{1}{T_2}\right)$$

the differential form of which is

$$\frac{d \ln p}{dT} = \frac{\Delta H^\ominus_{vap}}{RT^2}$$

For the small change in boiling point, ΔT_b, that occurs when a little solute is dissolved in a solvent, the equation can be rewritten:

$$\Delta T_b = \frac{RT^2}{\Delta H^\ominus_{vap}} \times \frac{(p_0 - p)}{p_0}$$

where p_0 is the vapour pressure of the solvent and p is the vapour pressure of the solution. According to Raoult's Law, as discussed in Chapter 7:

$$\frac{(p_0 - p)}{p_0} = \frac{n_{solute}}{n_{solute} + n_{solvent}}$$

Substitution in the equation above yields:

$$\Delta T_b = \frac{RT^2}{\Delta H^\ominus_{vap}} \times \frac{n_{solute}}{n_{solute} + n_{solvent}}$$

For dilute solutions in which $n_{solvent}$ is much greater than n_{solute}

$$\Delta T_b = \frac{RT^2}{\Delta H^\ominus_{vap}} \times \frac{n_{solute}}{n_{solvent}}$$

The ebullioscopy constant, K_b, is the elevation of boiling point that would be observed, *provided the mixture continued to behave ideally at the higher concentrations*, when one mole of solute is dissolved in 1000 g of solvent. Thus for a solvent of relative molar mass M_r:

$$K_b = \frac{M_r RT^2}{\Delta H^\ominus_{vap} \times 1000}$$

A statistical approach to entropy

For a fuller appreciation of the concept of entropy it must be realized that a substance can have entropy due to other factors than the location in space of its particles. In addition to such translational entropy, it has entropy due to the distribution of energy between the available electron energy levels of its particles. This contribution to the total entropy is known as the system's internal entropy. Consider an assembly of 36 atoms as shown in Figure 11.17. If 36 quanta of energy are to be distributed among these atoms, three of the many possible ways this can be done are shown:

(*a*) One quantum is allotted to each atom; there is only one way this can be done, W = 1.
(*b*) All the quanta can be given to one of the atoms, this can be done in 36 ways.
(*c*) This is the most probable distribution if the quanta are allocated at random; there are 2×10^8 ways this can be done.

Table 11.4 shows the contributions made by the translational and internal entropies to the total entropy for selected substances at 1 atom and 298 K. As expected, similar substances have much the same values for translational entropy and the internal entropy increases sharply with molecular complexity because more electron energy levels are available.

Table 11.4 Translational and internal entropies of some substances at 298 K

Substance	*Entropy/J mol^{-1} K^{-1}*		
	Translational	*Internal*	*Total*
$N_2O(g)$	156	64	220
$NO_2(g)$	157	83	240
$N_2O_4(g)$	164	141	305

Boltzmann proposed a logarithmic relationship between entropy and W, the ways the energy can be distributed (for internal entropy) or the ways the particles can be distributed in space (for translational entropy); the entropy per particle being:

$$S = k \ln W = \frac{R}{L} \ln W$$

The Boltzmann constant, k, is the gas constant per molecule.

Imagine two identical flasks connected together by a stopcock. One flask is evacuated and the other contains 1 mole of gas. Let the total number of ways the translational energy can be distributed initially be W. If the stopcock is opened the gas will diffuse into the vacuum. W increases because each molecule now has twice the original volume in which to find itself.

The mole of gas consists of L molecules each having this choice of compartments, so the new number of ways of distribution is $2^L W$. The entropy change on expansion for each molecule can thus be calculated using Boltzmann's equation:

$$\begin{aligned} S_2 - S_1 &= k \ln 2^L W - k \ln W \\ &= k \ln 2^L \end{aligned}$$

thus

$$\begin{aligned} \Delta S/\text{J mol}^{-1}\text{ K}^{-1} &= R \ln 2 \\ &= 5.77 \end{aligned}$$

This statistical mechanics approach to entropy is presented in more detail in *Nuffield Advanced Science—Physics Unit 9: Change and Chance* (London: Penguin Books, 1972).

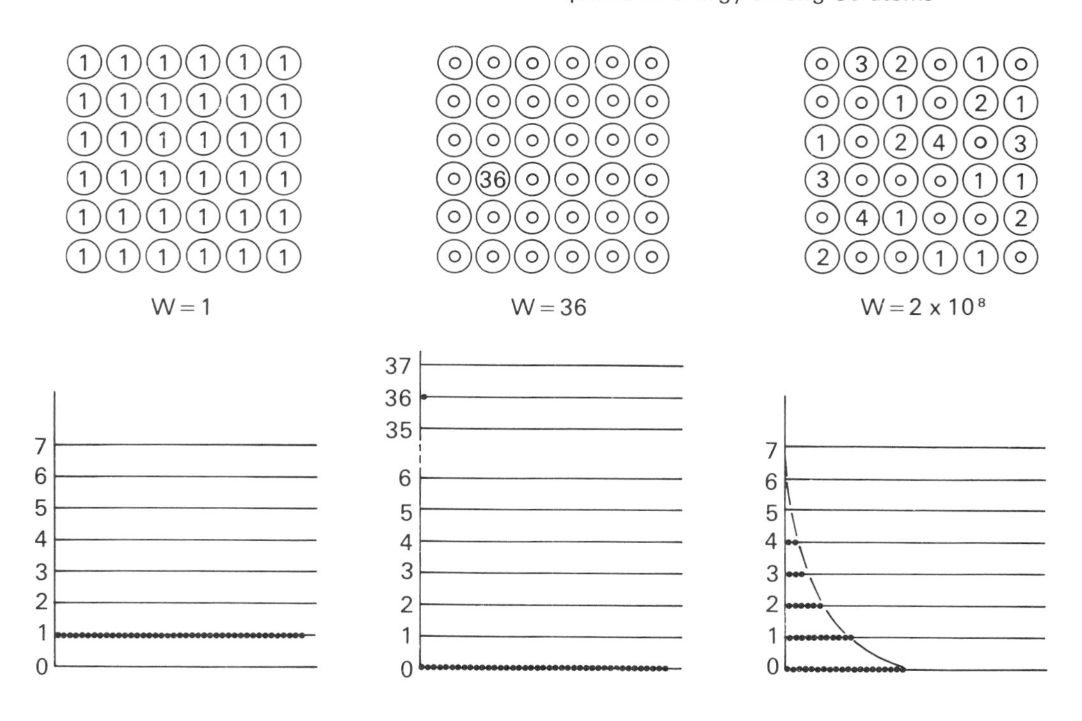

Figure 11.17 Some ways of distributing 36 quanta of energy among 36 atoms

Suggestions for further reading

SMITH, BRIAN, *Basic Chemical Thermodynamics* (Oxford: Oxford University Press, 1973)
ANGRIST, S. W. and HEPLER, L. G., *Order and Chaos* (Harmondsworth: Penguin Books, 1973)
JOHNSTONE, A. H. and WEBB, G., *Energy, Chaos, and Chemical Change* (London: Heinemann Educational Books, 1977)

Problems

1. Using data from Table 9.1 (page 126):

(*a*) Calculate $E^{\ominus}$ for the cell:

$$Sn(s) \mid Sn^{2+}(aq) \vdots Pb^{2+}(aq) \mid Pb(s)$$

(*b*) Write the equation for the reaction that would occur if the two electrodes of the cell were connected by a conductor.

(*c*) Calculate (i) $\Delta G^{\ominus}$ and (ii) K_c for this reaction at 298 K.

2.

$$CdCO_3(s) \rightarrow CdO(s) + CO_2(g)$$

Use the data in Table 6.4 (page 87) to determine $\Delta H^{\ominus}$ and $\Delta S^{\ominus}$ for the reaction above. The standard enthalpies of formation/kJ mol^{-1} for cadmium carbonate, cadmium oxide and carbon dioxide are -748, -255, and -394 respectively. Use these in a suitable energy cycle to make an independent check on $\Delta H^{\ominus}$ for the reaction.

3. The enthalpies of combustion of rhombic and monoclinic sulphur/kJ mol^{-1} are -296.606 and -297.063 respectively. Determine $\Delta H^{\ominus}$ for the change:

$$S(\text{rhombic}) \rightarrow S(\text{monoclinic})$$

The two allotropes are in equilibrium at 369 K, the transition temperature. Calculate $\Delta S^{\ominus}$ for the change.

4.

Substance	$\Delta G^{\ominus}_{f298}$/kJ mol^{-1}
$N_2O(g)$	104
$NH_3(g)$	-16.5
$H_2O(g)$	-228

Use the data above to calculate (i) $\Delta G^{\ominus}$ and (ii) K_p for the following reaction at 298 K.

$$N_2O(g) + 4H_2(g) \rightleftharpoons 2NH_3(g) + H_2O(g)$$

5. What do you understand by the terms *free energy*, *enthalpy*, and *entropy*?

What is the relationship between the standard free energy change, the standard enthalpy change and the standard entropy change in a chemical reaction?

In the reaction

$$C_2H_4 + H_2 = C_2H_6$$

the activation energies of the forward and reverse reactions are 171 kJ mol^{-1} and 305 kJ mol^{-1} respectively. Explain how these two quantities are related to the enthalpy change of the reaction. (Oxford and Cambridge 'S')

6. What do you understand by the 'standard molar entropy of a substance at 298 K'? What patterns are there in the entropy of substances and what is a possible interpretation of the patterns? (Nuffield 'S')

7. Discuss the elevation of the boiling point of a solvent by an involatile solute in terms of the Clausius–Clapeyron equation

$$\frac{d \ln p}{dT} = \frac{L}{RT^2}$$

Calculate the boiling point elevation for a solution containing 0.1 mole of naphthalene (regarded as involatile) in 50 moles of benzene using the following vapour pressure data for benzene

p/mmHg	400	518	628	760
T/K	333.7	341.3	347.2	353.2

(Cambridge Entrance and Awards)

8. A relationship known as Trouton's Rule that, for many liquids, the enthalpy change on evaporation per mole divided by the normal boiling temperature (at a pressure of one atmosphere) is approximately 88 J mol^{-1} K^{-1}. What interpretation, in terms of molecular behaviour, can you suggest for Trouton's Rule? Convert a selection of the data in the table below into an appropriate form to test Trouton's Rule. Justify your choice of liquids and comment critically on the validity of Trouton's Rule.

Liquid	M/g mol^{-1}	T_b/K	Δh_b/ J g^{-1}
Water	18.0	373.2	2261
Pentane	72.2	309.2	359
Hexane	86.2	341.9	332
Heptane	100.2	371.6	325
Octane	114.2	398.8	292
Dichloromethane	84.9	313.4	330
Trichloromethane	119.4	334.9	249

Liquid	M/g mol^{-1}	T_b/K	Δh_b/ J g^{-1}
Tetrachloromethane	153.8	349.7	195
Methanol	32.0	337.9	1103
Ethanol	46.1	351.4	839
Propan-1-ol	60.1	370.4	687
Propan-1,2,3-triol	92.1	563.2	—
Benzene	78.1	353.3	394
Cyclohexane	84.2	353.9	357
Methylbenzene	92.1	383.8	359
1,2-dimethylbenzene	106.2	417.6	347
1,3-dimethylbenzene	106.2	412.3	343
1,4-dimethylbenzene	106.2	411.5	339
Ethanoic (acetic) acid	60.1	391.1	394
Propanone	58.1	329.4	522
Ethoxyethane (ether)	74.1	307.4	372

where

M = Molar mass
T_b = Normal boiling temperature (1 atm)
Δh_b = Specific enthalpy change on evaporation (at T_b)

(Nuffield 'S')

9. What do you understand by the term *entropy*? Study the table of thermodynamic data given below and comment on any features of the values of $\Delta H_f^\ominus$ and $S^\ominus$.
In a low temperature fuel cell having an aqueous electrolyte, hydrazine reacts with oxygen (from the air) to produce nitrogen and water. Using the relevant data from the table below, calculate, for the cell reaction at 298 K (25 °C):

(*a*) the standard enthalpy change, $\Delta H^\ominus$;
(*b*) the standard entropy change, $\Delta S^\ominus$;
(*c*) the standard free energy change, $\Delta G^\ominus$.

(The heat of solution of hydrazine in water may be ignored.)

Substance	Standard enthalpy of formation at 298 K $\Delta H_f^\ominus$ kJ mol^{-1}	Standard entropy at 298 K $S^\ominus$ J K^{-1} mol^{-1}
$CaH_2(s)$	−186.9	41.8
$H_2O(l)$	−285.8	69.9
$H_2S(g)$	−20.6	+205.9
$O_2(g)$	0	+205.0
$N_2(g)$	0	+191.5
$NH_2 . NH_2(l)$ (hydrazine)	+50.6	+121.2

(Oxford and Cambridge 'S')

10. If chemical change always proceeds in the direction of greater stability, why is it that endothermic reactions occur at all? Illustrate your answer with some specific examples.
(Oxford Colleges Entrance and Awards)

11. *Either*, Explain why the heat capacities of substances are of interest to chemists.
Or, Discuss the significance of the concept of *energy levels* in chemistry.
(Oxford Colleges Entrance and Awards)

12. The Periodic Classification

IN 1789, Lavoisier classified the elements known at that time into metals and non-metals. His main criterion for this distinction was the acidic or basic properties of the oxides of the approximately thirty elements whose existence was known. In the early years of the last century an approximately equal number of other elements were isolated, many using the then novel method of electrolysis. The chemists of that time were faced with the problem of systematizing the properties of these new elements. In 1829, the German chemist Johann Döbereiner observed that the recently discovered element bromine had properties half-way between those of chlorine and iodine. Moreover, its atomic mass (80) was virtually the average of the atomic masses of chlorine (35.5) and iodine (127). Other 'triads' which seemed related in this way in terms of properties and atomic masses were later discovered by Döbereiner and others. These include:

calcium (40)	sulphur (32)	lithium (7)
strontium (87.6)	selenium (79)	sodium (23)
barium (137.4)	tellurium (127.6)	potassium (39)

Döbereiner's work did not meet general acceptance, largely because over 80 per cent of the then known elements could not be fitted into triads.

In 1864, the English chemist John Newlands read a paper to the Chemical Society of London in which he claimed that if the elements are arranged in order of atomic mass, there are similarities in the properties of each element and each eighth element coming after or before it. By analogy with the musical scale, Newlands called his generalization 'The law of octaves'. Part of his scheme is shown below:

H	Li	Be	B	C	N	O
F	Na	Mg	Al	Si	P	S
Cl	K	Ca	Ti	Cr	Mn	Fe

Newlands' law works well for the lighter elements but, unfortunately, with increasing atomic mass it gets progressively out of tune! His paper was rejected, a Dr. Foster suggested sarcastically that he arrange the elements in alphabetical order and then look for patterns in properties. With historical hindsight, Newlands' failure can be seen to be due to (*a*) some incorrect values for atomic masses being accepted at that time, (*b*) the existence of what are now known as the 'transition elements' and (*c*) not all the elements required for the full sequence having been discovered.

In 1870, the German chemist Lothar Meyer calculated the volume occupied by one mole of each element in the solid state. This quantity is known as the atomic volume. When he plotted these values against the atomic masses of the elements he obtained a curve having sharp peaks and broad minima. A graph of this type is shown in Figure 12.1. This shows data for some elements discovered since 1870 and atomic number rather than atomic mass is plotted as abscissa (the justification for this will be seen later). It will be observed that elements exhibiting similar properties occupy corresponding positions on the peaks and troughs.

One year previously, in Russia, Dmitri Mendeléev first propounded his *Periodic Table of the Elements*. This great generalization was able to contain and extend the discoveries discussed above. He constructed a table by arranging the elements in horizontal rows (known as periods) in order of atomic mass and in vertical columns (known as groups) each containing elements with similar properties. He improved upon Newlands' classification in several important respects:

(*a*) by leaving gaps for elements which he claimed would be subsequently discovered;

(*b*) by adding an eighth group in which he placed triads of elements which followed sequentially in the list of elements in order of atomic mass but which seemed most closely to resemble each other in properties, e.g. the magnetic metals iron, cobalt, and nickel;

(*c*) by reversing the order of certain pairs of elements, e.g. iodine and tellurium, when this was necessary for consistency in the table;

(*d*) by adopting an A and B sub-group structure on a similarly pragmatic basis.

Figure 12.2 shows a periodic table based on Mendeléev's principles but to which has been added some elements discovered since 1869. In particular a completely new vertical column, labelled Group O, contains the noble gases discovered around the turn of our century. A major cause of the acceptance of Mendeléev's table was the success with

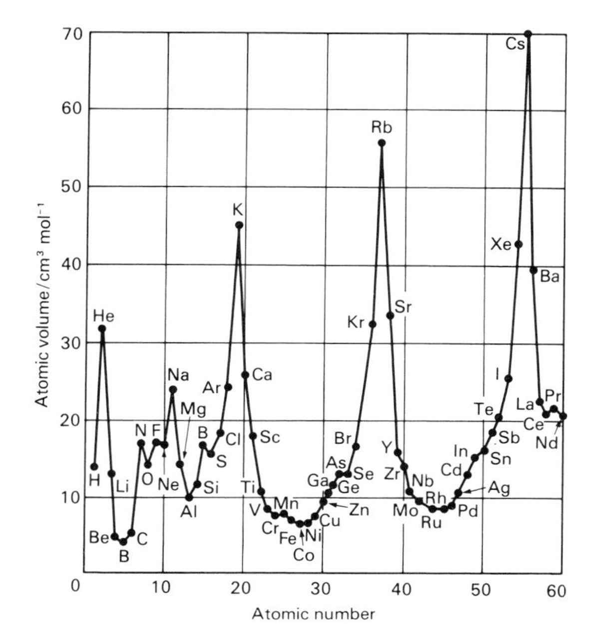

Figure 12.1 Atomic volume plotted against atomic number for some elements

which he predicted the properties of the elements for which he left blanks in the table. On their subsequent discovery his predictions were seen to be remarkably accurate. In 1871 he predicted the existence of an element to fall into Group IVB between silicon and tin. He named this eka-silicon and his predictions of its properties with those of the element germanium, discovered in 1886, and found to fill the gap, are given in Table 12.1.

Mendeléev's predictions were based principally on graphical interpolation of the properties of neighbouring elements in the table. It should be recognized that Mendeléev's achievement was essentially a pragmatic one, fitting elements into positions in the table that correctly reflected their properties. By the time of his death in 1907, only the first steps had been taken in the quest to understand atomic structure. Work in this area was eventually to explain the reason why elements obey a periodic law in their properties.

Figure 12.2 A Periodic Table of the Elements based on Mendeléev's principles (the Rare Earths are now known as the Lanthanides and the Actinons as the Actinides)

period	group															
	0	I A	I B	II A	II B	III A	III B	IV A	IV B	V A	V B	VI A	VI B	VII A	VII B	VIII
1		H 1														
2	He 2	Li 3		Be 4		B 5		C 6		N 7		O 8		F 9		
3	Ne 10	Na 11		Mg 12		Al 13		Si 14		P 15		S 16		Cl 17		
4	A 18	K 19	Cu 29	Ca 20	Zn 30	Sc 21	Ga 31	Ti 22	Ge 32	V 23	As 33	Cr 24	Se 34	Mn 25	Br 35	Fe 26 Co 27 Ni 28
5	Kr 36	Rb 37	Ag 47	Sr 38	Cd 48	Y 39	In 49	Zr 40	Sn 50	Nb 41	Sb 51	Mo 42	Te 52	Tc 43	I 53	Ru 44 Rh 45 Pd 46
6	Xe 54	Cs 55	Au 79	Ba 56	Hg 80	La 57 Rare Earths 58-71	Tl 81	Hf 72	Pb 82	Ta 73	Bi 83	W 74	Po 84	Re 75	At 85	Os 76 Ir 77 Pt 78
7	Rn 86	Fr 87		Ra 88		Ac 89		Th 90		Pa 91			U 92 Actinons 93—			

Table 12.1 Comparison of Mendeléev's predictions for eka-silicon (1871) with the properties of germanium (1886)

Property	*Eka-silicon (Es)*	*Germanium (Ge)*
Atomic mass	72	72.6
Density/g cm^{-3}	5.5	5.469
Melting point	high	958 °C
Atomic volume/cm^3 mol^{-1}	13	13.2
Appearance	Dirty-grey metal	Greyish-white metal
Oxide	EsO_2, a white solid of high melting point and density 4.7 g cm^{-3}; amphoteric	GeO_2 a white solid melting at 1100 °C and density 4.70 g cm^{-3}; amphoteric
Extraction	Reduction of oxide or K_2EsF_6 with sodium metal	Reduction of K_2GeF_6 with sodium metal
Chloride	$EsCl_4$, will boil below 100 °C and have a density of 1.9 g cm^{-3}	$GeCl_4$ boils at 83 °C and has a density of 1.89 g cm^{-3}
Metallo-organic compound formation	Like silicon and tin it will readily form such compounds as $Es(C_2H_5)$ which will boil at 160 °C and have a density a little less than that of water	$Ge(C_2H_5)$ readily forms which boils at 169 °C and has density of 0.97 g cm^{-3}

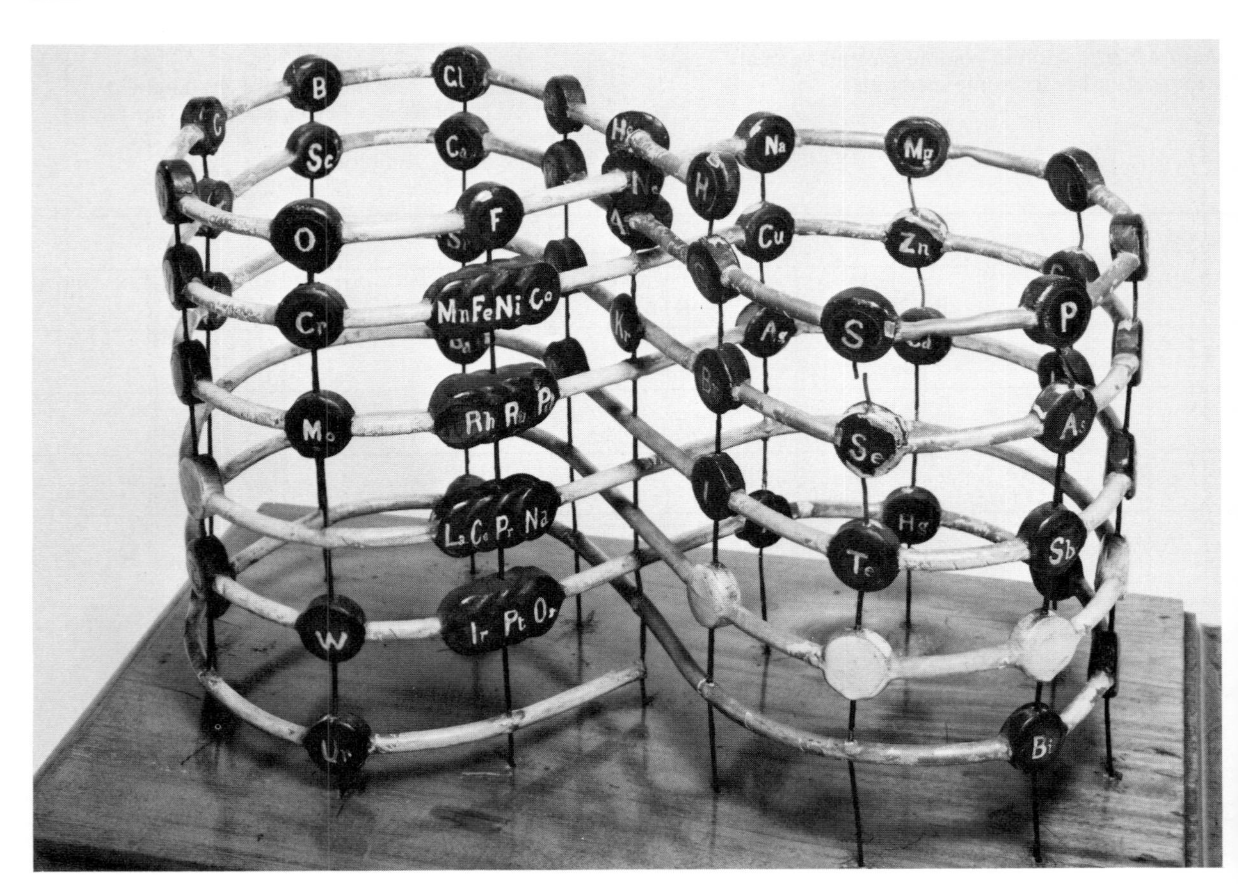

Figure 12.3 Sir William Crooke's Spiral of the Elements (1888)

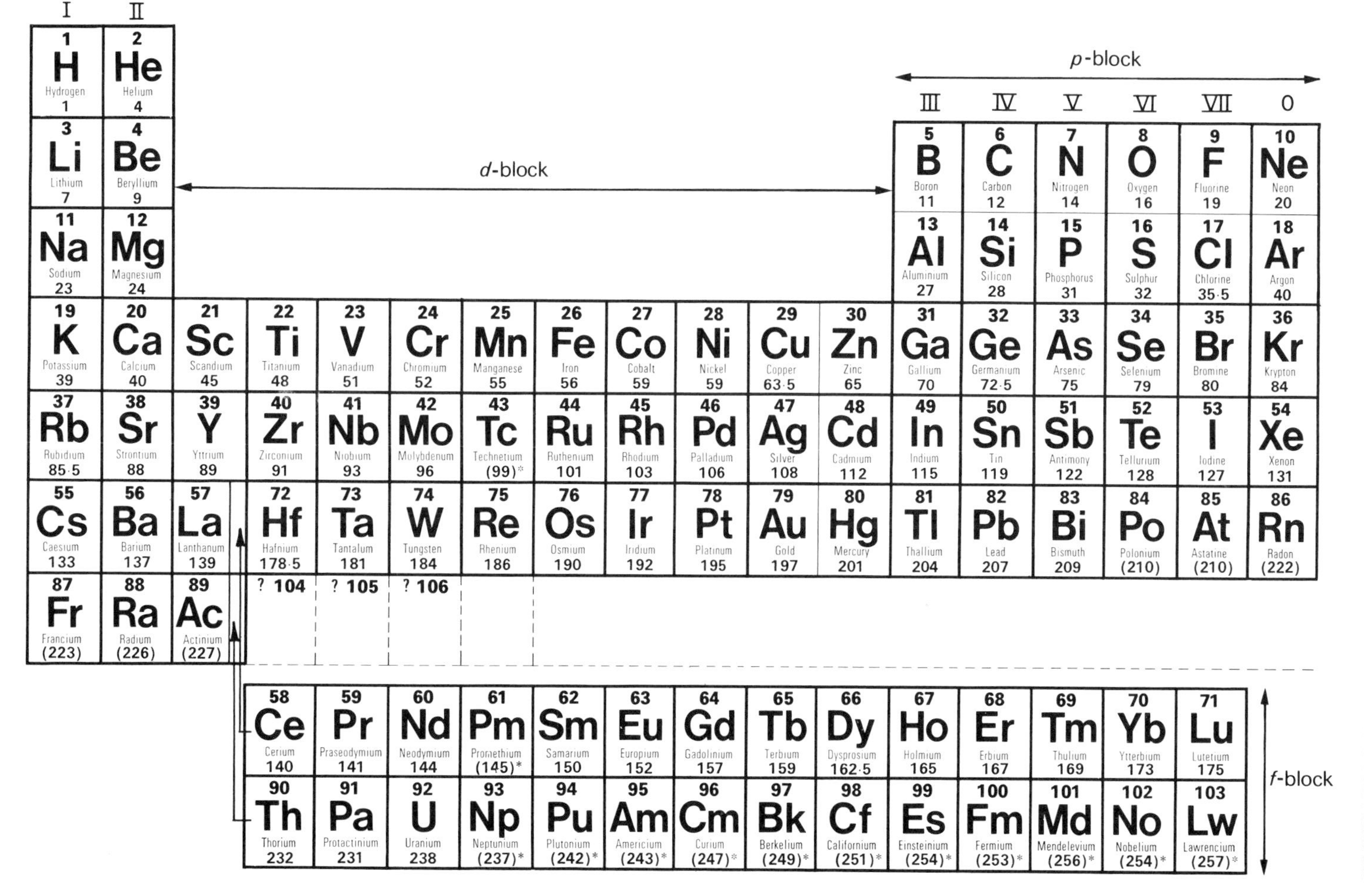

Figure 12.4 Periodic table based on electron structures

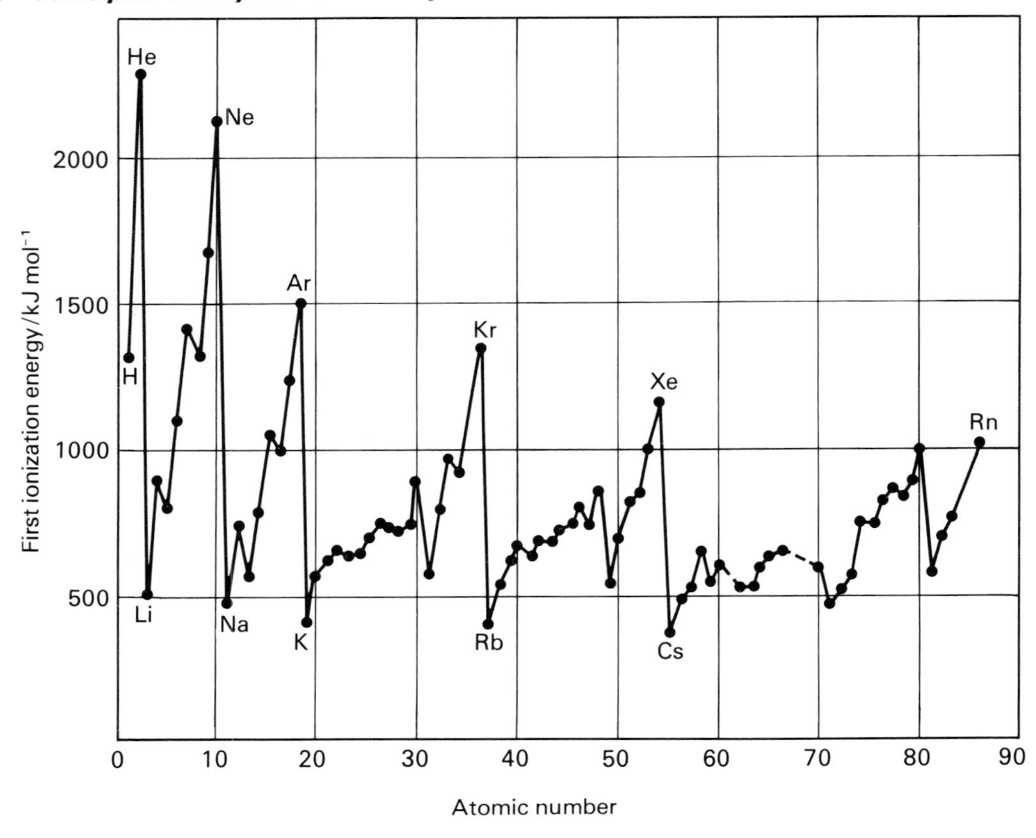

Figure 12.5 Variation of first ionization energy with atomic number

Towards the end of the nineteenth century, many attempts were made to present the information concerning periodicity in different forms to give a more consistent pattern in properties. One, Sir William Crooke's 'Spiral of the Elements' of 1888, is shown in Figure 12.3. The next substantial progress came, however, with Moseley's work on the X-ray spectra of elements in 1913. As discussed in Chapter 2, this led to the establishment of the atomic number of an element as more than its ordinal number on the list in order of atomic mass. When Lothar Meyer's curve and Mendeléev's table are redrawn on the basis of atomic number instead of atomic mass, several difficulties are resolved, as, for example, the need to reverse the order of such elements as iodine and tellurium. With the discovery of the fine electron structures of the elements, it became possible to use this as a basis for the construction of a periodic table. Sodium and potassium, for example, have similar physical and chemical properties and hence fall into the same group of the periodic table because of the similarity of their outer electron structures. All the Group IA metals have the structure:

$$[\text{noble gas}] \times s^1$$

Figure 12.4 shows a periodic table based on the electron structures of the elements. In such a table the elements fall into four blocks according to the orbital type which hold their most energetic electrons. These blocks are the *s*-block, *p*-block, *d*-block, and *f*-block. The *s*-block and *p*-block together contain elements often referred to as the typical elements. The *d*-block and *f*-block elements are known as the transition elements. Subsets within these blocks are recognized, e.g. the *s*-block consists of Group I (the alkali metals) and Group II (the alkaline earth metals). Similarly within the *p*-block, Group VI are known as the chalcogens and Group VII as the halogens. Recognition of subsets is not confined to groups, e.g. the first and second periods of *f*-block metals are known as the lanthanides and actinides respectively from their first members.

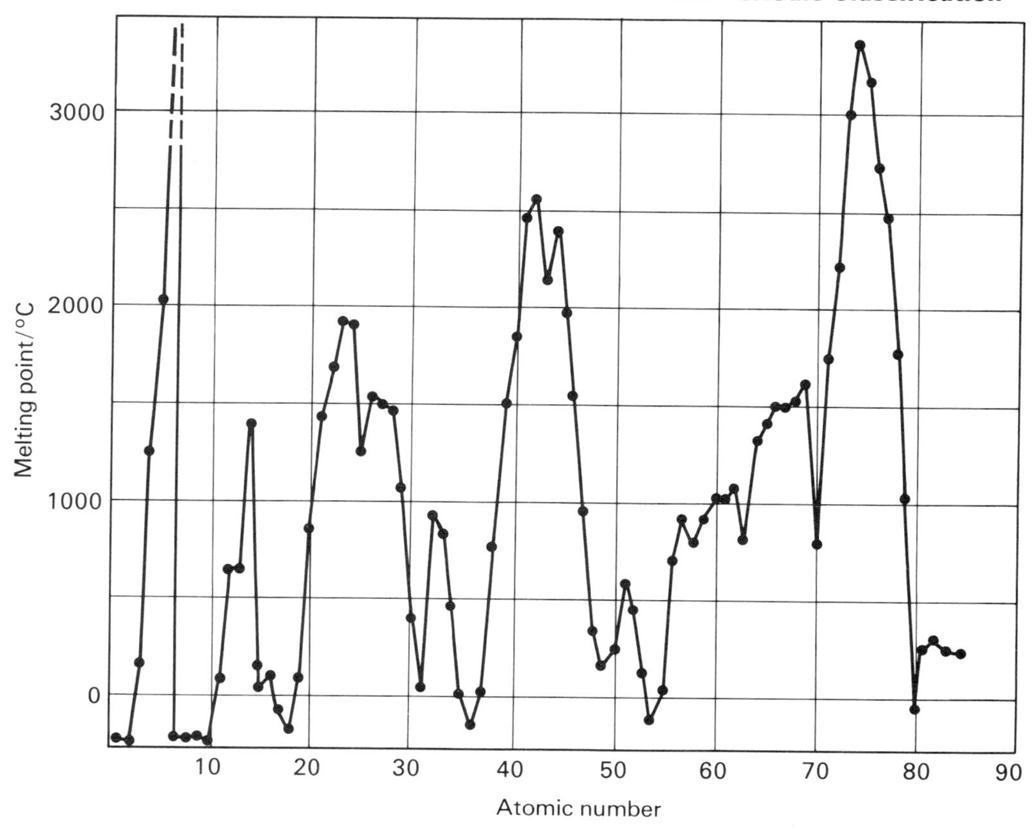

Figure 12.6 Melting points of the elements as a periodic function of atomic number

Trends in physical properties

When numerical values for a variety of physical properties of the elements are plotted against their atomic number a graph showing a periodically repeating pattern is often obtained with elements in the same group occupying corresponding positions on the curves. In Figure 12.5 the first ionization energies of some elements is plotted in this way. Inspection shows that the peaks of the curve are occupied by the noble gas elements and troughs by the Group I, alkali metals. This results from the situation of the most energetic electron (i.e. the most easily removable one) in an orbital of the same type within any particular group.

Their electron structures dictate the type of bonding that occurs in each element and thus physical properties which reflect bonding type also exhibit periodic trends. This is shown in Figures 12.6 and 12.7 where the melting points and the molar latent heats of fusion of the elements are plotted against atomic number.

Trends in chemical properties of the typical elements

I	II	III	IV	V	VI	VII
NaCl	$MgCl_2$	$AlCl_3$	$SiCl_4$	PCl_3 PCl_5	S_2Cl_2	—

In the formulae of the chlorides of the third period given above, the oxidation number of each element is often seen to be equal to the group number or to (8 − group number). In Chapter 4 the factors controlling the stoichiometry of compound formation and hence the oxidation numbers of the elements, is discussed. The metals of the *s*-block invariably have oxidation numbers in their compounds equal to their group numbers. The *p*-block elements are rather more complex than this, although when they form simple compounds they often follow this rule. The simple ions of Group VI elements, for example, are O^{2-}, S^{2-}, and so on. Similarly the simple ions of Group VII are F^-, Cl^-, Br^-, and I^-. With increasing group number the elements in

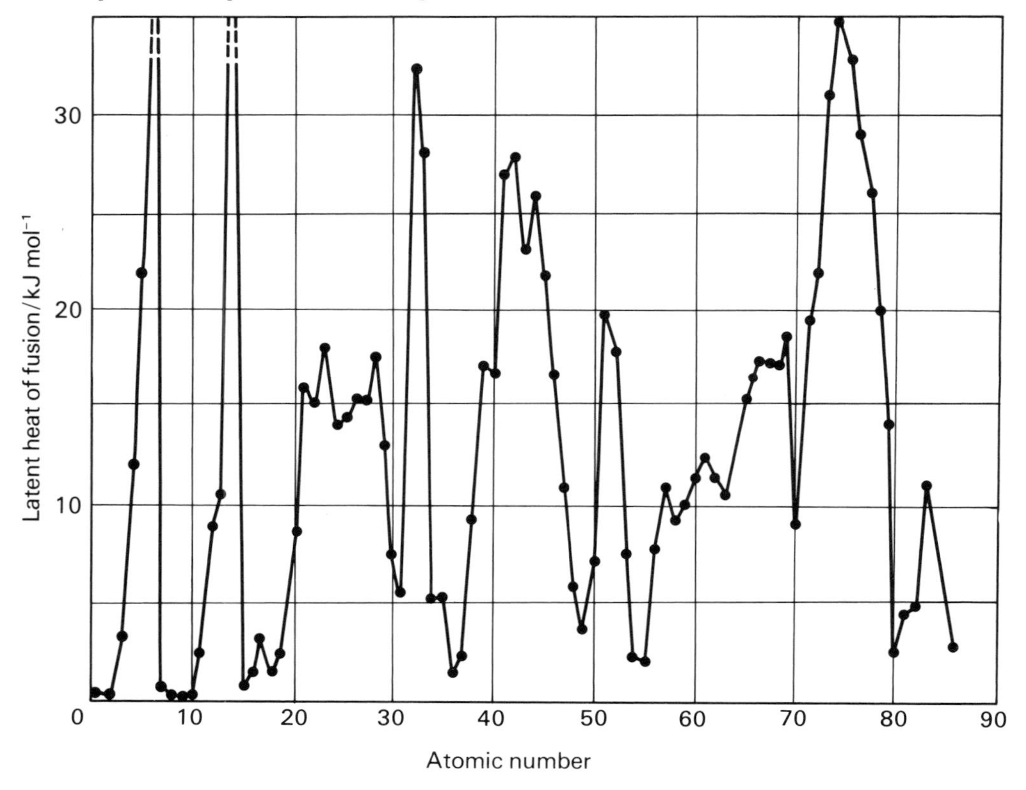

Figure 12.7 Molar latent heat of fusion as a periodic function of atomic mass

any one period tend to become more electronegative and thus exhibit increasing non-metallic behaviour. With increasing atomic number, the elements within any one group become more electropositive and show increasing metallic behaviour. Thus the only Group VII element to form a stable monocharged positive ion is iodine. The more metallic elements of the *p*-block form complex ions such as $Al(H_2O)_6^{3+}$ in aqueous solution, this behaviour which is only shown to any appreciable extent by the less electropositive *s*-block metals such as magnesium. The non-metallic elements of the *p*-block exhibit oxidation numbers other than those suggested by their group numbers by the formation of oxy-anions such as SO_4^{2-} and ClO_3^-. More details concerning this will be found in a suitable textbook of inorganic chemistry.

Na_2O MgO	Al_2O_3	SiO_2
basic	amphoteric	very weakly acidic
P_2O_5	SO_3 Cl_2O_7	
weakly acidic	strongly acidic	

The trend in acid/base behaviour of the oxides across the third period is shown above. This follows the expected pattern as the elements become less metallic and more non-metallic in character.

Patterns in the chemical properties of the transition metals

Transition metals are characterized by the ability to form compounds in a range of oxidation numbers. One aspect of this is the formation of more than one stable compound with a particular non-metallic element. Iron, for example, when heated reacts with chlorine gas to give iron(III) chloride:

$$Fe(s) + 1\tfrac{1}{2}Cl_2(g) \rightarrow FeCl_3(s)$$

When hydrogen chloride gas is passed over heated iron, however, the reaction results in the formation of iron(II) chloride:

$$Fe(s) + 2HCl(g) \rightarrow FeCl_2(s) + H_2(g)$$

This is very different behaviour from that of the

s-block metals which invariably form a single stable compound with any non-metal. The reasons for this are discussed in Chapter 4. When the appropriate Born–Haber cycles are drawn for transition metal compounds, it is found that more than one compound is possible with sufficiently negative values of $\Delta H^\ominus$ to ensure its thermodynamic stability. (Strictly speaking $\Delta G^\ominus$ values should be used in the construction of Born–Haber cycles. However, enthalpy values used for this purpose are generally a reliable guide to thermodynamic stability.)

The transition metals and their compounds often exhibit catalytic activity. This can be due to their ability to form compounds in a range of oxidation numbers or, in the case of heterogeneous catalysis, to their having suitable interatomic distances between the atoms on their surfaces. This is discussed in Chapter 10.

Copper(II) salts in aqueous solution contain the metal in the form of a complex cation which can be represented by the formula $Cu(H_2O)_6{}^{2+}$. The arrangement of the water molecules around the copper is shown in Figure 12.8. The complex ion is held together by bonds formed when each water molecule provides an electron pair which is inserted into an empty *d*-orbital around the copper. Substances that form complex ions with metals in this manner are known as *ligands*. The octahedral shape is a common one for the complex ions of transition metals. If a series of solutions of a copper salt are taken and varying amounts of sodium chloride are dissolved in them, ligand displacement reactions occur which can be represented by the equations:

$$Cu(H_2O)_6{}^{2+}(aq) + Cl^-(aq) \rightleftharpoons [Cu(H_2O)_5Cl]^+(aq) + H_2O(l) \qquad \lg K_c = 2.80$$

$$[Cu(H_2O)_5Cl]^+(aq) + Cl^-(aq \rightleftharpoons [Cu(H_2O)_4Cl_2](aq) + H_2O(l) \qquad \lg K_c = 1.60$$

$$[Cu(H_2O)_4Cl_2](aq) + Cl^-(aq) \rightleftharpoons [Cu(H_2O)_3Cl_3]^-(aq) + H_2O(l) \qquad \lg K_c = 0.49$$

$$[Cu(H_2O)_3Cl_3]^-(aq) + Cl^-(aq) \rightleftharpoons [Cu(H_2O)_2Cl_4]^{2-}(aq) + H_2O(l) \qquad \lg K_c = 0.73$$

The equilibrium constants for these reactions, known as the stability constants for the complex ion formed, can be determined from measurements of partition coefficients (Chapter 6, page 88), redox potentials (Chapter 9, page 125) and spectroscopy (Chapter 3, p. 37). In solutions of copper(II) chloride of differing concentrations the equilibria above are established and this results in differences in the visible absorption spectra of such solutions as can be seen in Figure 12.9. Changing the ligands is seen to result in differences not only in the intensity of absorption but also in the spectral region in which absorption occurs. The colour of many of

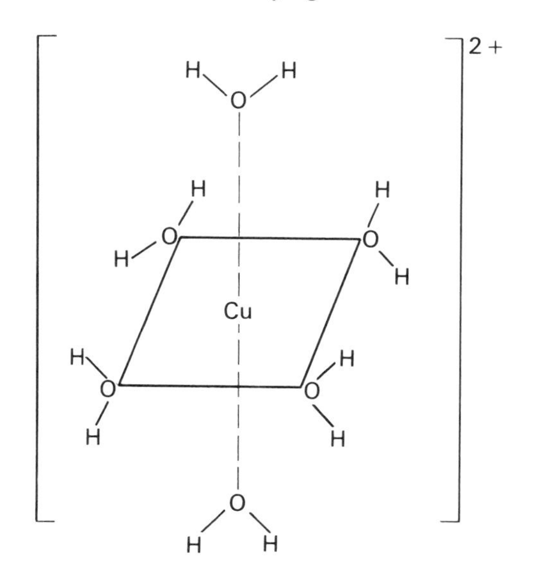

Figure 12.8 Structure of the hydrated copper(II) ion

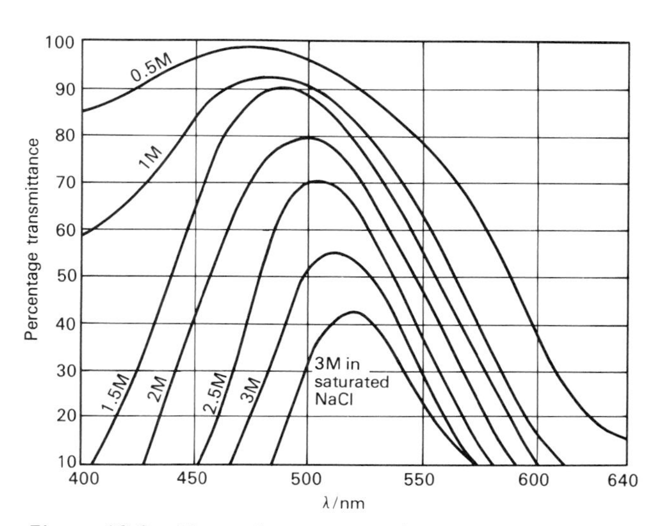

Figure 12.9 Absorption spectra of copper(II) chloride solutions of different concentrations

the compounds of the transition metals is regarded as one of their characteristics.

Salt hydrolysis

As discussed in Chapter 8, page 112, aqueous solutions of salts do not always have neutral pH values. The chlorides of the *d*, *f*, and *p*-block metals dissolve in water to give acidic solutions. This is due to reactions of the type:

$$Fe(H_2O)_6{}^{3+}(aq) + H_2O(l) \rightleftharpoons [Fe(H_2O)_5(OH)]^{2+} + H_3O^+(aq)$$

A proton from one of the water molecules acting as a ligand around the metal ion is donated to an outside water molecule, thus increasing the concentration of hydronium ions. Further donation of protons from the remaining water molecule ligands is also thought to occur. The ions of the more electropositive metals of the *s*-block do not behave in this way because they form much weaker links with the water molecules acting as solvent. The chlorides of the less electropositive *s*-block metals, such as magnesium, do, however, dissolve to give acidic solutions.

The colour of transition metal compounds

The colours of the hydrated metal ions in aqueous solution of the first row of the *d*-block transition metals together with details of their electron structures is given in Table 12.2. This information refers to ions of the type $M(H_2O)_6{}^{2+}$ (aq) present as their sulphate.

Table 12.2 Colours of some transition metal hydrated ions

Ion	*Colour*	*Total number of electrons in d-orbitals*	*Number of unpaired electrons in d-orbitals*
Sc^{3+}	Colourless	0	0
Ti^{3+}	Pink-violet	1	1
V^{3+}	Green	2	2
Cr^{3+}	Violet	3	3
Cr^{2+}	Blue	4	4
Mn^{2+}	Pale pink	5	5
Fe^{3+}	Pale violet		
Fe^{2+}	Green	6	4
Co^{2+}	Red	7	3
Ni^{2+}	Green	8	2
Cu^{2+}	Blue	9	1
Cu^{+}, Zn^{2+}	Colourless	10	0

Inspection of the table shows that no colour is shown when the central ion either has no electrons in the *d*-orbital or when the *d*-orbital is full. The extremely pale colour of Mn^{2+} and Fe^{3+} is associated with the presence of one electron in each of the five *d*-orbitals. The shapes of *d*-orbitals are shown in Figure 2.27 (b), page 33. In an isolated, non-complexed transition metal ion, the five *d*-orbitals have identical energies, they are said to be degenerate. In the case of Ti^{3+}, for example, the single *d*-electron has the choice of five orbitals of equal energy and transfer between orbitals involves no overall energy change. It can be shown that if six ligands (either negative ions or polar molecules) are disposed octahedrally around it, because the *d*-orbitals are differently orientated in

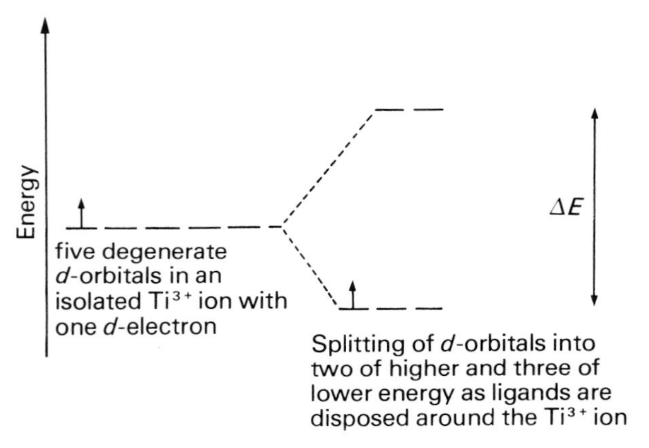

Figure 12.10 Effect of disposing ligands around an isolated Ti^{3+} ion on the energy levels of its *d*-orbitals

space, the electron will be repelled from an orbital close to a ligand. The energy of such orbitals thus becomes greater than that of the others and the degeneracy is lost. This situation is depicted in Figure 12.10. Two of the *d*-orbitals have greater energy associated with them than the other three. An electron in an orbital of lower energy must absorb the energy difference (ΔE) to be promoted to one of the higher energy orbitals. From Planck's relationship:

$$\Delta E = h\nu$$

such an electron will absorb energy of a particular frequency if presented with a continuous spectrum of radiation. The value of ΔE depends on the particular transition metal ion and on the ligands involved. For transition metal ion complexes, the values of ΔE generally correspond to frequencies in the visible region of the electromagnetic spectrum. Thus when white light is viewed through copper(II) sulphate solution, for example, light in the red portion of the spectrum is absorbed and the solution appears to be blue. This is a simplified explanation. A more advanced analysis must take into account so-called 'forbidden' electron transitions. It can be shown that certain transitions of electrons between orbitals are not possible and thus no energy adsorption due to them can occur. A detailed treatment of this restriction is beyond the scope of this book.

Suggestions for further reading

PUDDEPHATT, R. J., *The Periodic Table of the Elements* (Oxford: Oxford University Press, 1972)

13. Surface Chemistry and Colloidal Systems

SURFACE chemistry is concerned with chemical phenomena occurring at the interface or boundary between two phases. In Chapter 10 the role of finely divided nickel as a catalyst in the hydrogenation of ethene is discussed. The attachment of the hydrogen molecules to the metal surface, known as *adsorption*, is a typical surface phenomenon. Adsorption should not be confused with absorption, this latter being the penetration of molecules into the bulk of the absorbing medium as when gases dissolve in liquids or liquids soak into porous solids.

Activated wood or animal charcoal can be used to adsorb certain solute molecules from aqueous solution. This effect can be conveniently investigated using a suitable acid as solute, so allowing the equilibrium proportions between adsorbed and non-adsorbed material to be determined by titration. In such an experiment 100 cm^3 portions of 0.30, 0.20, 0.10, 0.05, 0.01, and 0.005 M solutions of ethanedioic (oxalic) acid were placed in separate stoppered bottles. A 5 g portion of activated charcoal was added to each and the mixtures left for some hours to allow equilibrium to be established. Each was separately filtered and the filtrates titrated against a standard solution of sodium hydroxide using phenolphthalein indicator. From the titration the moles of solute adsorbed and the concentration remaining in the solution with which it is in equilibrium were calculated. The results are shown in Table 13.1.

In 1906, Freundlich proposed the equation:

$$\frac{\text{moles of solute adsorbed}}{\text{mass of adsorbent/g}} = k\,[\text{solute(aq)}]^{n}_{\text{eqm}}$$

where k and n are constants that depend on the particular solvent and adsorbent. The validity of this equation can be verified from the experimental results. On taking logarithms

$$\lg\left\{\frac{\text{moles of solute adsorbed}}{\text{mass of adsorbent/g}}\right\} = n\lg\,[\text{solute(aq)}]_{\text{eqm}} + \lg k$$

The result of plotting

$$\lg\left\{\frac{\text{moles of solute adsorbed}}{\text{mass of adsorbent/g}}\right\}$$

against $\lg\,[\text{solute(aq)}]_{\text{eqm}}$ is shown in Figure 13.1. The linear relationship confirms the validity of the Freundlich equation. If the constants n and k are required, these can be determined from the gradient and the intercept respectively.

Applications of surface chemistry

Detergency

This is the most important commercial application of surface chemistry. Consider the situation when a grease particle that is attached to a clothing fibre is removed during laundering. The surface between the particle and the fibre is broken down and two new surfaces, between particle and solution and between fibre and solution, are formed. For this to occur the appropriate free energy change must be negative:

$$\Delta G = G_{(\text{solution/fibre})} + G_{(\text{solution/particle})} - G_{(\text{fibre/particle})}$$

Table 13.1 Adsorption of ethanedioic acid on to activated charcoal

Moles of solute adsorbed	*Moles of solute adsorbed / Mass of adsorbent/g*	$[Solute(aq)]_{eqm}$
0.11	0.022	0.011
0.14	0.028	0.022
0.32	0.064	0.034
0.61	0.122	0.083
0.84	0.168	0.110
1.20	0.240	0.280

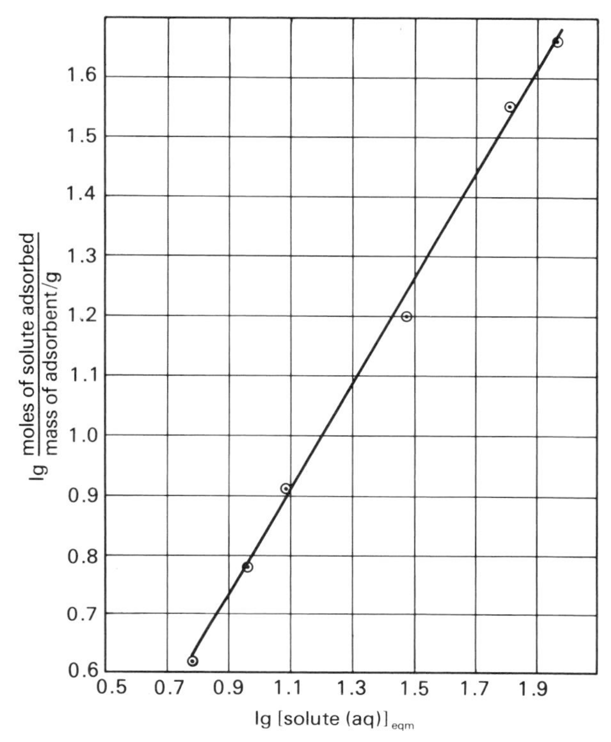

Figure 13.1 Verification of the Freundlich equation

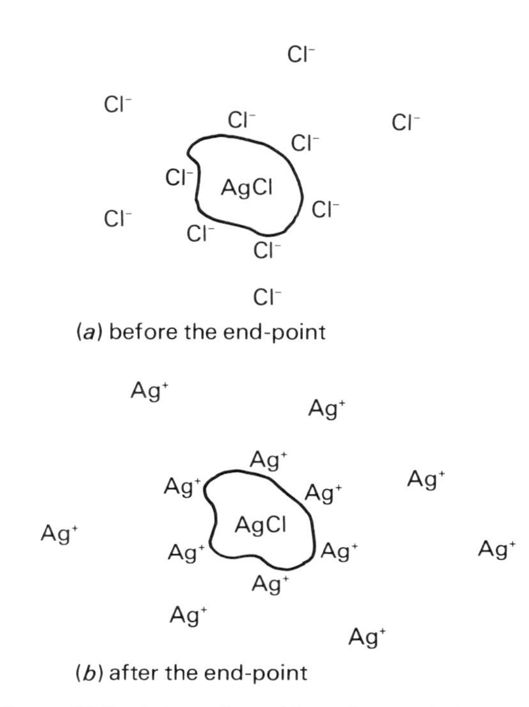

Figure 13.2 Adsorption of ions from solution onto a silver chloride surface

The surface tension of a liquid is the free energy per unit area of surface with which the liquid is in contact. Thus for a liquid to be a good detergent it must have a low surface tension at the surfaces of both particle and fibre. Substances added to water as detergents work by producing a solution of lower surface tension than the water alone. Soap, the traditional detergent, contains ions such as stearate ($C_{17}H_{35}CO_2^-$) and oleate ($C_{17}H_{33}CO_2^-$) which accomplish this. Synthetic detergents have structures which involve a long hydrocarbon tail with an ionic head.

'Spot' reagents

The freshly precipitated hydroxides of many metals selectively adsorb on to their surface specific organic dyes. This provides a useful and versatile method of determining the presence of many metal ions in aqueous solution. The sensitivities of such tests are among the most sensitive without involving physical methods. Details will be found in suitable textbooks of analytical chemistry.

Adsorption indicators

During titrations involving precipitation, the solid as it forms generally adsorbs ions from the surrounding solution on to its surface, thus acquiring an electrical charge. When silver nitrate solution, for example, is run into a chloride solution the silver chloride solid as it is first formed becomes negatively charged by adsorption of excess chloride ions as shown in Figure 13.2 (*a*). As the end-point is approached this charge is reduced as the chloride ions are reacted; and at the end-point the solid is electrically neutral. After the end-point there is an excess of silver ions which are adsorbed as shown in Figure 13.2(*b*), giving the solid a positive charge. The reaction can be monitored by adding some fluorescein solution to the chloride solution. Fluorescein is a weak acid and its anion, fluoresceinate, is coloured. Prior to the end-point the solid and fluoresceinate have the same negative charge and so the fluoresceinate anions are not adsorbed. After the end-point, however, the now positive solid attracts the fluoresceinate ions which form a distinctive secondary layer on its surface.

Colloids

The work of Thomas Graham during the early part of the nineteenth century on the diffusion of gases is discussed in Chapter 1. He extended his studies to include the diffusion of solutes from aqueous solution through a parchment membrane. He found that sodium chloride, for example, diffused about four hundred times faster through the membrane than did gum arabic. Solutions of both these substances appear homogeneous and they cannot be separated by filtration. Graham called substances showing rapid diffusion *crystalloids* from their obviously crystalline appearance. For those showing slow diffusion he introduced the name *colloid* (from the Greek and meaning glue-like). X-ray diffraction studies have shown that this distinction is not always valid, and that many substances he classified as colloids are crystalline. A colloidal solution is essentially a heterogeneous two-phase system with a large area of boundary between the phases. It is thus intermediate between a true solution and a suspension. Table 13.2 shows a comparison of solutions, colloids, and suspensions.

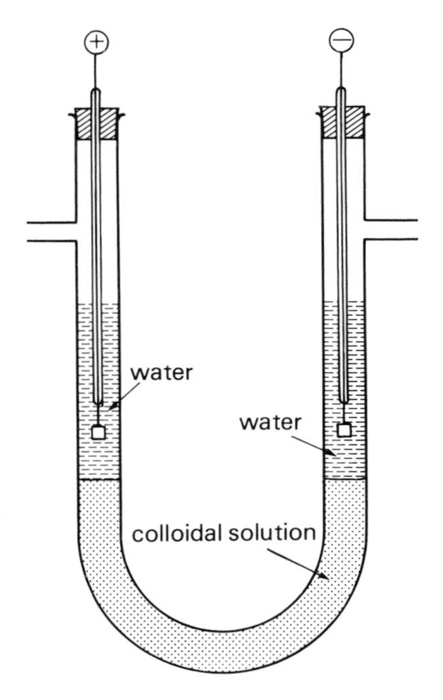

Figure 13.3 Apparatus for electrophoresis

The Tyndall effect

The *Tyndall effect* is the name given to the scattering observed when a beam of light is passed through a colloidal solution held in a transparent container. An ultracentrifuge operates in the region of 60 000 revolutions per minute. Apparatus suitable for electrophoresis is shown in Figure 13.3. When the colloid migrates to the cathode this is called cataphoresis and when it migrates to the anode, anaphoresis.

Graham used apparatus of the type shown in Figure 13.4 to separate the colloidal materials from true solutions. The technique is known as

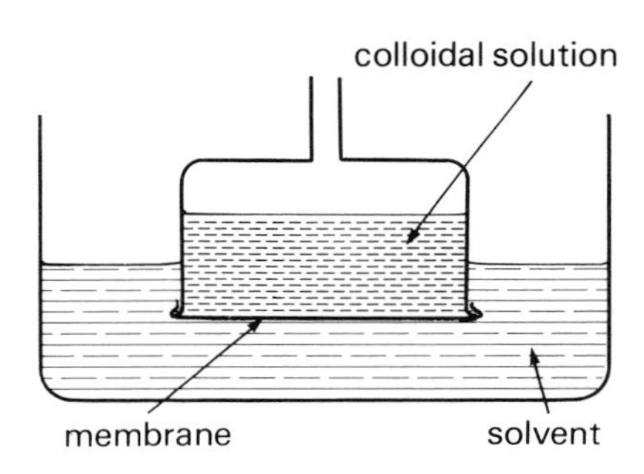

Figure 13.4 Apparatus of the type used by Graham for dialysis

Table 13.2 Comparison of solutions, colloidal systems, and suspensions

Solution	*Colloidal system*	*Suspension*
Homogeneous	Heterogeneous	Heterogeneous
Particles less than 10^{-9} m	Particles between 10^{-9} and 10^{-6} m	Particles greater than 10^{-6} m
Does not scatter light	Scatters light (Tyndall effect)	Scatters and absorbs light
Never separates by sedimentation	No sedimentation under gravity but possibly by ultracentrifuge	Sedimentation under gravity
An electric current sometimes results in electrolysis	An electric current can cause the migration of colloidal particles to one electrode (electrophoresis)	An electric current has no effect

dialysis. Patients suffering from kidney failure may be connected at intervals to a kidney machine as shown in Figure 13.5. Their blood is first passed through the instrument in the centre of the photograph where it is monitored for electrolyte content and temperature. The blood then passes into the dialyser shown in Figure 13.5 with an arrow. Here the blood is passed along one side of a dialysing membrane along the other side of which there is a counter-current flow of water. Non-colloidal excretory products, such as urea, pass through the membrane but the colloidal components of blood do not.

Types of colloids

The dispersion of solids in a liquid is only one type of colloid, known as a *sol* or *suspensoid*. Other types of colloidal system are also recognized. In each case it is possible to distinguish between a continuous phase, known as the disperse medium, and a phase consisting of discrete particles contained in it, known as the *disperse phase*.

disperse phase + disperse medium = colloidal system

Some common colloidal systems are listed in Table 13.3.

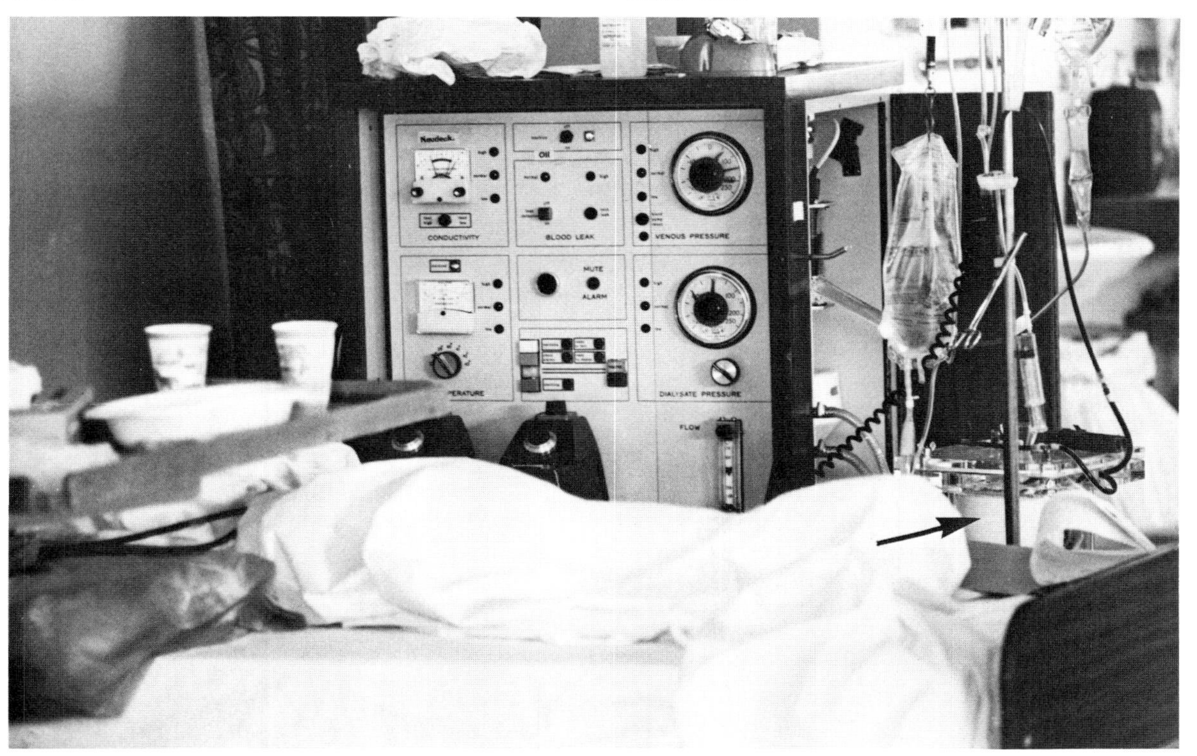

Figure 13.5 A kidney machine and dialyser

Table 13.3 Some common colloidal systems

Disperse phase	*Disperse medium*	*Colloidal system*
Gas	Liquid	Foam (e.g. whipped cream)
Liquid	Gas	Mist or cloud
Liquid	Liquid	Emulsion (e.g. hair cream)
Liquid	Solid	Gel (e.g. jelly)
Solid	Gas	Smoke
Solid	Liquid	Sol or suspensoid

Lyophilic and lyophobic sols

Lyophilic sols (from the greek meaning 'release-loving') have considerable mutual affinity between the two phases. Such systems include sols of the proteins albumen and gelatin with water. *Lyophobic sols* (meaning 'release-hating') have little affinity between the two phases and tend to be unstable. Examples include arsenic(III) sulphide and gold with water. The major characteristics of these two types of sols are summarized in Table 13.4.

Preparation of sols

The discussion will be limited to hydrosols in which the disperse medium is water. The methods available fall into two categories:

(1) dispersion methods in which larger particles are broken down to colloidal size, and
(2) aggregation methods in the colloidal sized particles are built up from smaller particles.

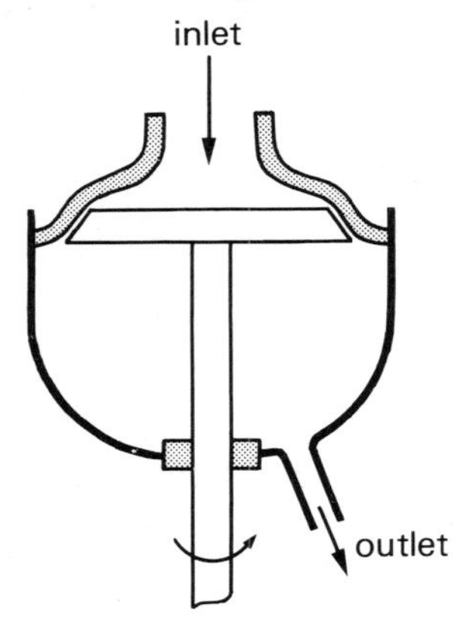

Figure 13.6 A colloid mill

Table 13.4 Characteristics of lyophilic and lyophobic sols

Lyophilic sols	*Lyophobic sols*
Individual particles cannot be detected by ultramicroscope	Individual particles can be detected by ultramicroscope
Do not migrate under an applied electrical potential	Exhibit electrophoresis under an applied electrical potential
Not readily precipitated by addition of electrolyte solutions	Readily precipitated by the addition of electrolyte solutions
Have high viscosities	Viscosity similar to that of disperse medium
Have low surface tensions	Surface tension similar to that of disperse medium
On evaporation or cooling form gels which reform sol on addition of extra disperse medium	On evaporation or cooling the disperse phase coagulates and the sol cannot readily be re-formed

Dispersion methods

(*a*) The colloid mill: the disperse phase is shredded between high-speed plates until the particle size is of colloidal proportions. This is shown in Figure 13.6.
(*b*) Peptization: when certain precipitates are well shaken with water, some of the solid undergoes the reverse of coagulation. A suitable reagent, known as a peptizing agent, may be added to assist in the process. The peptizing agent generally contains an ion in common with the substance precipitated.
(*c*) Bredig's method: hydrosols of certain metals such as platinum and gold can be obtained by striking an electric arc between electrodes of the metal placed under water. The apparatus is illustrated in Figure 13.7.

Aggregation methods

(*a*) Hydrolysis: if a concentrated solution of certain metal salts is slowly added to a large volume of boiling water, hydrolysis occurs which results in the formation of a hydrosol with the metal hydroxide as the disperse phase, e.g. when iron(III) chloride solution is treated in this way:

$$Fe(H_2O)_6{}^{3+}(aq) \rightarrow Fe(OH)_3(s) + 3H_3O^+(aq)$$

(*b*) Precipitation: when suitable mixtures of ions that give insoluble solids are made, the precipitate formed may be partly or completely colloidal, e.g. when hydrogen sulphide gas is bubbled into a solution of an arsenic(III) compound:

$$2As^{3+}(aq) + 3S^{2-}(aq) \rightarrow As_2S_3(s)$$

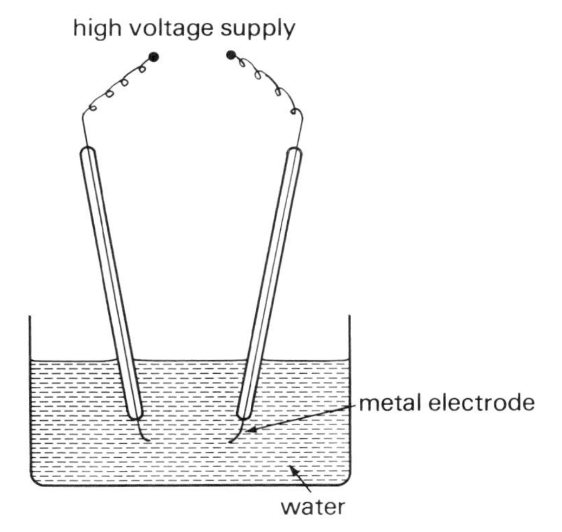

Figure 13.7 Bredig's method of hydrosol formation

(*c*) Oxidation and reduction: reactions of this type between substances in aqueous solution can give solid products of colloidal proportions, e.g. when solutions of hydrogen sulphide and sulphur dioxide react:

$$2H_2S(aq) + SO_2(aq) \rightarrow 2H_2O(l) + 3S(s)$$

When certain metal ions react with suitable reducing agents, colloidal sized metal particles may be formed.

(*d*) Change of solvent: when a solution of resin in ethanol is added to an excess of water, the solid resin precipitates in colloidal form.

Table 13.5 Precipitation of sols by selected electrolyte solutions

	Critical concentration/10^{-3} mol dm^{-3}	
Electrolyte	*Iron(III) hydroxide (positive)*	*Arsenic(III) sulphide (negative)*
NaCl	9.3	51
KCl	9.0	50
K_2SO_4	0.20	
$MgSO_4$	0.22	
$K_3Fe(CN)_6$	0.096	
$MgCl_2$		0.72
$ZnCl_2$		0.69
$Al(NO_3)_3$		0.095
$Ce(NO_3)_3$		0.092

In those methods described above where the production of the sol results in the presence of unnecessary spectator ions, the sol can be separated from them by the technique of dialysis described previously.

Precipitation of sols

As discussed previously, the individual particles of the disperse phase of a sol can adsorb a layer of ions from the disperse medium and thus acquires a positive or negative charge. The sign of this charge can be deduced by carrying out electrophoresis of the sol. As all the particles of the disperse phase carry a charge of the same sign, the mutual repulsion that results between them contributes to the stability of the sol. If electrolyte solutions are added to the sol, the ions of opposite charge electrically neutralize the particles and thus tend to precipitate them. Ions carrying multiple charges are more efficient than single charged ions in this respect and cause precipitation at lower concentrations. Table 13.5 shows the minimum concentrations of selected electrolytes that are required to precipitate iron(III) hydroxide, a typical anaphoretic sol, and arsenic(III) sulphide, a typical cataphoretic sol.

Emulsions

These are colloidal systems in which the disperse phase and disperse medium are a pair of immiscible liquids. An emulsion may thus be stabilized by adding an emulsifying agent which lowers the surface tension between the phases.

Applications of colloids

Many natural colloidal systems are of importance in everyday life, of which blood and milk are two of the most obvious examples. Some man-made uses of colloidal systems include:

(*a*) paints: all paints are colloidal systems and consist of a solid disperse phase in a liquid disperse medium. 'Non-drip' or thixotropic paints are special examples. At rest these have high viscosity as they contain long chains of polymer molecules which coil and entrap much disperse medium. When a shearing stress is applied with a paint-brush, the coils straighten (as when a coiled spring is pulled) resulting in a sharp drop in viscosity.

(*b*) Dairy products: milk has already been discussed as a natural colloidal material. It is an aqueous emulsion of fat stabilized by the protein casein. Cream is formed on the top of milk by the less dense fat particles rising to the surface of the aqueous phase. On the vigorous agitation of cream, fat particles congeal forming lumps of butter, and the excess water is drained off. When cream is whipped, air is beaten into it as a disperse phase and a foam is produced.

(*c*) Cooking: meringues are made by beating egg whites until a stiff foam with air as disperse phase and albumen as disperse medium is produced. Mayonnaise is made by beating egg yolks and olive oil into an emulsion.

(*d*) Pharmaceutical preparations: many ointments for application to the skin consist of the physiologically active component dissolved in an oil and made into an emulsion with water. Materials of high molar mass such as insulin and penicillin are produced in colloidal form suitable for injection.

Suggestions for further reading

SHAW, D. J. *Introduction to Colloid and Surface Chemistry*. (London: Butterworth Group, 1970)

Problems

1. 1-g portions of activated charcoal were allowed to equilibrate with a series of aqueous solutions of propanone. The mixtures were filtered and from analysis of the filtrate the results show in the table at the foot of the page were obtained.

Show that these results are in agreement with the Freundlich equation and evaluate the constants k and n.

2. Give two examples of lyophobic sols and two examples of lyophilic sols.

Describe briefly three experiments which could be carried out in order to decide which of two sols is lyophobic and which is lyophilic.

Explain why both lyophobic and lyophilic sols do not coagulate in normal conditions.

Why are car headlights almost ineffective in fog? (Associated Examining Board 'A')

3. A silver iodide sol is visible under the ultramicroscope, and is more readily precipitated by ammonium sulphate solution than by potassium chloride solution. It is less readily precipitated if some gelatin is added to the system. Explain these observations and suggest a method of preparation for the sol. How would you expect the particles to move in an electric field?

Describe how the charge on the sol might be reversed and explain the changes that take place during the reversal. (Cambridge 'S')

4. Give a concise account of the colloidal state of matter (preparative methods are *not* required). Include in your account (*a*) the factors which determine the stability of a lyphobic colloid, (*b*) a brief description of how you would demonstrate three characteristics of the colloidal state. (Oxford and Cambridge 'S')

Equilibrium concentration of propanone/mol dm^{-3}	0.0147	0.0410	0.0886	0.178
Amount propanone adsorbed/10^{-3} mol	0.618	1.08	1.51	2.08

14. Nuclear and Radiochemistry

In Chapter 2, the contribution made by discoveries concerning radioactivity to modern theories of atomic structure are discussed. The student is advised to revise this chapter before considering the further development of the subject that follows.

The stability of the nucleus

The mass of an atom might be expected to equal the sum of the masses of the protons, neutrons, and electrons of which it is composed. When this calculation is made for specific atoms it is found that this is not exactly true. The modern standard of atomic mass is based on the isotope $^{12}_{6}C$. This is defined as having a relative atomic mass of 12.000 00. This isotope is formed from six protons with six neutrons in the nucleus together with six extranuclear electrons. The six protons and six electrons are equivalent to six hydrogen atoms.

Mass of six hydrogen atoms	=	6.046 80
Mass of six neutrons	=	6.051 78
Total	=	12.098 58
Observed mass of $^{12}_{6}C$	=	12.000 00
Difference (mass defect)	=	0.098 58

A mass defect is observed when the corresponding calculation is made for all other isotopes. It is explained in terms of Einstein's relationship discussed previously:

$$E = mc^2$$

The energy equivalent to the mass defect is the nuclear binding energy of the isotope. The greater the mass defect, therefore, the greater the nuclear binding energy and the more stable the isotope. This stability is conveniently measured in terms of the '*packing fraction*' of the nucleus. The packing fraction was defined by Aston as:

$$\text{packing fraction} = \frac{(\text{mass of isotope} - \text{mass number}) \times 10^4}{(\text{mass number})}$$

The packing fraction of the most abundant isotope of selected elements is plotted against atomic number in Figure 14.1. The most stable atoms are those with the most negative values of packing fraction and thus the most stable atoms are those close to iron of atomic number 26.

Nuclei become progressively less stable with increasing atomic number because the electrostatic repulsion between the protons more and more outweighs the nuclear binding energies. Figure 14.2 shows that the proportion of neutrons to protons steadily increases with atomic number in the case of stable nuclei. Radio-isotopes of the heavier elements decay by α-emission (involving the loss of 2 protons and 2 neutrons) and the daughter product formed occupies a position on the graph closer to the line representing nuclear stability. With the lighter elements β-decay and positron decay result in daughter products also closer to this line.

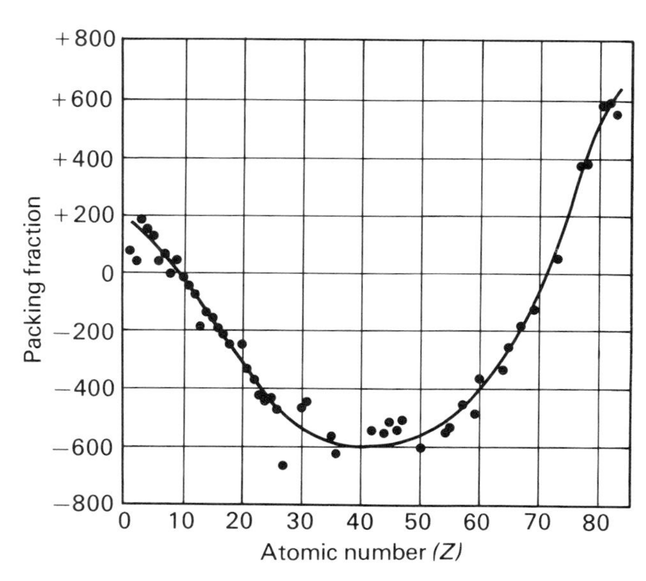

Figure 14.1 Packing fraction of the most abundant isotope of selected elements plotted against atomic number

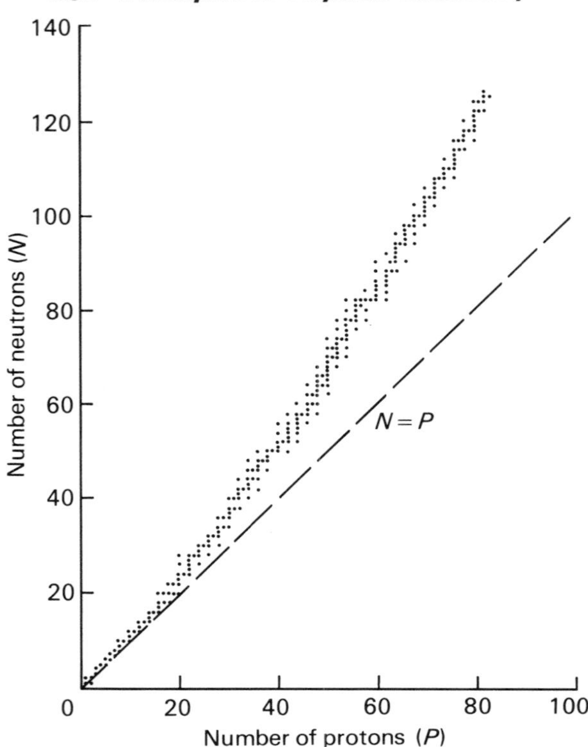

Figure 14.2 Proportion of neutrons to protons in stable naturally occurring isotopes

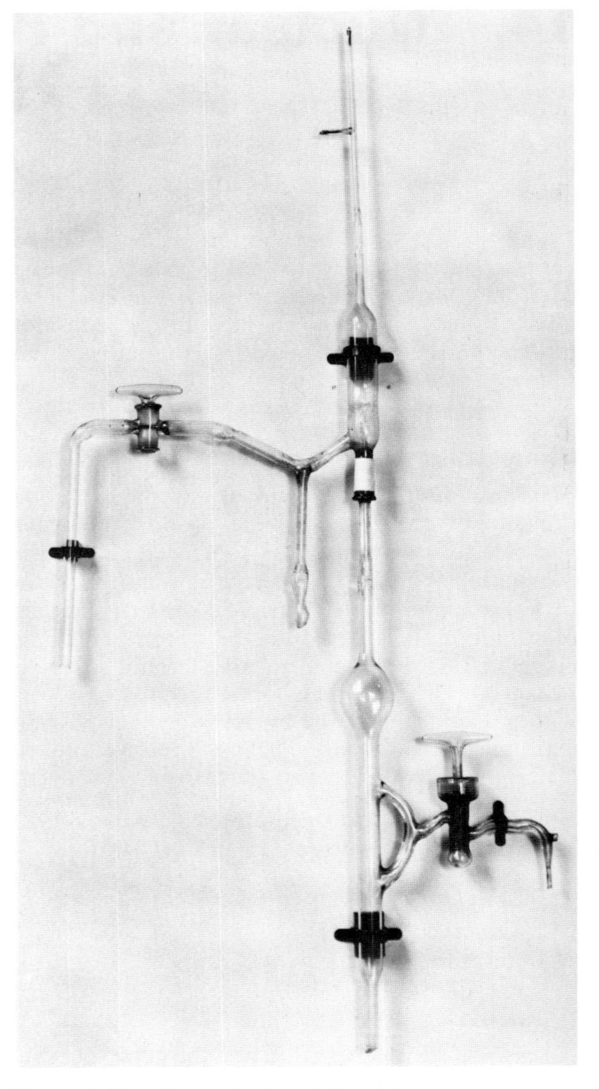

Figure 14.3 Rutherford and Royds' apparatus for the identification of α-particles

Modes of radioactive decay

The discovery that the radiation from radium could be split into three components, α, β, and γ is discussed in Chapter 2. In 1932, Carl Anderson discovered the existence of positive particles with the same mass as an electron in cosmic radiation. It was later shown that these particles, known as positrons, can be emitted during radioactive decay.

α-emission

Proof that α-particles are the nuclei of helium atoms came from an experiment by Rutherford and Royds in 1909. Their apparatus is shown in Figure 14.3. A schematic drawing of the apparatus is shown in Figure 14.4. A quantity of the α-emitting gas radon-220 was placed in the thin-walled tube A surrounded by the concentric outer tube B which is attached to the spectrum tube S fitted with electrodes through which an electric discharge could be passed. B was evacuated and after several days gas had collected in it. This gas was compressed by raising the mercury level and, on passing an electrical discharge, the characteristic spectrum of helium was observed. Previous experiments showed that helium gas could not penetrate the walls of A and thus the helium must have been formed, as a monatomic gas, from α-particles passing from A into B. As each α-particle gives one

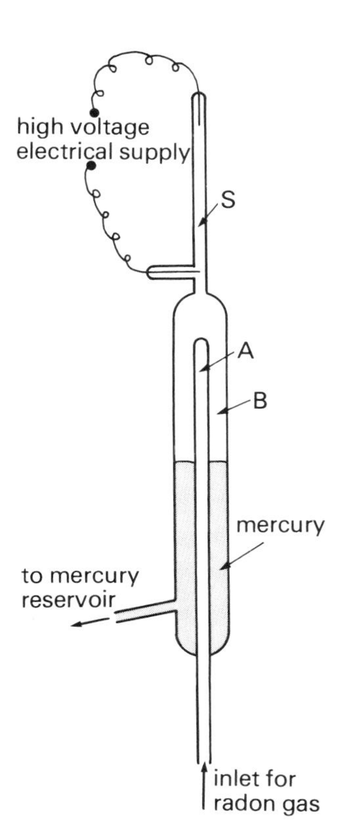

Figure 14.4 Schematic diagram of Rutherford and Royds' apparatus

helium atom, this can be made the basis of a determination of Avogadro's number.

Rutherford and Geiger showed that radium emits α-particles at a rate of $3.40 \times 10^{10}\ \mathrm{s^{-1}\ g^{-1}}$. Rutherford and Royds found that radium produces helium at a rate of $1.27 \times 10^{-9}\ \mathrm{cm^3\ g^{-1}\ s^{-1}}$ at s.t.p.

Therefore 3.40×10^{10} α-particles give 1.27×10^{-9} $\mathrm{cm^3}$ helium at s.t.p.

$3.40 \times 10^{10} \times \dfrac{22\,400}{1.27 \times 10^{-9}}$ α-particles give 22 400 $\mathrm{cm^3}$ of helium at s.t.p.

1 mole of helium contains

$$3.40 \times 10^{10} \times \frac{22\,400}{1.27 \times 10^{-9}} \text{ atoms,}$$

i.e. 1 mole of helium contains 6.00×10^{23} atoms.

Using the apparatus shown in Figure 14.5, the charge to mass ratio, e/m, for α-particles was determined. Knowing the mass of helium atoms, this confirms that a double positive charge is carried by the particles.

When an isotope undergoes α-decay it loses two units of atomic number and four units of mass number. The decay of radon-220, for example, can be represented in two alternative manners:

$$^{220}_{86}\mathrm{Rn} \rightarrow {}^{216}_{84}\mathrm{Po} + {}^{4}_{2}\alpha$$

$$^{220}_{86}\mathrm{Rn} \rightarrow {}^{216}_{84}\mathrm{Po} + {}^{4}_{2}\mathrm{He}$$

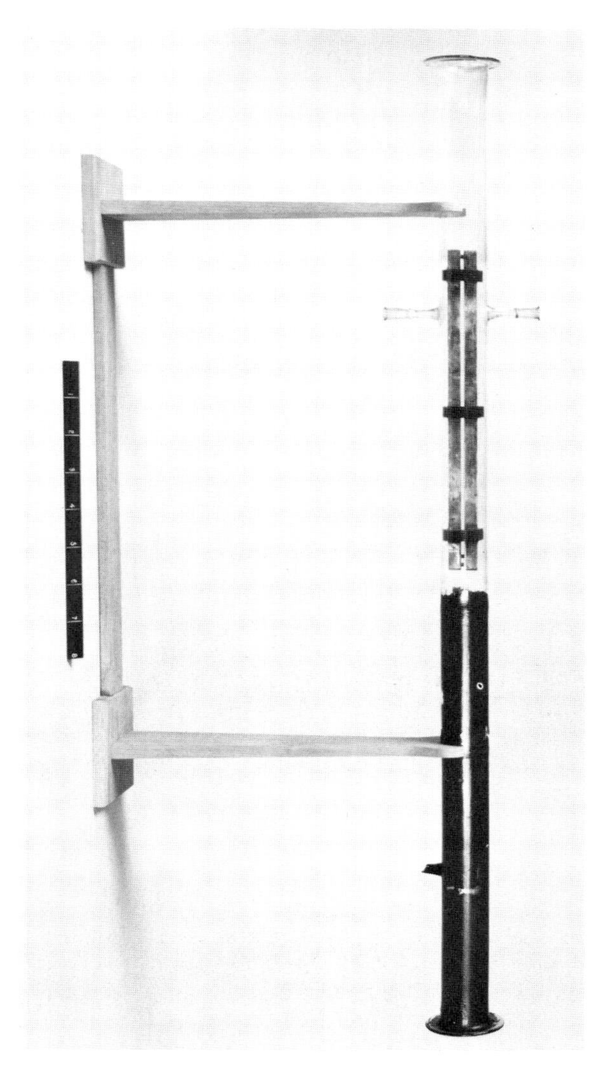

Figure 14.5 Apparatus for the determination of e/m for α-particles

β-emission

β-particles were identified as electrons by measurement of their mass and charge-to-mass ratio. They are emitted from the nucleus and this does not contain free electrons. It is considered that β-particles are formed by the change of a neutron into a proton:

$$^{1}_{0}\mathrm{n} \rightarrow {}^{1}_{1}\mathrm{p} + {}^{\;\,0}_{-1}\mathrm{e}$$

When an isotope undergoes β-decay its atomic number increases by one unit and its mass number

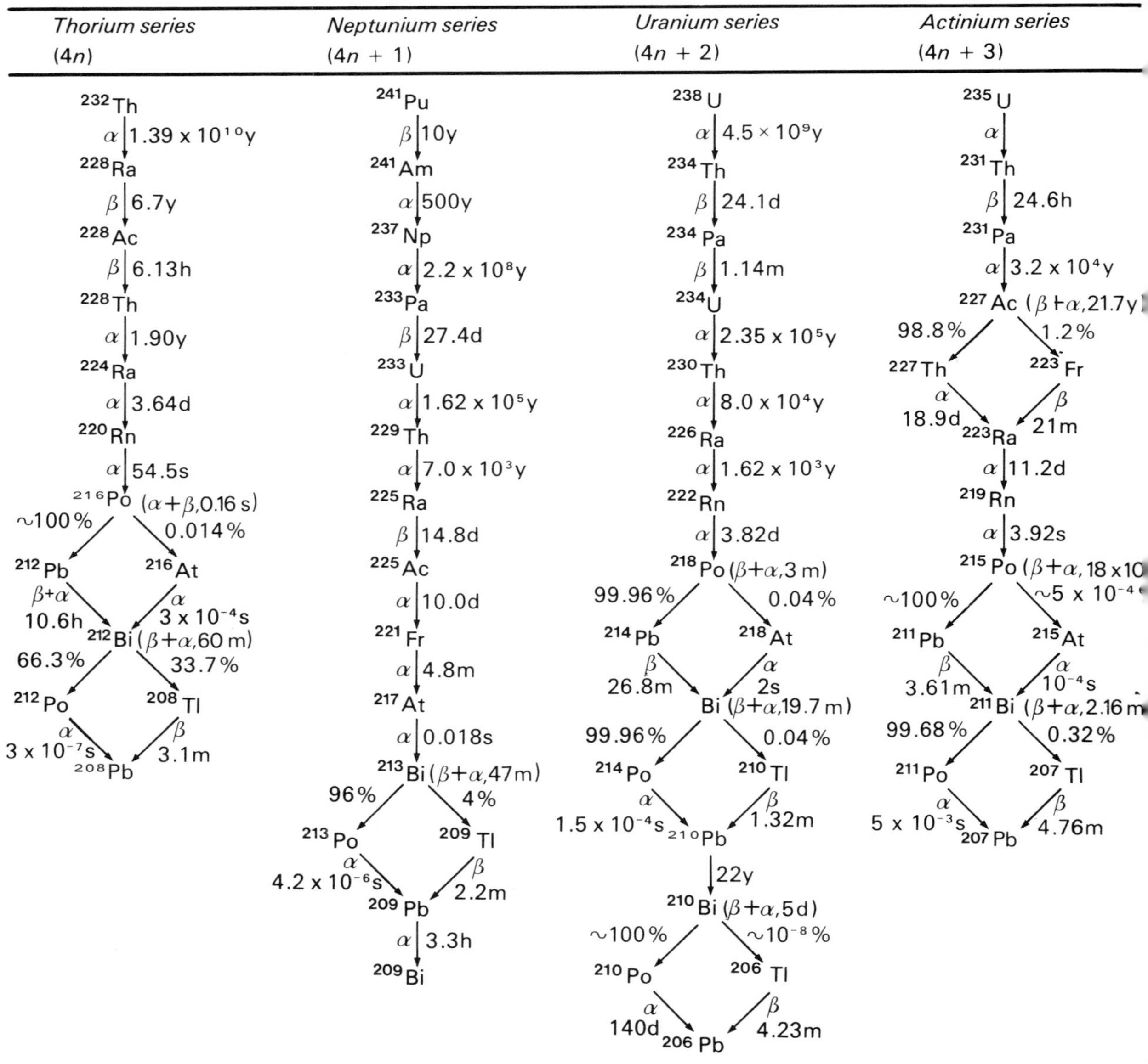

Thorium series ($4n$)

^{232}Th
α | 1.39×10^{10} y
^{228}Ra
β | 6.7y
^{228}Ac
β | 6.13h
^{228}Th
α | 1.90y
^{224}Ra
α | 3.64d
^{220}Rn
α | 54.5s
^{216}Po ($\alpha+\beta$, 0.16 s)
~100% → ^{212}Pb; 0.014% → ^{216}At
^{212}Pb → ^{212}Bi: $\beta+\alpha$, 10.6h
^{216}At → ^{212}Bi: α, 3×10^{-4} s
^{212}Bi ($\beta+\alpha$, 60 m)
66.3% → ^{212}Po; 33.7% → ^{208}Tl
^{212}Po → ^{208}Pb: α, 3×10^{-7} s
^{208}Tl → ^{208}Pb: β, 3.1m

Neptunium series ($4n + 1$)

^{241}Pu
β | 10y
^{241}Am
α | 500y
^{237}Np
α | 2.2×10^{8} y
^{233}Pa
β | 27.4d
^{233}U
α | 1.62×10^{5} y
^{229}Th
α | 7.0×10^{3} y
^{225}Ra
β | 14.8d
^{225}Ac
α | 10.0d
^{221}Fr
α | 4.8m
^{217}At
α | 0.018s
^{213}Bi ($\beta+\alpha$, 47m)
96% → ^{213}Po; 4% → ^{209}Tl
^{213}Po → ^{209}Pb: α, 4.2×10^{-6} s
^{209}Tl → ^{209}Pb: β, 2.2m
^{209}Pb
α | 3.3h
^{209}Bi

Uranium series ($4n + 2$)

^{238}U
α | 4.5×10^{9} y
^{234}Th
β | 24.1d
^{234}Pa
β | 1.14m
^{234}U
α | 2.35×10^{5} y
^{230}Th
α | 8.0×10^{4} y
^{226}Ra
α | 1.62×10^{3} y
^{222}Rn
α | 3.82d
^{218}Po ($\beta+\alpha$, 3 m)
99.96% → ^{214}Pb; 0.04% → ^{218}At
^{214}Pb → Bi: β, 26.8m
^{218}At → Bi: α, 2s
Bi ($\beta+\alpha$, 19.7 m)
99.96% → ^{214}Po; 0.04% → ^{210}Tl
^{214}Po → ^{210}Pb: α, 1.5×10^{-4} s
^{210}Tl → ^{210}Pb: β, 1.32m
^{210}Pb
| 22y
^{210}Bi ($\beta+\alpha$, 5d)
~100% → ^{210}Po; ~10^{-8}% → ^{206}Tl
^{210}Po → ^{206}Pb: α, 140d
^{206}Tl → ^{206}Pb: β, 4.23m

Actinium series ($4n + 3$)

^{235}U
α |
^{231}Th
β | 24.6h
^{231}Pa
α | 3.2×10^{4} y
^{227}Ac ($\beta+\alpha$, 21.7y)
98.8% → ^{227}Th; 1.2% → ^{223}Fr
^{227}Th → ^{223}Ra: α, 18.9d
^{223}Fr → ^{223}Ra: β, 21m
^{223}Ra
α | 11.2d
^{219}Rn
α | 3.92s
^{215}Po ($\beta+\alpha$, 18 x 10
~100% → ^{211}Pb; ~5 x 10^{-4} → ^{215}At
^{211}Pb → ^{211}Bi: β, 3.61m
^{215}At → ^{211}Bi: α, 10^{-4} s
^{211}Bi ($\beta+\alpha$, 2.16 m
99.68% → ^{211}Po; 0.32% → ^{207}Tl
^{211}Po → ^{207}Pb: α, 5×10^{-3} s
^{207}Tl → ^{207}Pb: β, 4.76m

Figure 14.6 The radioactive series

remains unchanged. When lead-212 undergoes β-decay, for example, the change can be represented in the two equivalent ways:

$$^{212}_{82}Pb \rightarrow {}^{212}_{83}Bi + {}^{0}_{-1}\beta$$

$$^{212}_{82}Pb \rightarrow {}^{212}_{83}Bi + {}^{0}_{-1}e$$

Positron decay

As a consequence of positron decay, an isotope loses one unit of atomic number while its mass number remains unchanged. When phosphorus-30 undergoes positron decay the change can be represented as:

$$^{30}_{15}P \rightarrow {}^{30}_{14}Si + {}^{0}_{+1}\beta$$

The radioactive series

Three families of radioactive elements are found in nature. One starts with thorium-232 which undergoes a sequence of decays and finishes with stable lead-203. A second starts with uranium-238 and finishes with stable lead-206. A third starts with uranium-235 and finishes with stable lead-207. The mass numbers of the thorium-232 series are all divisible by 4 and can be represented as $4n$. Those of uranium-238 and uranium-235 series can be similarly represented as $(4n + 2)$ and $(4n + 3)$ respectively. This suggests the existence of a fourth series of which the mass numbers can be represented by $(4n + 1)$. This series is not found in nature but its members have been produced artificially. These four families of radioactive isotopes are shown in Figure 14.6.

Many artificial radioactive isotopes have now been prepared, not all of which are members of the radioactive series discussed above. At least one radioactive isotope of every element in the periodic table is now known.

Artificial transmutation of the elements

In 1919, Lord Rutherford observed the first artificial transmutation of an element. His apparatus is shown in Figure 14.7. Nitrogen gas was bombarded with α-particles and oxygen and a proton were the products:

$$^{14}_{7}N + {}^{4}_{2}He \rightarrow {}^{17}_{8}O + {}^{1}_{1}H$$

Oxygen-17, the product of this reaction, is stable. His achievement was quickly followed by other scientists using protons and neutrons as other bombarding particles to interact with atoms of some lighter elements.

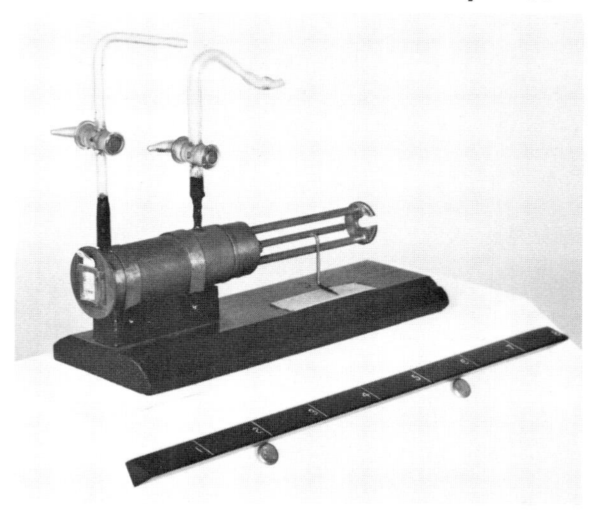

Figure 14.7 Apparatus in which Rutherford achieved the transmutation of nitrogen into a stable isotope of oxygen

The daughter of Marie and Pierre Curie, Irène Joliot-Curie, together with her husband Frédéric, were the first to induce radioactivity into previously non-radioactive material. In 1934 they bombarded magnesium with α-particles. After the bombardment ceased, the target was found to continue to radiate. The radiation induced is due to the formation of an isotope of silicon which is itself radioactive:

$$^{24}_{12}Mg + {}^{4}_{2}\alpha \rightarrow {}^{27}_{14}Si + {}^{1}_{0}n$$

The silicon-27 decays by positron emission:

$$^{27}_{14}Si \rightarrow {}^{27}_{13}Al + {}^{0}_{+1}\beta$$

A useful alternative notation for such processes is:

$$^{24}_{12}Mg\ (\alpha, n) \rightarrow {}^{27}_{14}Si \rightarrow {}^{27}_{13}Al + {}^{0}_{+1}e$$

in which α represents the bombarding particle and n the particle emitted. In 1932, John Cockcroft and Ernest Walton working at Harwell produced the first device to accelerate particles and use these to bombard atomic targets. Their apparatus is shown in Figure 14.8. They used it to bombard lithium with accelerated protons when the following nuclear reaction took place:

$$^{7}_{4}Li + {}^{1}_{1}p \rightarrow 2{}^{4}_{2}He$$

Figure 14.8 Cockcroft and Walton's apparatus

Cockcroft and Walton's was a linear accelerator. Ernest Lawrence, an American physicist, realized the advantage of using an accelerator in which the particles are forced to travel in a slowly expanding spiral. Lawrence built the first of these accelerators, known as *cyclotrons*, and today's versions of this, over a kilometre in circumference, are being used to learn more about the nature and function of the plethora of sub-atomic particles that have recently been discovered.

No element is known to occur naturally with an atomic number greater than 92, the atomic number of uranium. In 1940 Edwin McMillan and Philip Abelson bombarded uranium with neutrons and obtained element number 93 and named this neptunium. McMillan in collaboration with Glenn Seaborg produced element 94, plutonium, in 1941. Seaborg's team at the University of California during the next decade produced americium (95), curium (96), berkelium (97), californium (98), einsteinium (99), and fermium (100). With the discoveries of mendelevium (101) in 1955, nobelium (102) in 1957, and lawrencium (103) in 1964, the actinides are all known in at least one isotope. To these transuranic elements must be added element number 104, claimed by a team of Russian scientists and provisionally named kirchotovium.

Methods of detecting and measuring radioactivity

As discussed in Chapter 2, radioactivity was first detected through its capacity to cause fogging of a photographic plate wrapped to exclude visible light. Although this can be developed to give semi-quantitative measurements, as for example in the lapel badges worn by workers in establishments using potentially dangerous levels of radioactivity, more accurate techniques were soon evolved. Radioactive emissions will ionize the air in a gold-leaf electroscope, causing them to discharge. Early experimenters used the rate of discharge of such an electroscope to measure radioactivity, but the technique was cumbersome and unreliable and was largely discarded after the invention of the Geiger–Müller (G–M) tube. A particularly useful form of the G–M tube for radiochemical investigations, known as a liquid counter, is shown in Figure 14.9. In this modification, the G–M tube proper is surrounded concentrically by a tube to hold the liquid under test. The arrangement may be optionally surrounded by a 'lead castle' to absorb background radiation. The G–M tube contains a gas, often a halogen or an organic compound, under reduced pressure. It is fitted with two electrodes which are maintained at a voltage just below that required to cause an electric current to pass between the electrodes. When radiation enters the tube, it causes the gas to ionize and pass an electric current. The G–M tube is con-

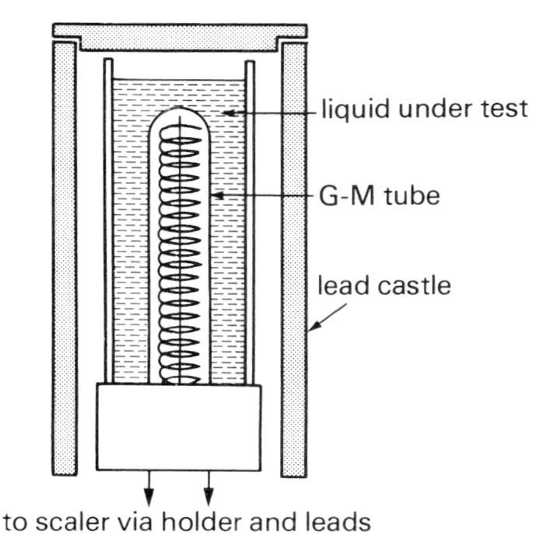

Figure 14.9 A liquid counter with lead castle

nected either to a 'scaler' which counts the number of times this ionization occurs during the time interval for which it is switched on or to a 'ratemeter' which differentiates this information and presents it as a count rate with respect to time.

A valuable device that makes the paths of radioactive emissions visible is the 'cloud chamber' developed by C. T. R. Wilson in 1911. This consists of a chamber fitted with a piston containing dust-free air saturated with water vapour. The piston is operated to expand the air causing the temperature to fall and condensation of the water to occur. If a charged particle passes through the air it ionizes the molecules in its path and, because water droplets tend to condense on the ionized air molecules, the trace of the particle appears as a misty line. The track of an α-particle in a cloud chamber is shown in Figure 14.10. The cloud chamber played an important role in the discovery of the positron and, together with a modern development known as a 'bubble chamber', it is still a useful technique today for scientists investigating interactions between radioactive particles and matter.

Investigating the rate of decay of a radioactive isotope

For the rate of decay of a radioactive isotope to be investigated, it must be obtained separate from other radioactive atoms so that its activity alone is counted. Various techniques, such as ion-exchange, selective extraction by a solvent, and

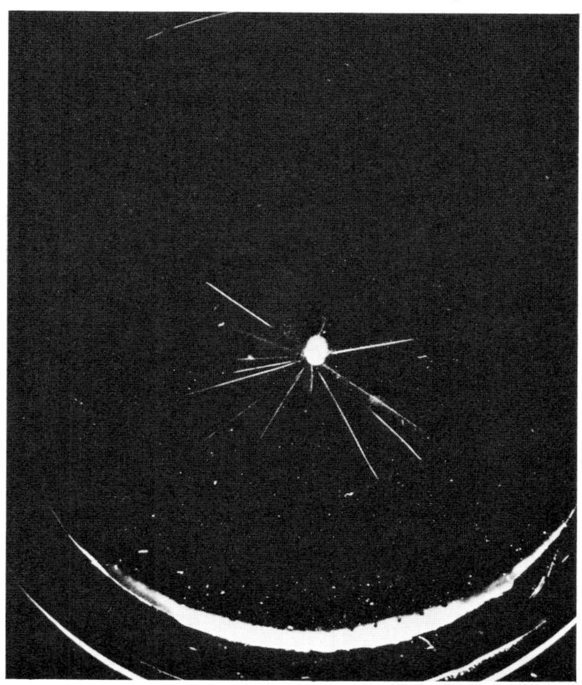

Figure 14.10 Cloud chamber photograph of an α-particle track

the use of solubility differences, are used. In the example discussed below extraction by a suitable solvent of protoactinium-234 depends on the property of the protoactinium to form largely unionized H_2PaCl_6 in a solution with a high concentration of both hydrogen and chloride ions.

Uranium-238 undergoes sequential decay until it becomes the stable isotope lead-206. In the first step of this sequence α-emission is involved:

$$^{238}_{92}U \rightarrow {}^{234}_{90}Th + {}^{4}_{2}\alpha$$

This is followed by the β-decay of the thorium-234:

$$^{234}_{90}Th \rightarrow {}^{234}_{91}Pa + {}^{0}_{-1}\beta$$

The protoactinium-234 decays by β-emission:

$$^{234}_{91}Pa \rightarrow {}^{234}_{92}U + {}^{0}_{-1}\beta$$

The half-life is the time taken for half any given quantity of a radioactive isotope to decay. Uranium-234 has a half-life of 2.35×10^5 years. As the half-life of protoactinium is very much shorter than this, if it can be extracted from a solution of a uranium salt containing the full range of daughter products, the activity of the extract can be considered to be entirely due to the decay of protoactinium-234.

In an experiment of this type, carried out with the full precautions essential when working with radioactive materials, 3 cm^3 of 35% aqueous uranyl(VI) nitrate plus 7 cm^3 of concentrated hydrochloric acid were shaken with 10 cm^3 of 4-methylpentan-2-one in a separating funnel. The ketone is not miscible with the aqueous layer and the protoactinium-234 is selectively absorbed into it. The layers were allowed to separate and the lower aqueous layer was discarded. The upper layer was quickly transferred to a liquid counter, as illustrated in Figure 14.9, and counted over 20-second intervals with a 10-second gap between each to read and re-set the scaler. Ideally the count rate at an instant is required. A compromise must be made in these cases, however, and it is necessary to assume that the count rate over the counting period is that at the time of the start of the period. The results of the experiment are shown in Figure 14.11. From this decay curve it can be deduced that 72 seconds are required before the initial count rate of 920 drops to 460. This value of the

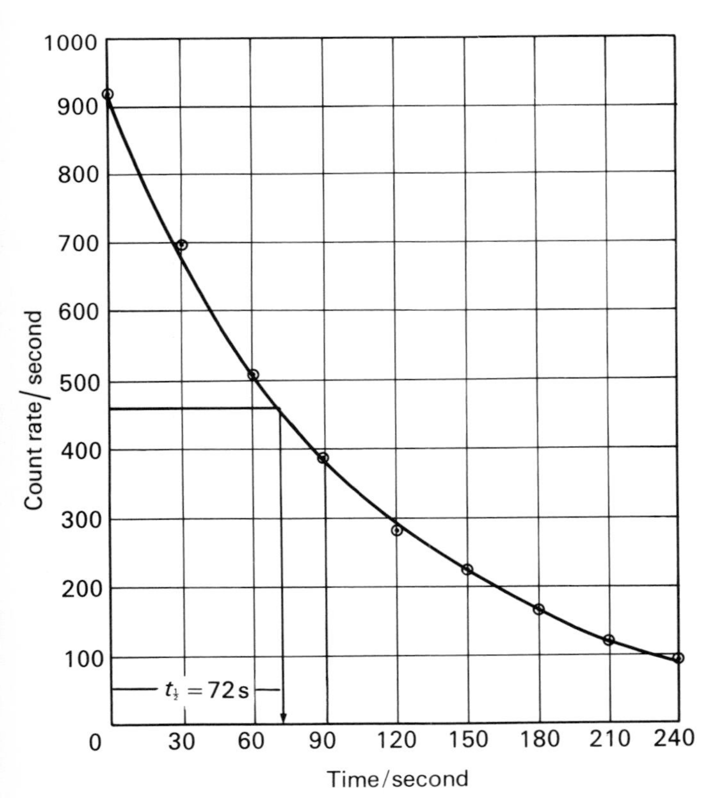

Figure 14.11 Decay curve for protoactinium-234

half-life of protoactinium must be confirmed by choosing one or two other count rates and finding how long it takes for this to be reduced to half.

Although the same proportion of radioactive atoms decay at any one instant, as the sample becomes used up it is this proportion of a diminishing number of atoms that decays. The actual number of atoms decaying, therefore, decreases with time and this explains the shape of the decay curve. The rate of decay is proportional to the number of radioactive atoms that remain, i.e.:

$$-\frac{dN}{dt} = \lambda N$$

where N is the number of radioactive atoms present and λ, the proportionality constant, is known as the decay constant.

This can be rearranged in a form suitable for integration as follows:

$$-\frac{dN}{N} = \lambda t$$

On integration this gives

$$-\ln N = \lambda t + C$$

C, the integration constant, can be evaluated by observing that when $t = 0$, the number of radioactive atoms present is the original number, given the symbol N_0. Thus

$$C = -\ln N_0$$

and

$$-\ln N = \lambda t - \ln N_0$$

or

$$\ln \frac{N_0}{N} = \lambda t$$

This relationship, known as the radioactive decay law, can also be stated in the exponential form:

$$N = N_0 \exp(-\lambda t)$$

For the special case of the half-life, $t_{\frac{1}{2}}$, $N = N_0/2$. In this case the radioactive decay law takes the form:

$$\ln 2 = \lambda t_{\frac{1}{2}}$$

or

$$t_{\frac{1}{2}} = \frac{\ln 2}{\lambda} = \frac{0.693}{\lambda}$$

This relationship can be used to determine the decay constant for protoactinium-234 from its half-life:

$$\lambda = \frac{0.693}{t_{\frac{1}{2}}} = \frac{0.693}{72 \text{ s}} = 9.6 \times 10^{-3} \text{ s}^{-1}$$

The validity of the radioactive decay law can be tested by plotting logarithms of the count rates of a radioactive isotope against time. This is done for protoactinium-234 in Figure 14.12. The linear plot obtained confirms the validity of the law.

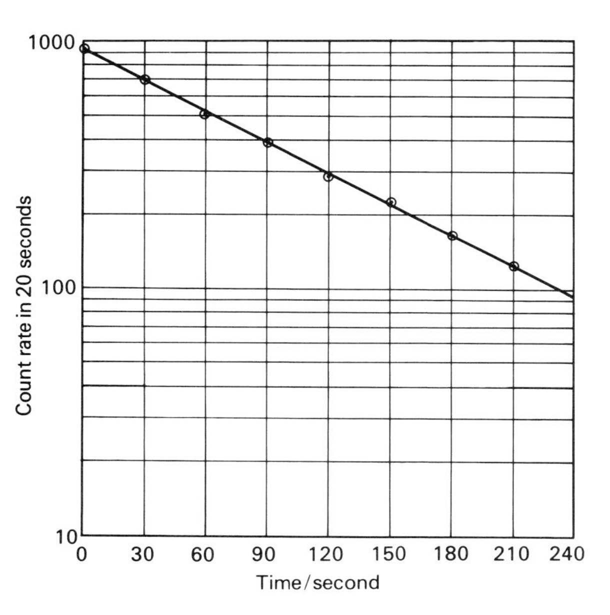

Figure 14.12 Lg count rate plotted against time for protoactinium-234

Nuclear fission

In 1939, Otto Hahn and Fritz Strassman were working in Berlin on the neutron bombardment of uranium. They were surprised to detect as products of the process, elements such as barium and krypton from the middle of the periodic table. It was recognized that this must be due to uranium atoms undergoing fission to produce elements with about half its atomic number. It was shown that of the isotopes present in natural uranium, only uranium-235 undergoes fission in this manner. A typical fission reaction of uranium-235 is:

$$^{235}_{92}\text{U} + {}^{1}_{0}n \rightarrow {}^{142}_{56}\text{Ba} + {}^{91}_{36}\text{Kr} + 2{}^{1}_{0}n$$

It must be stressed that this is just one possible fission process, some 300 isotopes of elements ranging from zinc (atomic number 30) to gadolinium (atomic number 64) have been detected. As each bombarding neutron results in the formation of two neutrons, provided the sample is above a critical size which ensures that most neutrons do not escape from the sides of the sample before initiating a further reaction themselves, there is the possibility of a chain reaction. When this occurs the rate of fission will escalate exponentially. As the total nuclear binding energy of the fission products is less than that of the uranium-235, such a chain reaction will be accompanied by a large evolution of energy. Hahn and Strassman's discovery came on the eve of World War II. The American Government launched an immediate research project to develop a bomb using the vast energy available from the fusion of uranium-235. One problem faced by the American team, led by Enrico Fermi, an Italian émigré, was to separate uranium-235 (only present at 0.7%) from the bulk of non-fissionable uranium. This was achieved by converting the uranium to uranium(VI) fluoride and utilizing the difference in the rate of diffusion of molecules of this gas due to the different masses of the uranium atoms they contained. The neutrons had to be moderated (slowed down) to increase the probability of their absorption by uranium-235 nuclei. The first atomic bomb was tested in July 1945 in the Alamogordo desert of New Mexico. In August of the same year, similar devices were exploded over the Japanese cities of Hiroshima and Nagasaki and brought World War II to an end.

Nuclear fission, properly controlled, can be used to produce energy to meet the power requirements of home and industry. Britain's present annual energy requirements could be met by the annihilation of only about 12 kg of matter. Since 1950 nuclear power stations have been constructed all over the globe. In these the fission process has to be slowed down to produce the energy over a much longer time scale. This is achieved by placing in the nuclear reactor control rods whose function it is to absorb a proportion of the free neutrons. Raising or lowering these control rods provides a means of adjusting the rate of fission and hence the rate of energy production. Figure 14.13 shows fuel rods being loaded into a reactor at Hinkley Nuclear Power Station in Somerset.

Inspection of Figure 14.1 shows that energy can be expected from the fission of elements of atomic number greater than that of iron. It also shows that energy will be evolved by the fusion of nuclei of the lighter atoms.

Figure 14.13 Fuel being loaded into a nuclear reactor

Figure 14.14 Examination of a jet engine intake by γ-radiography

Nuclear fusion

Life on Earth completely depends on the vast quantities of energy continually radiated by the sun. This energy comes from the fusion of hydrogen atoms into helium:

$$4{}^1_1H \rightarrow {}^4_2He + 2{}_{+1}^{0}\beta$$

In the hydrogen bomb, deuterium (so-called heavy hydrogen) is made to undergo fusion to helium. A standard atomic bomb is used to generate the very high temperatures necessary before this will happen:

$$2{}^2_1H \rightarrow {}^4_2He$$

The long-term energy requirements of mankind are likely to be met from fusion rather than from fission and much current research is being undertaken into these processes to provide knowledge from which it is hoped that the power stations of the future can be designed.

Uses of radioactive isotopes

γ-radiography

γ-rays fall into the same portion of the electromagnetic spectrum as X-rays. A source of γ-rays is very much more compact and portable than an X-ray source and γ-radiography is a useful alternative to X-radiography when such considerations are important. In Figure 14.14 a γ-source is shown being used to examine the metal structure of a jet engine intake for faults.

Medicinal uses

Cancer therapy Although it is known that exposure to radiation can induce cancer, malignant cells are particularly sensitive to radioactive emissions, particularly γ-rays. In Figure 14.15 a patient is seen under treatment from a cobalt-60 γ-source. The machine allows the accurate focusing of the radiation onto the malignant region.

Radiopharmaceuticals A wide range of radioactive material is used in modern medicine both for diagnostic and treatment purposes. The dispensing of such preparations must be carried out under sterile conditions and via remote handling because of the radiological hazard. Figure 14.16 shows work of this type in progress at the Radiochemical Centre, Amersham. In Figure 14.17 the kidneys of a patient are being monitored for radioactivity after an injection of iodine-131 in a form which concentrates it in these organs. From the rate at which it is eliminated, the efficiency of the kidneys can be ascertained. Iodine is also concentrated in the thyroid gland. The rate of uptake of iodine-131 can be used as a measure of thyroid efficiency. Stronger emitting doses of iodine-131 can be administered to destroy a portion of thyroid tissue and thus treat an overactive gland without the need for surgery.

Nuclear pacemaker Contraction of the heart is controlled by nervous impulses originating in that part of the heart known as the pacemaker. Sometimes the pacemaker is inefficient and the patient suffers from weak and irregular heartbeats. An artificial pacemaker can be surgically implanted into the heart to stimulate regular contractions electrically.

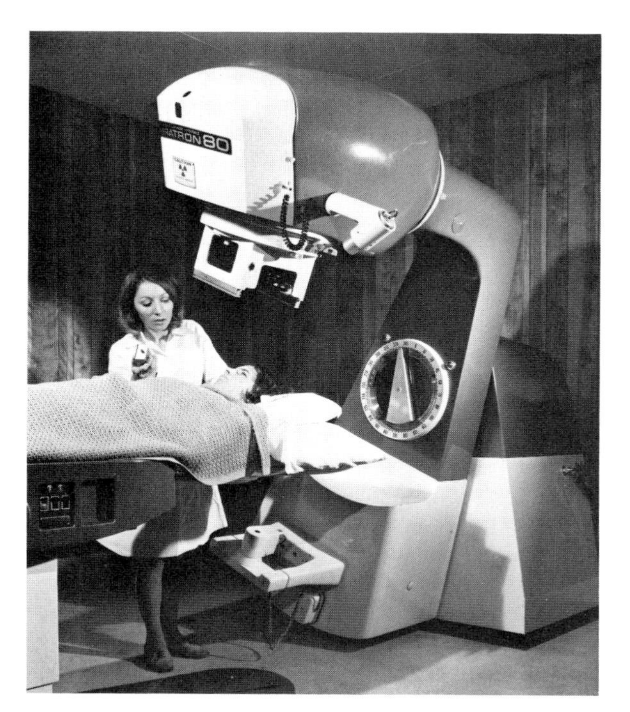

Figure 14.15 A patient undergoing radiation treatment from a cobalt-60 γ-source

Conventional artificial pacemakers which are battery-powered require surgical replacement every two years. The nuclear-powered pacemaker shown in Figure 14.18 will operate for ten years before replacement.

Tracer techniques

Some uses of radioactive isotopes as tracers in medicine have been described above. Tracer techniques are now widely used in many branches of science and technology. Radioactive phosphorus, as phosphate(V) ion, can be used in agricultural research to follow the uptake of this element by plants. Figure 14.19 shows the use of a radioactive solution to trace the path of liquid flow in the pumping system of a hydroelectric power station. In biochemistry, much knowledge of the metabolic pathways of processes such as tissue respiration and photosynthesis has been gained using substrates labelled with radioactive carbon-14. In chemical research, information concerning equilibrium and kinetics of reactions can be obtained using a suitable radioactive isotope as a tracer.

Figure 14.16 Remote handling of radiopharmaceutical preparations

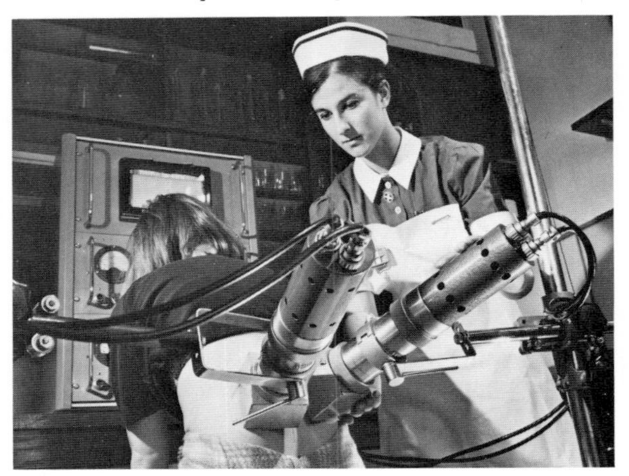

Figure 14.17 Monitoring the uptake of iodine-131 by the kidneys

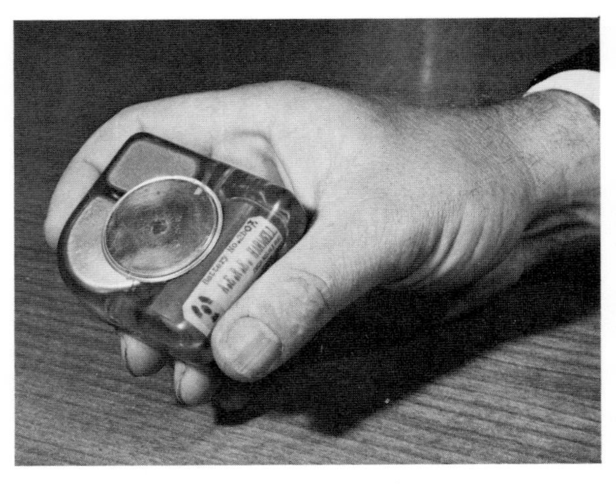

Figure 14.18 A nuclear-powered heart pacemaker

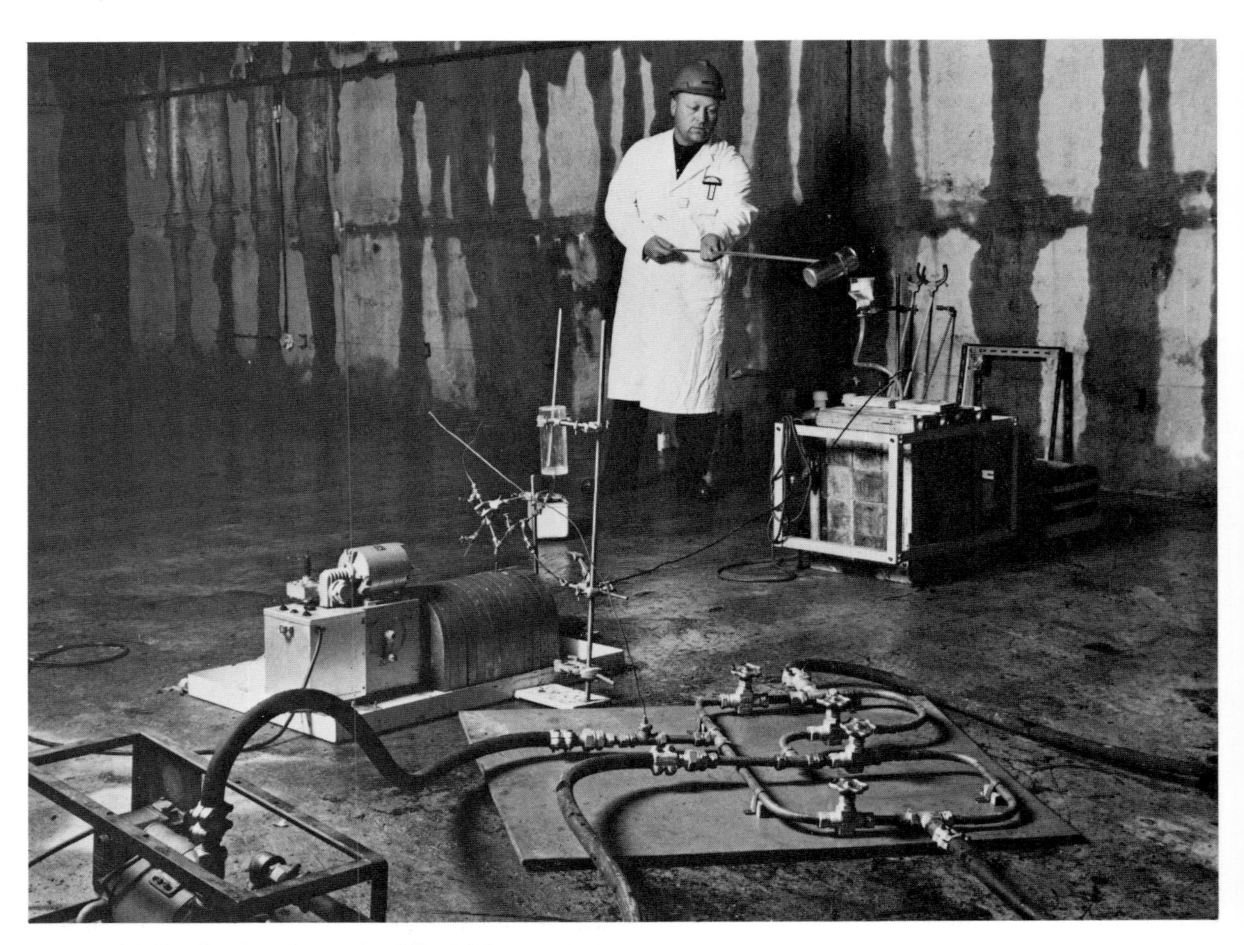

Figure 14.19 Tracing the path of liquid flow through the pumping system of a hydroelectric power station

Dilution analysis

This technique can best be explained using a medical application as an example. In a diagnostic investigation to find the total blood volume of a patient, 10 cm^3 of a solution of saline containing a proportion of radioactive sodium-24 was injected intravenously. The count rate of 10 cm^3 of this saline solution was found to be 1256 per minute. After sufficient time for the saline to become thoroughly mixed with the patient's blood, a 10 cm^3 sample of blood was withdrawn, counted and found to have a count rate of 2.314 per minute. (The counting time was adjusted to make an accurate determination of this latter count rate possible.)

$$\text{Total blood volume} = 10 \times \frac{1256}{2.314}\ \text{cm}^3 = 5428\ \text{cm}^3 = 5.43\ \text{dm}^3$$

The technique is not, of course, limited to medical procedures.

Sterilization

γ-radiation in powerful doses will kill bacteria and other micro-organisms. Attempts have been made to preserve food by sealing it in an air-tight container and irradiating with a γ-source. Commercial success has not yet been achieved for the technique but research continues. The method has been successfully used to sterilize hypodermic syringes and surgical instruments.

Thickness gauging

The thickness of a wide-ranging series of materials can be gauged using either the absorption or the back-scatter of γ-radiation. Figure 14.20 shows a back-scatter radiation gauge used to determine the wall thickness of pipework affected by internal corrosion.

Radioactive dating

Radioactive carbon-14 is produced in the upper atmosphere by the bombardment of nitrogen atoms with neutrons formed by cosmic radiation:

$$^{14}_{7}\text{N} + ^{1}_{0}n \rightarrow ^{14}_{6}\text{C} + ^{1}_{1}\text{H}$$

This radioactive carbon is distributed over the earth's surface in the carbon dioxide in the air and as hydrogencarbonate ions in the sea. It is therefore incorporated into growing plants as they photosynthesize; on the death of the plant this ceases and the radioactivity decays with the half-life of carbon-14, which is 5720 years. If it is assumed that C-14 has been produced over many centuries at a constant rate, this can be used to date articles of historical and archaeological interest. A sample of wood from an Egyptian sarcophagus gives a count rate of 9.50 per minute. The count rate from the same mass of a similar piece of modern wood is 15.30 per minute. From this information the age of the sarcophagus can be deduced:

$$\ln \frac{N_0}{N} = \lambda t$$

$$\lambda = \frac{0.693}{t_{\frac{1}{2}}} = \frac{0.693}{5720\ \text{y}} = 1.21 \times 10^{-4}\ \text{y}^{-1}$$

If the number of atoms is proportional to the

Figure 14.20 Determining the wall thickness of pipework by γ-radiation back-scatter

count rate, these values can be substituted in the radioactive decay law above:

$$\ln \frac{15.30}{9.50} = (1.21 \times 10^{-4}\ \mathrm{y}^{-1})t$$

$$t = \frac{\ln \dfrac{15.30}{9.50}}{1.21 \times 10^{-4}\ \mathrm{y}^{-1}}$$

$$= \frac{0.477}{1.21 \times 10^{-4}}\ \mathrm{y}$$

$$= 3940\ \mathrm{y}.$$

In addition to its use by archaeologists and historians, radiodating can be used by geologists to determine the ages of rocks. Rocks from the earth and from the moon have been investigated in this manner.

As shown in Figure 14.6, uranium-238 undergoes sequential decay and finishes as stable lead-206. Lead-206 cannot have been formed in the earth in any other way. A sample of the uranium bearing ore phouranylite was found to have a lead-206/uranium-238 ratio of 0.053. (The uranium-238 was determined by counting its radiation and the lead-206 was assayed using a mass spectrometer.) Because it has by far the greatest half-life of any member of its series, the decay constant for uranium-238 can be taken as that for the series as a whole; $t_{\frac{1}{2}} = 4.5 \times 10^9$ y.

$$\lambda = \frac{0.693}{4.5 \times 10^9\ \mathrm{y}} = 1.54 \times 10^{-10}\ \mathrm{y}^{-1}$$

The 0.053 moles of lead-206 must come from 1.053 moles of uranium-238 to give the ratio observed. The 1.053 moles of uranium-238 originally will be reduced to 1.000 moles today.

$$\ln \frac{N_0}{N} = \lambda t$$

$$\ln 1.053 = (1.54 \times 10^{-10}\ \mathrm{y}^{-1})t$$

$$t = \frac{\ln 1.053}{1.54 \times 10^{-10}\ \mathrm{y}^{-1}}$$

$$= \frac{0.0516}{1.54 \times 10^{-10}\ \mathrm{y}^{-1}}$$

$$= 4.97 \times 10^8\ \mathrm{y}$$

The age of rocks can be determined by the alternative method of comparing the relative abundances of two radioactive isotopes it contains. Consider the data in Table 14.1 for the two isotopes of uranium that almost exclusively account for the uranium content of a certain mineral.

Table 14.1 Data for the major isotopes of uranium

Isotope	$t_{\frac{1}{2}}/\mathrm{y}$	λ/y^{-1}	*Present abundance*
Uranium-235	7.1×10^8	9.8×10^{-10}	0.72%
Uranium-238	4.5×10^9	1.4×10^{-10}	99.28%

The radioactive decay law applied to each of these isotopes can be written in the form:

$$\ln N_0(\text{U-235}) - \ln N(\text{U-235}) = \lambda(\text{U-235})t$$
$$\ln N_0(\text{U-238}) - \ln N(\text{U-238}) = \lambda(\text{U-238})t$$

Assuming that when they were originally formed the two isotopes were present in equal amount, i.e.

$$N_0(\text{U-235}) = N_0(\text{U-238})$$

the two statements of the decay law can be combined to give:

$$\lambda(\text{U-235})t + \ln N(\text{U-235}) = \lambda(\text{U-238})t + \ln(\text{U-238})$$

Substitution of appropriate values into this equation yields

$$(9.8 \times 10^{-10}\ \mathrm{y}^{-1})\ t + \ln 0.72 = (1.4 \times 10^{-10}\ \mathrm{y}^{-1})\ t + \ln 99.28$$

$$(9.8 \times 10^{-10}\ \mathrm{y}^{-1})\ t - (1.4 \times 10^{-10}\ \mathrm{y}^{-1}) = \ln \frac{99.28}{0.72} = \ln 138$$

$$(8.4 \times 10^{-10}\ \mathrm{y}^{-1})\ t = \ln 138$$
$$= 4.93$$

$$t = \frac{4.93}{(8.4 \times 10^{-10}\ \mathrm{y}^{-1})}$$
$$= 5.9 \times 10^9\ \mathrm{y}$$

Developments in mass spectrometry

The electrons that cause ionization of the sample in a mass spectrometer have an energy due to the accelerating potential to which they are subjected. A typical accelerating potential is 70 V. Each of the bombarding electrons can be said to have an energy of 70 electron volts (70 eV). To convert this to the normal energy units used in chemistry this value must be multiplied by the charge on one electron and by Avogadro's number:

Electron energy

$= 70\ V) \times (1.602 \times 10^{-19}\ C) \times (6.023 \times 10^{23}\ mol^{-1})$

$= 6754\ kJ\ mol^{-1}$

Ionization energies

Even if a pure monisotopic and monatomic material is examined in a mass spectrometer using 70 eV electrons, the spectrum still shows several peaks because of the formation of multicharged ions. By reducing the electron energy, the peaks due to such species gradually disappear as eventually does that of the monocharged atomic ion X^+. If the instrument is focused to transmit X^+ ions, the variation of ion intensity with electron energy can be investigated. A typical result is shown graphically in Figure 14.21. (The curve in the initial portion of the graph is due to small energy differences between individual electrons.) Extrapolation as shown enables the ideal intercept to be found which corresponds to the first ionization energy of the material:

$$X(g) + e^-(fast) \rightarrow X^+(g) + 2e^-(slow)$$

For example, when hydrogen gas, which is virtually monisotopic, is examined in this manner, the intercept obtained is 15.41 eV or 1487 kJ mol^{-1}. It must be stressed that this is the first ionization energy of the hydrogen molecule, not the isolated hydrogen atom. By focusing on each of the multicharged ions in turn, the higher ionization energies can be determined.

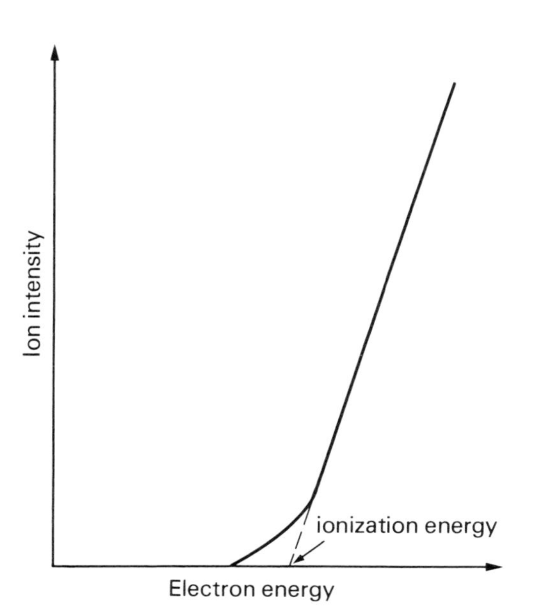

Figure 14.21 Determination of ionization energy by mass spectrometry

Compound analysis

When polyatomic molecules are examined in a mass spectrometer, the electron energy is often sufficient to produce fragmentation. The fragment ions so obtained are a valuable clue to the arrangement of the atoms in the original molecule. Figure 14.22 shows the mass spectrum of propane using 70 eV electrons. Table 14.2 shows the relative ion intensities in the mass spectrum of propane. The intensities have been normalized, this means that the most abundant ion is given a relative intensity of 100 and the intensities of the other ions are scaled proportionately.

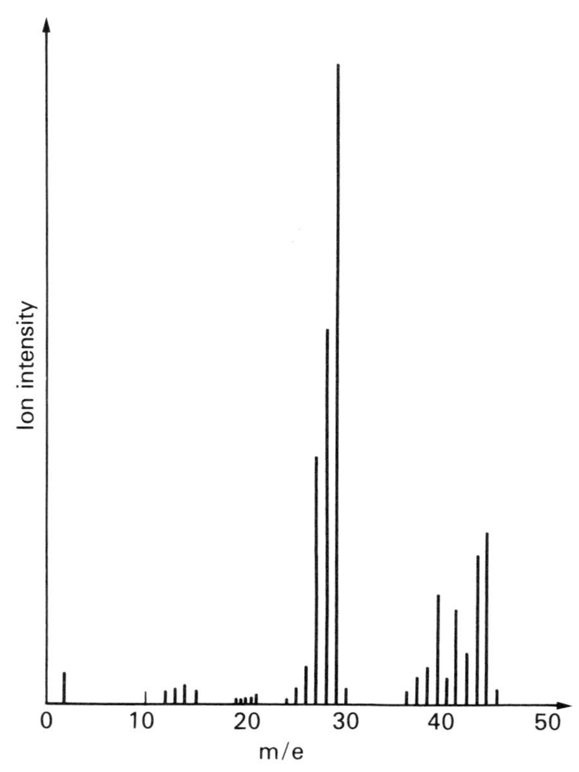

Figure 14.22 The mass spectrum of propane

Table 14.2 Ion intensities in the 70 eV mass spectrum of propane

m/e	*Relative intensity*	*m/e*	*Relative intensity*
2	6.87	27	37.9
12	0.35	28	59.1
13	0.48	29	100.0
14	2.48	30	2.09
15	3.9	36	0.43
16	0.20	37	3.06
19	0.82	38	4.91
19.5	0.48	39	16.2
20	0.94	40	2.8
20.5	0.25	41	12.4
21	0.02	42	5.15
24	0.11	43	22.3
25	0.71	44	26.2
26	7.58	45	0.81

With 70 eV electrons the most intense peak is seen to be that corresponding to the fragment ion $C_2H_5^+$ and not the parent ion $C_3H_8^+$. Ions of m/e 28 ($C_2H_4^+$) and m/e 27 ($C_2H_3^+$) are also more intense. The largest m/e value obtained might be expected to correspond to the relative molar mass of propane, 44, but there is an additional low intensity ion at m/e 45. This arises from the small proportion of propane molecules containing one carbon-13 atom. The natural abundance of carbon-13 is approximately 1% and since such an atom could occur as any one of the three carbon atoms in propane, it would be expected that:

$$\frac{\text{Relative intensity m/e 45}}{\text{Relative intensity m/e 44}} = 0.03$$

The data in Table 14.2 confirms this. Such intensity ratios, therefore, provide an indication of the number of carbon atoms in each molecule.

High resolution mass spectrometry

The resolving power of a mass spectrometer is a measure of its ability to distinguish and separate ions with similar m/e values. An improvement in resolution is obtained by using instruments in which electrostatic as well as magnetic deflection of the ions occurs. A photograph of such an instrument is shown in Figure 14.23 and its schematic diagram is shown in Figure 14.24. The electrostatic analyser allows the transmission of all ions with the same kinetic energy regardless of their m/e value. The ions thus enter the magnetic analyser with the same kinetic energy and correspondingly sharper focusing is possible. Such instruments are capable of distinguishing between ions with m/e values differing only by a few parts per million.

As discussed at the beginning of this chapter, the mass defect varies from element to element. Thus the masses of ions with the same nominal mass show small differences if precise measurements are made. In Table 14.3 the precise masses of some ions of nominal mass 28 are given.

Figure 14.23 A high resolution mass spectrometer

Table 14.3 Precise masses of some ions with nominal m/e 28

Ion	*Precise mass*
CO^+	27.994 914
N_2^+	28.006 148
CH_2N^+	28.018 724
$C_2H_4^+$	28.031 300

Tables listing the precise masses of the parent ions of a wide range of compounds are available and these can be used to rapidly identify substances.

Suggestions for further reading

NEWTON, G. W. A. and ROBINSON, V. J. *Principles of Radiochemistry* (Basingstoke and London: Macmillan Education Limited, 1971)

PUTMAN, J. L. *Isotopes* (Harmondsworth: Penguin Books Ltd., 1965)

Problems

1. Use the information in Figure 14.2 to predict the likely mode of decay of the isotopes: (*a*) $^{101}_{48}Cd$; (*b*) $^{93}_{36}Kr$; (*c*) $^{216}_{90}Th$.
2. The results in the table below were obtained during the investigation of the radioactive disintegration of sodium-24 by β^- decay.

Time/h	0	2	5	10	20	30
Count rate/s	751	683	594	472	298	188

(*a*) Write an equation for the decay of sodium-24.

(*b*) Deduce the half-life and the decay constant for this isotope.

3. Complete the following radioactive decay equations:

(*a*) $^{214}_{82}Pb \rightarrow {}^{0}_{-1}\beta + X$

(*b*) $^{210}_{84}Po \rightarrow {}^{4}_{2}\alpha + Y$

(*c*) $^{54}_{27}Co \rightarrow {}^{0}_{+1}\beta + Z$

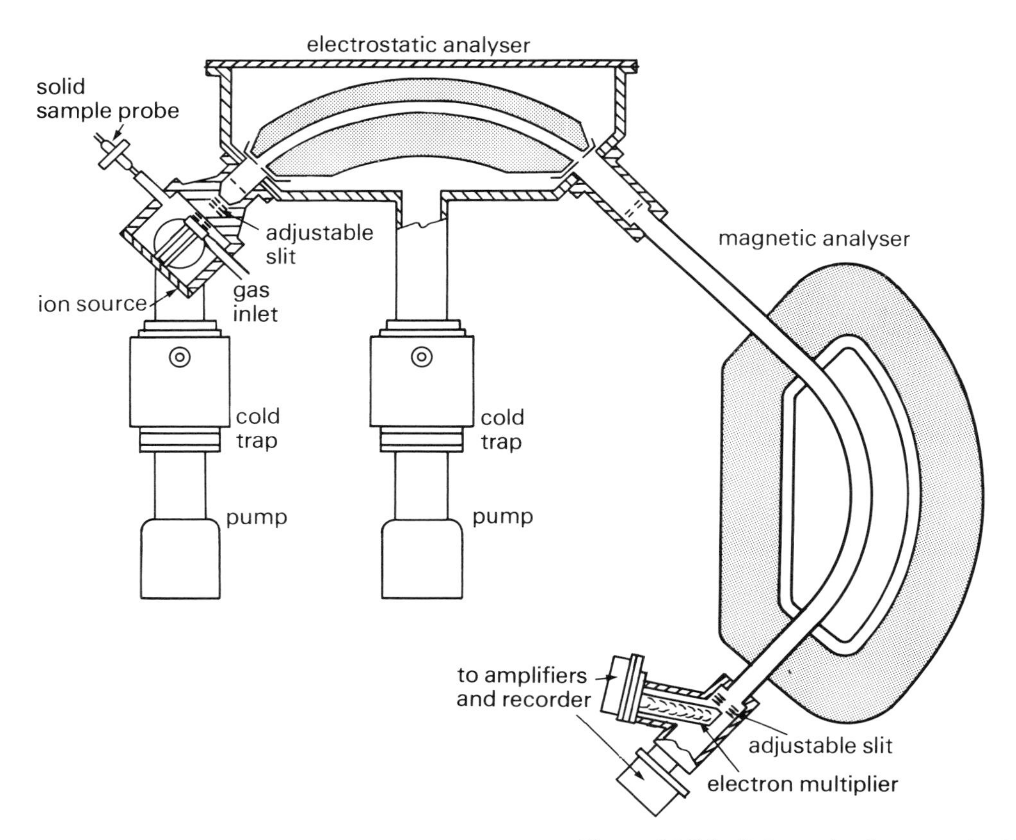

Figure 14.24 Schematic diagram of a high resolution mass spectrometer

4. 10 g of a protein was hydrolysed by refluxing with hydrochloric acid. 0.100 g of pure deuterium substituted alanine ($CD_3CHNH_2CO_2H$) was added and after thorough mixing, alanine crystallized out which had a deuterium content of 1.03% by mass. Calculate the percentage of alanine in the protein. (Deuterium is a non-radioactive isotope of hydrogen which has mass number 2 because its nucleus contains one proton plus one neutron.) How might the percentage of deuterated alanine in the determination described above be obtained?

5. When iodine is examined in a mass spectrometer adjusted to focus the monocharged ion, I_2^+, by graphical extrapolation it is found that the minimum electron accelerating voltage that results in its appearance is 9.27 V.

(*a*) Calculate the first ionization energy of the iodine molecule in kJ mol^{-1}.

(*b*) Use your answer in (*a*) together with a suitable values for the first ionization energy of the iodine atom and the bond energy of the iodine molecule to construct an energy cycle to deduce the enthalpy change for the reaction:

$$I(g) + I^+(g) \rightarrow I_2^+(g)$$

6. Write an essay on radioactivity. Include in your answer reference to the following points, giving experimental details where appropriate. (*a*) The nature and origins of radioactive disintegrations. (*b*) The importance of this topic for an understanding of atomic structure. (JMB 'A')

7. (*a*) What are α-rays and β-rays and how may they be detected?

(*b*) Briefly describe how α-rays were used to show that atoms have a nuclear structure.

(*c*) Write a nuclear equation (in terms of mass numbers and atomic numbers) to show the decay of radio-carbon, ${}^{14}_{6}C$.

If this nuclide has a half-life of 5600 years, calculate the age of wood from an ancient tomb, if this wood gave 10 counts per minute per gram of carbon compared with the 15 counts that are given by the carbon from new wood.

(Oxford 'A')

8. (*a*) Carbon consists of 99% of the isotope ${}^{12}C$ and 1% of the isotope ${}^{13}C$. One way of estimating the number of carbon atoms in the molecule of a compound is to examine the mass spectrometer peaks which the molecule gives. The peak corresponding to the second highest mass is caused by the molecule ion having only ${}^{12}C$ in it but the highest mass peak is due to a molecule ion with one ${}^{13}C$ atom in it.

A hydrocarbon X gives a peak at mass M and a smaller peak at mass $M + 1$. There are no significant peaks at higher mass than this. The peak at mass M is 12.5 times as intense as the peak at mass $M + 1$.

When a quantity of X was completely oxidized, 0.11 g of CO_2 and 0.023 g of water were formed. X decolourizes bromine water and 1 mole of X reacts with 1 mole of bromine molecules.

What is the molecular formula and probable structure of X?

(*b*) Bromine consists of a mixture of isotopes ${}^{79}Br$ and ${}^{81}Br$. Assuming no bonds are broken in the mass spectrometer and that hydrogen has only one significant isotope, how many peaks will be given by the compound CH_2Br_2 and what will each peak be due to? (Nuffield 'S')

9. Write brief notes on (i) radioactive decay, (ii) the production of energy by nuclear fission.

The age of a mineral containing ${}^{238}U$ and ${}^{206}Pb$ formed by decay of the uranium isotope may be estimated from the atomic ratio ${}^{206}Pb:{}^{238}U$ if it is assumed that all the lead has arisen by decay of the uranium and that all the half-lives in the decay series from ${}^{238}U$ to ${}^{206}Pb$ are small in comparison with that of ${}^{238}U$. The half-life of ${}^{238}U$ is 4.51×10^9 years, and the atomic ratio ${}^{206}Pb:{}^{238}U$ in a sample of pitchblende is 0.231:1.00. Estimate the age of the pitchblende. (Cambridge Entrance and Awards)

Table of atomic masses

Element	Symbol	Atomic number	Atomic mass	Element	Symbol	Atomic number	Atomic mass
Actinium	Ac	89	227	Mercury	Hg	80	200.6
Aluminium	Al	13	27	Molybdenum	Mo	42	96
Americium	Am	95	243	Neodymium	Nd	60	144.2
Antimony	Sb	51	121.7	Neon	Ne	10	20.2
Argon	Ar	18	40	Neptunium	Np	93	237
Arsenic	As	33	75	Nickel	Ni	28	58.7
Astatine	At	85	210	Niobium	Nb	41	93
Barium	Ba	56	137.3	Nitrogen	N	7	14
Berkelium	Bk	97	249	Nobelium	No	102	254
Beryllium	Be	4	9	Osmium	Os	76	190.2
Bismuth	Bi	83	209	Oxygen	O	8	16
Boron	B	5	10.8	Palladium	Pd	46	106.4
Bromine	Br	35	80	Phosphorus	P	15	31
Cadmium	Cd	48	112.4	Platinum	Pt	78	195
Caesium	Cs	55	133	Plutonium	Pu	94	242
Calcium	Ca	20	40	Polonium	Po	84	210
Californium	Cf	98	251	Potassium	K	19	39
Carbon	C	6	12	Praseodymium	Pr	59	141
Cerium	Ce	58	140	Promethium	Pm	61	145
Chlorine	Cl	17	35.5	Protactinium	Pa	91	231
Chromium	Cr	24	52	Radium	Ra	88	226
Cobalt	Co	27	59	Radon	Rn	86	222
Copper	Cu	29	63.5	Rhenium	Re	75	186.2
Curium	Cm	96	247	Rhodium	Rh	45	103
Dysprosium	Dy	66	162.5	Rubidium	Rb	37	85.5
Einsteinium	Es	99	254	Ruthenium	Ru	44	101
Erbium	Er	68	167.3	Samarium	Sm	62	150.3
Europium	Eu	63	152	Scandium	Sc	21	45
Fermium	Fm	100	253	Selenium	Se	34	79
Fluorine	F	9	19	Silicon	Si	14	28
Francium	Fr	87	223	Silver	Ag	47	108
Gadolinium	Gd	64	157.2	Sodium	Na	11	23
Gallium	Ga	31	69.7	Strontium	Sr	38	87.6
Germanium	Ge	32	72.6	Sulphur	S	16	32
Gold	Au	79	197	Tantalum	Ta	73	181
Hafnium	Hf	72	178.5	Technetium	Tc	43	99
Helium	He	2	4	Tellurium	Te	52	127.6
Holmium	Ho	67	165	Terbium	Tb	65	159
Hydrogen	H	1	1	Thallium	Tl	81	204.4
Indium	In	49	115	Thorium	Th	90	232
Iodine	I	53	127	Thulium	Tm	69	169
Iridium	Ir	77	192.2	Tin	Sn	50	118.7
Iron	Fe	26	56	Titanium	Ti	22	48
Krypton	Kr	36	83.8	Tungsten	W	74	183.8
Lanthanum	La	57	139	Uranium	U	92	238
Lawrencium	Lw	103	257	Vanadium	V	23	51
Lead	Pb	82	207.2	Xenon	Xe	54	131.3
Lithium	Li	3	7	Ytterbium	Yb	70	173
Lutetium	Lu	71	175	Yttrium	Y	39	89
Magnesium	Mg	12	24.3	Zinc	Zn	30	65.4
Manganese	Mn	25	55	Zirconium	Zr	40	91.2
Mendelevium	Md	101	256				

Units and Symbols

THE Système International d'Unités (SI) now has almost world-wide recognition and is used, with very few exceptions, throughout this book. SI is based on seven fundamental physical quantities, of which the six listed in Table U.1 are required in physical chemistry at this level.

Table U.1 Basic SI units

Quantity	*Symbol*	*SI unit*	*Unit Symbol*
Length	l	metre	m
Mass	m	kilogram	kg
Time	t	second	s
Electric current	I	ampere	A
Temperature	T	kelvin	K
Amount of substance	n	mole	mol

Table U.2 Prefixes in common use

Multiples			*Submultiples*		
Prefix	*Symbol*	*Factor*	*Prefix*	*Symbol*	*Factor*
kilo	k	10^{3}	deci	d	10^{-1}
mega	M	10^{6}	centi	c	10^{-2}
giga	G	10^{9}	milli	m	10^{-3}
tera	T	10^{12}	micro	μ	10^{-6}
			nano	n	10^{-9}
			pico	p	10^{-12}

The basic SI units may be too large or too small to conveniently express particular measurements. In such cases they are used in association with a suitable prefix. SI prefers prefixes that are multiples or submultiples of 10^3 and these should be used if possible.

Some units derived from the basic SI units are given special names; Table U.3 contains some important examples of these. Table U.4 contains other derived SI units without special names.

Some other derived SI units encountered less frequently are defined in the text.

Use of the term 'specific'

This term should be used to prefix quantities only when it is intended to quote that quantity per unit mass of the material. E.g. the specific heat capacity of water is 4.184×10^3 J K^{-1} kg^{-1}.

Use of the term 'molar'

This term should be used to prefix quantities only when it is intended to quote that quantity per mole of the material. E.g. the molar latent heat of vaporization of water is 4.066×10^4 J mol^{-1}. It conflicts with this to describe a solution containing 1 mole of solute in 1 dm^3 of solvent as a molar solution, and it is recommended by the Royal Society that this usage be abandoned. It is, however, permissible to write 1 M, 2 M, 0.05 M, etc. to express the concentration of solutions in mol dm^{-3}.

Table U.3 Some derived SI units with special names

Quantity	*Symbol(s)*	*SI unit*	*Unit symbol*	*Definition*
Energy	E, U	joule	J	N m = kg m^2 s^{-2} = CV = VAs
Frequency	ν, f	hertz	Hz	s^{-1}
Force	F	newton	N	kg m s^{-2} = J m^{-1}
Power	P	watt	W	J s^{-1} = kg m^2 s^{-3} = VA
Electrical charge	Q	coulomb	C	As
Electrical potential	E	volt	V	J A^{-1} s^{-1} = kg m^2 s^{-3} A^{-1}
Electrical resistance	R	ohm	Ω	V A^{-1}
Electrical conductance	G	siemens	S	S = Ω^{-1}
Electrical conductivity		siemens m^{-1}	S m^{-1}	S m^{-1} = Ω^{-1} m^{-1}
Pressure	p	pascal	Pa	N m^{-2} = kg m^{-1} s^{-2}

Table U.4 Some derived SI units with no special names

Quantity	*Symbol(s)*	*Basic SI unit*	*Other useful SI acceptable units*
Area	A	m^2	
Volume	V	m^3	dm^3, cm^3
Velocity	v, c	$m\ s^{-1}$	
Density	ρ	$kg\ m^{-3}$	$g\ dm^{-3}$, $g\ cm^{-3}$
Molar mass	M	$kg\ mol^{-1}$	$g\ mol^{-1}$
Relative molar mass	M_r	dimensionless	
Gibbs free energy	G	$J\ mol^{-1}$	$kJ\ mol^{-1}$
Enthalpy	H	$J\ mol^{-1}$	$kJ\ mol^{-1}$
Entropy	S	$J\ mol^{-1}\ K^{-1}$	
Electrical field strength	E	$V\ m^{-1}$	
Magnetic field strength	B	$A\ m^{-1}$	
Wavelength	λ	m	μm, nm
Molar conductivity	Λ	$\Omega^{-1}\ m^2\ mol^{-1}$	

Some other derived SI units encountered less frequently are defined in the text.

Table U.5 Some important constants and their SI values

Constant	*Symbol*	*Value*
Avogadro's number	L	$6.023 \times 10^{23}\ mol^{-1}$
Bolzmann constant	k	$1.381 \times 10^{-23}\ J\ K^{-1}$
Planck constant	h	$6.626 \times 10^{-34}\ J\ s^{-1}$
Velocity of light	c	$2.998 \times 10^{8}\ m\ s^{-1}$
Gas constant	R	$8.314\ J\ K^{-1}\ mol^{-1}$
Charge on electron	e	$1.602 \times 10^{-19}\ C$
Faraday	F	$9.649 \times 10^{4}\ C\ mol^{-1}$
Molar volume of gas at s.t.p.		$2.241 \times 10^{-2}\ m^3 = 22.41\ dm^3$

Table U.6 Some non-SI units still in common use

Quantity	*Unit*	*Symbol*	*SI definition or equivalent*
Length	Angstrom	A	$10^{-10}\ m = 10^{-1}\ nm$
Volume	litre	l	$10^{-3}\ m^3 = 1\ dm^3$
Volume	millilitre	ml	$10^{-6}\ m^3 = 1\ cm^3$
Pressure	Atmosphere	atm	$1.013 \times 10^{5}\ Pa$
Pressure	millimetre of mercury	mm Hg	$1.333 \times 10^{2}\ Pa$
Temperature	degree Celsius	°C	K = °C + 273.2
Energy	kilocalorie	kcal	$4.184 \times 10^{3}\ J = 4.184\ kJ$
Energy	electronvolt	eV	$1.602 \times 10^{-19}\ J$
Gas constant	$0.08205\ dm^3\ atm\ K^{-1}\ mol^{-1}$	R	$8.314\ J\ K^{-1}\ mol^{-1}$

The solidus notation

This provides a useful alternative method of expressing relationships such as:

$$\text{Density of copper} = 8.93 \text{ g cm}^{-3}$$

The relationship is transposed so that a dimensionless number is alone on the right hand side of the equation:

$$\text{Density of copper/g cm}^{-3} = 8.93$$

This technique is very useful when constructing tables of data and when drawing graphs. Its use is illustrated throughout this book.

In tables and when plotting graphs where inverse relationships are involved, they should be treated as in the following example:

T/K	10^3K/T
273	3.66

The conversion of units

A useful general method of unit conversion is illustrated by the following example.

The density of copper is 8.93 g cm^{-3}; convert this into basic SI units.

$$x \text{ kg m}^{-3} = 8.93 \text{ g cm}^{-3} \times 10^{-3} \text{ kg g}^{-1} \times 10^{6} \text{ cm}^3 \text{ m}^{-3}$$

$$= 8.93 \times 10^3 \text{ kg m}^{-3}$$

Logarithmic relationships in chemistry

Logarithms can only be taken of pure numbers (i.e. without dimension). In a relationship such as $\Delta G^{\ominus} = -RT \ln K_p$ we are not strictly justified in taking logarithms of values of K_p for systems in which it has dimensions. If it is desired to observe the mathematical proprieties in such matters, the relationship can be rewritten

$$\Delta G^{\ominus} = -RT \ln K_{p/p^{\ominus}}$$

where each pressure term in the equilibrium expression is divided by unit pressure thus rendering it, and K_p, dimensionless. This sophistication is not stressed in the text.

The following symbols are used for logarithmic functions:

Logarithm to the base 10 of $x = \lg x$

Antilogarithm to the base 10 of $x = \lg^{-1} x$

Logarithm to the base e of $x = \ln x$

Antilogarithm to the base e of $x = \ln^{-1} x = \exp(x)$

Suggestions for further reading

McGLASHAN, M. C. *Physico-chemical Quantities and Units* (London: Royal Institute of Chemistry, 1968)

Chemical Nomenclature, Symbols and Terminology (Hatfield: Association for Science Education, 1975)

Index

ABCD rule, 125
Acids:
  acidity constants of, 135
  Arrhenius definition of, 110
  catalysis by, 150
  Lavoisier definition of, 110
  Lowry–Brønsted definition of, 111
  polyprotic, 141
Activation energy, 155–7
Activities, 143
Activity coefficients, 143
Adsorption, 194
Adsorption indicators, 195
Alpha particles, 21, 202–203
Amplitude, of waves, 26
Anticlockwise rule, 127
Aristotle, 15
Arrhenius, 108, 156
Aston, 23
Asymmetric carbon atom, 75
Atomic:
  absorption spectrophotometry, 35
  bomb, 209
  emission spectra, 26, 35
  mass, 24
  number, 22
  radius, 62
Atomicity, of gases, 9
Autocatalysis, 159
Avogadro's hypothesis, 6
Avogadro's number, 6, 64, 122
Azeotropic mixture, 102

BALMER, 27
Bases:
  Arrhenius definition of, 110
  basicity constants of, 136
  Lavoisier definition of, 110
  Lowry–Brønsted definition of, 111
Becquerel, 20
Benzene, structure of, 73
Beta particle, 21, 204
Binding energy (nuclear), 202
Body-centred cubic packing, 70
Boiling point, 95
Boltzmann, 13, 181
Bomb calorimeter, 48
Bond energy, 38–9, 52–3
Bonds:
  covalent, 65
  dative, 68
  delocalization of, 72
  hybridization, 78–80
  hydrogen, 71, 89, 176
  ionic, 62
  multiple, 68
  polarization of, 70
Born–Haber cycle, 54
Boyle's law, 1
Bragg, 61, 108
Brownian motion, 7
Buffer solutions, 137

CALOMEL ELECTRODE, 132
Calorimeter, 46–9
Cannizaro, 7
Capacity factors, 168
Catalysis, 145, 158–60
Cathode rays, 18, 20
Chain reaction, 160, 209
Charge cloud, of electrons, 65
Charge-to-mass ratio:
  of electrons, 18
  of positive ions, 23
Charles' law, 2
Chromatography, 90
Cleavage planes, in crystals, 61
Colligative properties, 98–9, 104–106
Collision factor, 157
Collision theory, of reaction rates, 155–7
Colloids:
  classification of, 196–8
  electrophoresis of, 196
  lyophilic, 198
  lyophobic, 198
  precipitation (coagulation) of, 199
Colorimetry, 36, 159
Combustion, enthalpy of, 48
Complex ion, 191
Conductimetric titrations, 117
Conductivity:
  electrolytic, 112
  molar, 113
Conjugate acids and bases, 111
Conservation of energy, law of, 50, 58
Conservation of mass, law of, 15
Constant composition, law of, 15
Continuous flow, for reaction rates, 162
Continuous flow reactor, 164
Coordinate bond, 68
Coordination number, 63
Coulometer, 109
Covalent bond, 65
Crooke, 17

Cryoscopy, 99
Crystal systems, 63
Cubic crystal systems, 77
Curie, Marie and Pierre, 21
Cyclotron, 206

DALTON'S ATOMIC THEORY, 16
Dalton's law, 9
Dative bond, 68
de Broglie, 26
Decay curve, of radio-isotopes, 208
Decomposition potentials, 120
Degenerative orbital, 192
Degree of dissociation, 86
Delocalization, of bonds, 72
Democritus, 15
Densities, of gases, 3
Dialysis, 196–7
Detergency, 194
Diffraction:
  of electrons, 66
  of X-rays, 60
Diffusion, of gases, 10
Dilution analysis, 213
Dipole moments, of molecules, 70
Disperse medium, 197
Disperse phase, 197
Distillation:
  fractional, 101
  simple, 99
  steam, 100
  under reduced pressure (vacuum), 100
Distribution coefficient, 88
Distribution of molecular energies, 13, 94, 155
Döbereiner's triads, 184
Dynamic equilibrium, 82

EBULLIOSCOPY, 98, 180
Effusion of gases, 10
Eigenfunctions and Eigenvalues, 32
Einstein, 58, 201
Electrode potential, 124
Electrolyte, 109
Electron:
  affinity, 54
  charge-to-mass ratio, 18
  density maps, 62, 65, 66, 73
  diffraction of, 66
  mass of, 19
  spin of, 30
  particulate/wave nature of, 26
Electronic configurations, of atoms, 29
Electronegativities, 71
Electron energy levels, 26–8
Ellingham diagrams, 171–2
Emission spectroscopy, 26, 35
Emulsion, 199
Electromagnetic spectrum, 35
Electrophoresis, 196

Enantiomer (enantiomorph), 75
Energy:
  activation, 155–7
  bond, 38–9, 52–3
  Gibbs free, 169–70
  internal, 45
  nuclear binding, 202
  profile, of reaction, 155
  rotational, 12
  translational, 12
  vibrational, 12
Entropy, 173, 178–81
Enzymes, 163
Equilibrium:
  constant, 82, 85, 91, 170, 176–7
  dynamic, 82
  heterogeneous and homogeneous, 87
  law, 82
Extraction, of metals, 171

FACE-CENTRED CUBIC PACKING, 76–7
Faraday, 109–10
First law, of thermodynamics, 50, 167
First-order reaction, 148
Fission (nuclear), 209
Flash photolysis, 163
Fractional distillation, 101
Free energy (Gibbs), 169–70
Free radical, 160, 163
Frequency, of waves, 26
Freundlich, 194
Fuel cells, 171
Fusion (nuclear), 210

GAMMA RADIATION, 21
Gamma radiography, 210
Gas constant, 3, 46, 91, 104, 128, 142, 170, 176
Gas equation (general), 2, 46, 104, 142
Gas laws:
  Boyle's, 1
  Charles', 2
  Dalton's, 9
  Gay-Lussac's, 7
  Graham's, 10
Gay-Lussac, 7
Gas phase reactions, rates of, 161
Geiger–Müller counter, 206
Gibbs free energy, 169–70
Goniometer, 62
Graham, 10, 196

HALF-LIFE, 147, 208
Hess, 50
Hund, 30
Hybrid orbital, 78–80
Hydrogen:
  bomb, 210
  bonding, 71, 89, 176
  electrode, 126–7

Hydrolysis:
of bromobutanes, 153
of salts, 112, 192
Hybridization, of orbitals, 78–80

ICE, structure of, 72
Ideal gas, 2
Ideal liquid mixtures, 97
Indicators:
acid/base, 140
Ostwald, theory of, 140
precipitation reactions, 132, 195
Infra-red, 27, 35, 39
Intensity factor, 168
Interference, of waves, 61
Intermolecular forces, 97
Ion exchange, 122
Ionic character, percentage of, 71
Ionic radius, 62
Ionization energy, 28–9, 55, 57, 188, 215
Isomerism, 64, 75
Isotonic solutions, 105
Isotopes, 23, 210–14

KELVIN TEMPERATURE SCALE, 2
Kinetic theory, of gases, 7
Kohlrausch, 113, 115

LATTICE ENERGY, 54–7
Laue, von, 60
Le Châtelier, 85
Ligand, 191
Limiting density, of gases, 11
Liquid counter, 206
Lone pair, of electrons, 67
Lyman, 27

MADELUNG CONSTANT, 54
Mass defect, 201
Mass–energy equivalence, 58, 201
Mass spectrograph, 24
Mass spectrometer, 24, 42, 214–17
Mechanism of reaction, 147, 154
Mendeléev, 184–7
Metals, structure of, 76
Michaelis–Menten theory, 163
Microbalance, 4
Migration, of ions, 117
Millikan, 19
Mitscherlich, 64
Molar volumes, of gases, 4
Molecularity, 154
Molecular mass, 4–6, 10–12, 43
Mole fraction, 95
Moseley, 22
Multiple proportions, law of, 15

NERNST EQUATION, 128, 142, 169
Neutralization (acid/base), 116
Neutron, 25
Nuclear binding energy, 202
Nuclear magnetic resonance (NMR), 40–43
Nucleus, 22

OPTICAL ROTATORY DISPERSION, 76
Ohm's law, 112
Orbital:
atomic, 31–3
molecular, 78–80
Order, of reaction, 148
Osmosis, 103–105
Ostwald, 115, 119, 140
Oxidation and reduction, 121

PACKING FRACTION, 201
Partial pressures, law of, 9
Partition coefficient, 88
Pauli, 30
Pauling, 71
Permittivity, of a vacuum, 54
pH concept, 135–41
pi bond, 79
Planck, 27, 38
Plane polarization, of light, 75
'Plasma' model, of metal structure, 77
Plasmolysis, of cells, 105
Polar bond, 70
Polarimeter, 76, 151–2
Polarography, 121
Positron, 20, 202, 205
Potentiometer principle, 134
Potentiometric titration, 134
Proton, 22
Proton acceptors/donors, 111
Pseudo 1st-order reaction, 151

QUANTUM NUMBER, 30
Quantum theory, 26

RADIOACTIVE DATING, 213–14
Radioactive series, 204
Raman, 40
Raoult, 95
Rate expression, 146, 150, 151, 161
Rate constant, 146
Rate determining step, 150
Reciprocal proportions, law of, 15
Reference electrodes, 132
Relativity, 58
Relaxation methods, 162
Reversibility, thermodynamic, 175
Roentgen, 20
Rotational energy, of molecules, 12
Rutherford, 21–2, 202–203, 205
Rydberg, 27

SALT HYDROLYSIS, 112, 192
Schrödinger, 31–2

Second law, of thermodynamics, 174
Second-order reaction, 148, 150
Semipermeable membrane, 103
Shell, of electrons, 30
Sigma bond, 78–80
Sols, 197–9
Solubility product, 129–32
Solvent extraction, 90
Spontaneous process, 167
'Spot' reagents, 195
Stability constants, of complex ions, 191
Standard:
  electrode potentials, 125
  enthalpy change, 49
  entropy, 178–9
  state, 49
  temperature and pressure (s.t.p.), 3
Steam distillation, 100
Stoichiometric equations, 147

Thomson, J. J., 18, 23
Third law, of thermodynamics, 178
Third-order reaction, 150
Titration curves, 134, 139, 141
Tracer techniques (radioactive), 211–12
Transition metals, 190–93
Transition state theory, of reaction rates, 156
Trouton's rule, 176
Tyndall effect, 196

Ultra-violet, 27, 35, 37
Unit cell, 63–4, 77

Valence bond, 65
van der Waals' equation, 10
Van't Hoff, 103
Vapour pressure, measurement of, 95–6
Vibrational energy, of molecules, 12
Vibrational modes, of molecules, 42
Visible spectrophotometry, 37, 151–2, 191
Volta, 17
Voltage standards, 133

Water composition of, 6
Wave function, 32
Wavelength, 26
Wave mechanics, 31–2
Wave number, 39
Wheatstone bridge, 112
Work done, by expanding gas, 45

X-rays, 20

Zartmann's apparatus, 12–13